BULGARIAN ACADEMY OF SCIENCES
NATIONAL EXPERT COUNCIL ON NANOTECHNOLOGIES

NANOSCIENCE & NANOTECHNOLOGY

NANOSTRUCTURED MATERIALS APPLICATION AND INOVATION TRANSFER

Edited by
E. Balabanova and I. Dragieva

HERON PRESS SCIENCE SERIES

Published by
Heron Press Ltd.,
18, Oborishte str., BG-1504 Sofia, Bulgaria; e-mail: heron_press@attglobal.net

NANOSCIENCE & NANOTECHNOLOGY
Proceedings of the Second Workshop *Nanostructured Materials Application and Innovation Transfer*, 23–24 November 2000, Sofia, Bulgaria

Edited by E. Balabanova and I. Dragieva

ISBN 954–580–097–6

Typeset by Heron Press Ltd.
Printed in Bulgaria by Heron Press Ltd.

PREFACE

Recently the nanotechnological research community in Bulgarian Academy of Sciences initiated the establishment of a National Expert Council on Nanotechnologies (March 1999), under the auspices of BAS, strongly supported by the Ministry of Education and Science. The National Center on Nanotechnologies of Bulgaria was created (October 1999) as a free association of partnering Academic and Academy institutes and scientific industrial research and development organizations, interested in the production and the application of nanosized materials in new high-tech and small-scale products and electronic devices (www.bas.bg/nano/).

The First Workshop NANOSCIENCE & NANOTECHNOLOGY was held at Sofia during 14–15 May 1999. Its proceeding contained 45 contribution of Bulgarian groups working with more than 65 partners from abroad and a round-table discussion on "Nanotechnologies and the EC 5th Frame Work Programme".

The Second Workshop NANOSCIENCE & NANOTECHNOLOGY "Nanostructured Materials Application and Innovation Transfer" was held at Sofia during 23–24 November 2000 and included 5 scientific sessions and a round-table discussion on "Participation of Bulgaria into COST 523 Action" with the personal presentation of Prof. H. Hofmann – President of COST Action 523.

The programme of the Second Workshop NS&NT consisted of invited talks given by Prof. W. Lorenz – Inst. Nanotechnology, Univ. Karlsruhe, Germany, Prof. H. Hofmann – Swiss Federal Inst. Tech. Laussane, Swiss, Prof. J. Schoonman – Delft Univ. Tech., The Netherlands, Prof. R. Hurly – Northern Ireland Semiconductor Research Centre, The Queen's Univ. of Belfast, UK and other contributed papers presented in 5 separate sessions on the following topics:

- Modeling and Fundamental Characterization of Nanoscaled Materials
- Synthesis and Processing (Optical Properties)
- Synthesis and Processing (Magnetic/Electrical Properties)
- Synthesis and Processing (Electrochemical/Chemical Properties)
- Synthesis and Processing (Mechanical Properties, Fullerenes).

The papers in this Proceeding are divided to topic sections following the programme of the Second Workshop NS&NT.

Special thanks to all participants – authors and institutions (Space Research Institute and small companies jointed with Dr. S. Stavrev, Central Lab. Electrochem. Power Sources – Director: Prof. Z. Stoynov, Central Lab. Photo processes – Director: Dr. I. Konstantinov, Institute of Electronics – Director : Dr. R. Enikov, Institute of Metal Science – Director: Prof. Y. Arsov belonging to the Bulgarian Academy of Sciences and State University of Sofia – Vice Rector: Prof. A. Popov) of the Second Workshop NS&NT – 2000 for their financial support.

The Editors
Sofia, February, 2001

Nanoscience & Nanotechnology
eds. E. Balabanova, I. Dragieva, Heron Press, Sofia, 2001

CONTENTS

SECTION 1: Modelling and Fundamental Characterization of Nanoscaled Materials

SECTION 2: Synthesis and Processing (Optical Properties)

SECTION 3: Synthesis and Processing (Magnetic/Electrical Properties)

SECTION 4: Synthesis and Processing (Electrochemical/Chemical Properties)

SECTION 5: Synthesis and Processing (Mechanical Properties, Fullerenes)

SECTION 1:
Modelling and Fundamental Characterization of Nanoscaled Materials

Nanoscience & Nanotechnology
eds. E. Balabanova, I. Dragieva, Heron Press, Sofia, 2001

SILICON ON INSULATOR (SOI) AND SILICON ON SILICIDE ON INSULATOR (SSOI) WITH "SMART-CUT" TECHNOLOGY FOR DEVICE APPLICATIONS RESEARCH

R. E. Hurley*, M. B. Armstrong, and H. S. Gamble

Northern Ireland Semiconductor Research Centre, School of Electrical and Electronic Engineering,
The Queen's University of Belfast, Ashby Building, Stranmillis Rd, Belfast BT9 5AH, UK

Abstract. The techniques of silicon on insulator (SOI), and silicon on silicide on insulator (SSOI), using wafer bonding are described. Various methods of producing the required thin layers are discussed with particular reference to "smart-cut" technology where the coalescence of nanocavities produced by H^+ implantation is used to accurately split a slice from a silicon wafer. Applications for the new generation of devices are described including low power, high speed bipolars, system-on-chip systems involving both digital and analogue devices and silicon based microwave circuits.

Keywords: Silicon, silicide, smart-cut, SOI, SSOI, bonding, bipolar, CMOS, SOC.

1. Introduction

The increasing demand in silicon microelectronics applications for faster devices, increased packing densities and more efficient use of power, continue to drive down the critical dimensions (CD = half-pitch between DRAM elements), with production of devices with CD $<$ 100 nm expected within the next few years. Front end processing requirements are for high quality defect-free thin films with precision doping and low thermal budgets with control over nanometric dimensions extending over 200 to 300 mm wafer diameters. Barrier layers for example, may have current thicknesses around 20 nm for 0.15 μm devices but this is expected to reduce to less than 10 nm as CDs shrink; gate-oxides would be only several nm thick. An important process for new generation devices is the production of thin, high crystalline purity silicon layers on insulating oxide substrates, silicon-on-insulator (SOI). SOI is especially suited to low power high speed bipolars and SiGe heterojunction bipolars, and for ultra-thin layers has application to CMOS devices. This paper examines the use of wafer-bonding with smart-cut and other technologies to produce the required thin SOI layers. A further development currently being studied incorporates thin tungsten-silicide layers which have applications for low resistivity collector contacts and dimensionally stable ground-plane layers.

2. Techniques

Wafer bonding is used to join an 'active' device wafer to an oxidised 'handle' wafer. Ultra-clean handling and RCA and/or plasma treatment is essential before the wafers are contacted together at room temperature to yield a hydrophilic bond of energy about 100 mJ/m^2. The wafers are then heated to a high temperature which for temperatures in excess of 1000°C increases the bond energy to around 2000 mJ/m^2; but for some applications lower temperatures must be used with a corresponding loss in bond strength. Several possibilities exist for removing the bulk of the active wafer to leave a thin SOI substrate: (1) Precision grinding and polishing will thin the active wafer surface to about 1.5 ± 0.3 μm and this is current technology used for linear ICs; (2) Chemical etching following grinding and polishing will reduce the surface layer to about 0.5 μm. In this case, before bonding, a SiGe etch-stop layer must be grown on a 'sacrificial' wafer followed by an epitaxial silicon layer of the required thickness which is then bonded to the oxidised wafer. Most of the sacrifical wafer is removed by grinding and this is followed by chemical etching which stops precisely on the SiGe etch-stop layer. The SiGe layer is then removed by selective etching [1]; (3) The polish stop process will produce SOI thicknesses down to 0.1 μm and is particularly suited to trench-patterned silicon layers. Following bonding and grinding and polishing, the silicon layer is trench-patterned and then coated with nitride which is patterned to lie in the trenches. Chemical-mechanical polishing (CMP) removes the excess silicon and stops on the nitride. The nitride, which has determined the final thickness of the silicon, is then removed selectively [2]; (4) The smart-cut process which avoids grinding and leaves a re-useable left-over wafer is currently used to produce SOI thicknesses in the range 0.1 to 0.2 μm. Research suggests that when combined with SiGe etch-stop technology, this can be reduced to 50 nm.

In the smart-cut process, the active wafer is implanted with hydrogen to the required depth and then bonded to the handle wafer. The bond is strengthened with a low temperature anneal and when the temperature is raised to $\sim$ 500°C the wafer splits along the projected range depth of the implant. A 'touch' polish is then usually required for device applications. The smart-cut process is applicable to other integrated substrate applications for example SiC or GaAs on oxidised silicon or Ge on glass [3], but although a potentially a powerful technique, research is needed to increase understanding and overcome problem areas. H^+ implantation creates Frenkel pairs in the silicon crystal matrix and H then passivates broken silicon bonds and also forms H_2

molecules. Vacancy-H complexes in close proximity coalesce to form platelets typically one or two atoms thick and 100 nm in length. During annealing, dissociation of Si-H complexes occurs with H diffusing to platelets forming H_2 and a build up of gas pressure that leads to microcracking and ultimate splitting along the line of the implant [4]. A major disadvantage for production applications is the high dose required which for silicon is $> 5 \times 10^{16}$ ions/cm^2, and such high doses can also lead to higher defect densities. Co-implantation with B [4], or He [5], are methods of reducing the required dose. He, implanted after H, acts physically on the existing platelets and microcracks driving the splitting process. The total combined dose may be several times less than using H alone [5]. Surface roughness following splitting ($\sim$ 10 nm), needs to be reduced, especially for ULSI applications. There is some evidence that the relatively low bond-anneal temperature (giving a low strength bond), used with the smart-cut process may be detrimental to surface roughness; when plasma or chemical surface treatment is applied to the wafer surface before bonding, the increased bond-strength acts as a stiffener with a resulting reduction in roughness from 10 to < 1 nm [6,7].

H^+ may be implanted through a silicon-germanium etch-stop layer and bonded to a handle wafer. Following annealing and splitting (releasing the re-useable wafer), grinding and polishing, etching of the remaining silicon to the etch-stop and selectively etching the SiGe layer can yield a final thickness of 50 nm. Current work is developing processes for silicon on silicide on insulator (SSOI) [8]. A silicide layer is formed on the active wafer by either sputtering and annealing or CVD. The surface is coated with polysilicon (and optionally TEOS which can aid bonding), and after planarisation, is bonded to a handle wafer. Grinding and polishing is used to produce the required thickness of silicon and an N^+ layer incorporated to provide ohmic contact to the silicide as shown in Figure 1. This process can be combined with smart-cut to produce ultra-thin SSOI substrates; alternatives exist here for H implantation into the silicon before silicide deposition or directly through the silicide. In the first case the deposited silicide must be suitably prepared to allow bonding to a handle wafer and this may involve coating with amorphous silicon and polishing this surface. Higher implantation energies are required to implant through silicide which may also be coated with a layer of polysilicon and TEOS. After implantation the TEOS surface is polished before bonding.

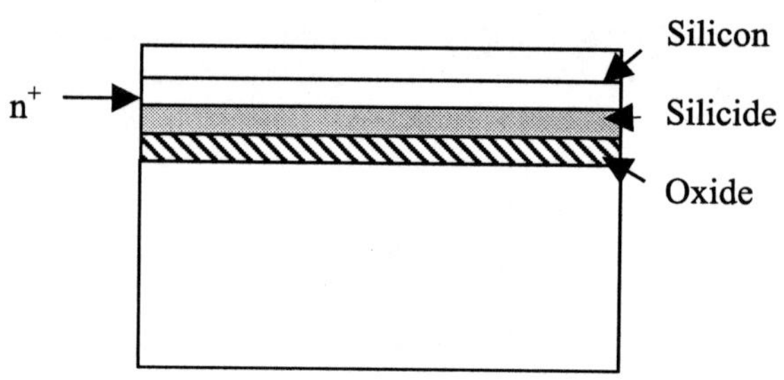

Figure 1. SSOI Substrate

3. Applications

SSOI substrates are ideal for complementary bipolar transistors and for ground-plane structures where the silicide is located under the oxide layer reducing cross-talk in system-on-chip (SOC), applications. These involve high-speed (CMOS), digital circuits and high frequency low noise analogue circuits on the same chip. The low resistivity substrates for CMOS can allow spurious signals to interfere with analogue circuits; SSOI, using a ground-plane (0.2 μm thick of $2\Omega/pi$) to pin the substrate potential, can suppress cross-talk [9]. Another application for SSOI is with sub-micron MOSFETs, being used to reduce the drain induced barrier lowering effect caused by field penetration into the buried oxide and silicon substrate. The silicon ground-plane provides an equipotential reference at the back gate interface which terminates fringing fields. Further applications include silicon MMIC substrates and coplanar waveguides (CPWs) as shown in Figure 2. SOI substrates are particularly advantageous for millimetre wave applications since low resistivity active silicon layers can be provided on the high resistivity silicon substrates required for low losses at high frequencies. However, positive charges in the oxide tend to induce a mobile layer of electron charge at the silicon surface with consequent transmission losses. An undoped layer of polysilicon deposited prior to oxidation will effectively immobilise this. The most recent work suggests that this technique can be used with low cost standard silicon wafer substrates with substantial economic benefits for certain applications [8]. Figure 3 shows the integration of a pin diode with CPW on an SSOI substrate where the silicide acts as a low resistance back-contact to the n^+ region. This device has reduced parasitic series resistance and easier interconnection technology.

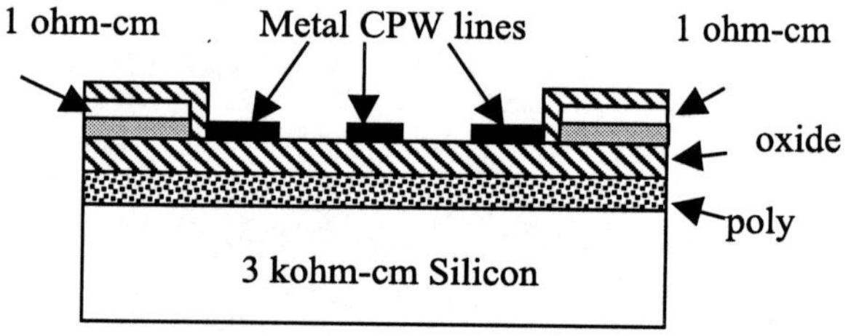

Figure 2. Coplanar wave-guide structure on SSOI substrate

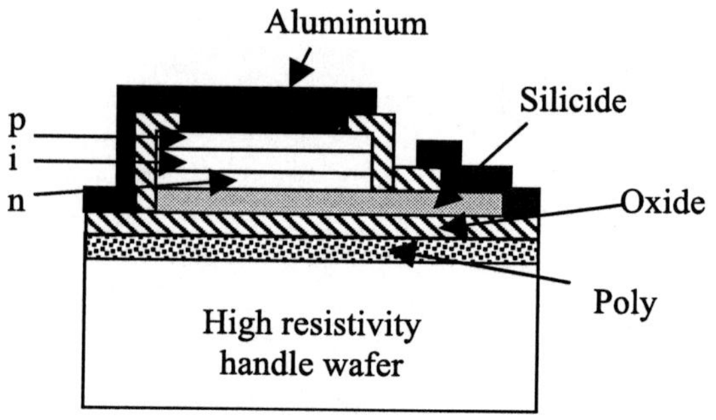

Figure 3. *pin* diode on SSOI for integration with CPW on SOI

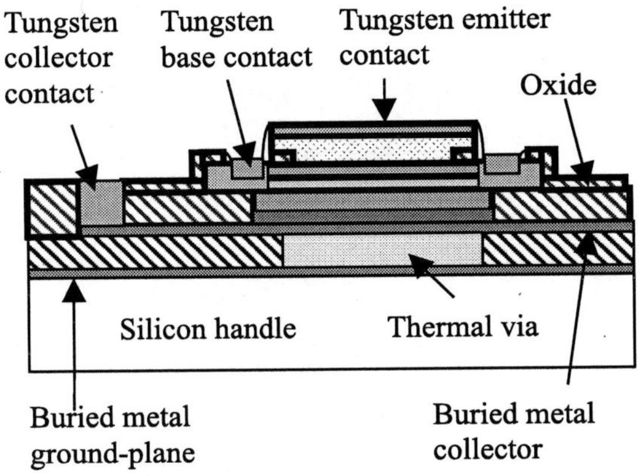

Figure 4. SSOI Transistor

Figure 4 shows an advanced transistor structure employing SSOI. Parasitic capacitance and resistance are reduced yielding low power and higher speed, ground-planes suppress chip noise and cross-talk, the buried metal collector and metallised base and emitter reduce parasitic resistance with faster performance and the thermal via conducts heat from the transistor while maintaining adequate electrical resistance.

4. Conclusions

SOI and SSOI technologies combined with the "smart-cut" wafer splitting process are valuable process technologies for certain new generation silicon-based devices. These techniques have the potential to produce the high quality thin layers essential to meet required performance specifications of modern devices. High speed low power bipolars and devices that require buried metal layers for collectors or ground-planes, as well as silicon-based microwave devices, are fruitful areas of application.

References

1. Uppal S, Gay D, Armstrong GA, McNeill DW, Baine P, Armstrong BM, Gamble HS, Yallup K. The Electrochemical Society 195th Meeting, Proceedings of the Ninth International Symposium on Silicon-on-Insulator Technology and Devices. Proc. Vol.. 99(3), 1999, pp. 219-224.
2. Baine PT, Gay DL, Armstrong BM, Gamble HS. J Electrochem Soc. 1998; 145(5):1738.
3. Tong Q-Y, Gutjahr K, Hopfe S, Gosele U, Lee T-H. Appl Phys Lett 1997;70(11):1390.
4. Tong Q-Y, Bower R W. MRS Bulletin 1998;December:40.
5. Agarwal A, Haynes TE, Venezia VC, Holland OW, Eaglesham DJ. Appl Phys Lett 1998;72(9):1086.
6. Li YA, Bower RW. Jpn J Appl Phys 2000;39(1):275.
7. Auberton-Herve AJ, Bruel M, Aspar B, Maleville C, Moriciau H. IEICE Trans Electron 1997;E80:C3.
8. Gamble HS, Armstrong BM, Baine P, Bain M, McNeill DW. SOI substrates with buried tungsten silicide layer. Submitted for publication to Solid State Electronics.
9. Hamel JS, Stefanou S, Bain M, Armstrong BM, Gamble HS. IEEE Microwave & Guided Wave Letters. 2000;10(4)134.

Nanoscience & Nanotechnology
eds. E. Balabanova, I. Dragieva, Heron Press, Sofia, 2001

THE ROLE OF METASTABLE STATES IN THE TEMPERATURE DRIVEN PHASE CHANGES OF MOLECULAR NANO-CLUSTERS

A. Proykova

University of Sofia, Faculty of Physics, 5 J. Bourchier Blvd., Sofia-1126, Bulgaria

Abstract. The thermal-equilibrium coexistence of two solid structures, a bcc orientationally-disordered phase and a strained monoclinic orientationally-ordered phase, is more difficult to observe in computer simulations of SF_6 clusters than of TeF_6 clusters. This study shows that the reason is rooted in the topography of the potential energy surface (PES). For the case of 89 SF_6 cluster the relative energies of most of the linked minima differ only slightly, the barriers between them are low and the competition between metastable states facilitates the system to escape fast from each minimum. At relevant temperatures, the fraction of the available phase space of the SF_6 89 cluster corresponding to a partially-ordered structure is smaller than it is for the TeF_6 89 cluster. The latter readily exhibits coexistence of the ordered and disordered forms. This results in better separation of the higher-energy local minima of the TeF_6 cluster, and longer residence times in the partially-ordered form. This makes the TeF_6 substance more promising for technological purposes.

Keywords: nano-sized systems, phase changes, theoretical analysis, simulations, Monte Carlo, Molecular Dynamics

1. Phase Transitions in Finite-size Systems

Among the most spectacular macroscopic events in nature are transformations between the various states of matter. Such transformations, or phase transitions, appear to be of both fundamental and technological importance, [1,2].

Abrupt changes, discontinuities, and strong fluctuations as a consequence of a cooperative phenomenon characterize phase transitions in bulk. The phase transitions are associated with singularities of the measurable quantities like heat capacity. These singularities are caused by fluctuations of the thermodynamic potentials, which cover all length scales up to macroscopic wavelengths. In finite systems, however, the range of fluctuations is limited by the system size. This limitation causes rounding of the measurable quantities.

The very first question to be answered in a computer study of the cooperative behavior of an interacting many-body finite-size system is the existence of a stable distinguishable phase at finite temperatures. Having established the existence and structure of the ordered phases, we turn to the question of kind of transition between the phases: discontinuous (first order) or continuous (higher order). Discontinuous transitions are characterized with a latent heat. Knowledge of the type of transition is important in order to distinguish between systems with and without memory capabilities. The ability of a material to memorize is one of the properties actively sought by the modern technology.

The special property of clusters, which we emphasize in this review, is that thermodynamically unfavored but locally stable phases occur in thermal equilibrium situations, *in observable amounts*. These are metastable states having a finite relaxation time. This means that some phases can be *stable* for clusters but only as minority species, and never be found in bulk systems in equilibrium. *Stable* are solutions of the master equation of the system, which move exponentially toward the origin with time and those that move exponentially away from the origin are *unstable*. The presence of metastable states in the transition region indicates a first-order phase transition, e.g. available memory. In this region the free energy has two minima and the system can be trapped in the upper metastable one for a finite time comparable to the relaxation time of the metastable state. The rate of separation of two phases of different structures can be radically altered by the presence of a metastable intermediate phase.

Systems trapped in such secondary minima are *effectively* in stable equilibrium as if the deeper minimum did not exist at all.

For example, clusters produced in a supersonic jet appear to be in metastable states [3,4]. Some states correspond to phases, which obey properties interesting for the industry and others do not. This is the reason that controlling the appearance of desired metastable states is very often the goal of phase transition study.

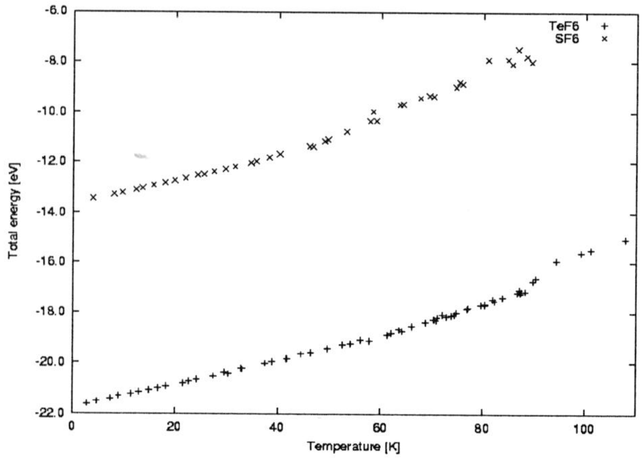

Figure 1

A good knowledge of the potential energy surface, which determines the structure, thermodynamics and dynamics of a cluster helps controlling the production conditions, [5]. By investigating the dependence of the topography of the potential energy surface on the molecular size, a microscopic explanation of the difference in transitions of TeF_6 and SF_6 clusters has been provided in [6]. The influence of the surface on the cluster transformation is investigated in [7]. The SF_6 cluster potential energy surface is shallow with saddle points (unstable steady states) while the topography of a TeF_6 cluster of the same size is characterized with a smaller number of saddle points and deeper wells. This means that TeF_6 clusters stay longer with a specific structure than the SF_6 clusters.

2. Canonical and Microcanonical Treatment of Plastic Nanoclusters

To be specific in our analysis, we have implemented a constant energy and a constant temperature Molecular Dynamics [8] to simulate the thermal behavior of nanometric clusters consisting of 89 TeF_6 and SF_6 molecules.

We describe a cluster as a classical aggregate of rigid, octahedral species, which are allowed to rotate and translate. The potential energy contains the information expressing the intermolecular interactions. The potential function of the system is a sum of pair-wise Lennard - Jones atom-atom potential and an atom-atom Coulomb interaction.

We simulated the thermal behavior of clusters consisting of 89 SF_6 or 89 TeF_6 molecules by solving the classical equation of motion in the Hamiltonian form [9]. The integration step is 5 fs in the case of microcanonical ensemble [5] and 10 fs for the case of canonical one [10]. Both steps are short in comparison with the vibrational characteristic time τt of the systems: for two TeF_6 molecules $\tau \sim 2.45$ ps; for two SF_6 molecules $\tau \sim 2.26$ ps. These characteristic times have been computed in the harmonic oscillator approximation, [5].

The caloric curve, Figure 1, shows that both an 89 SF6 cluster and an 89 TeF_6 cluster undergo phase transitions when the temperature changes. The simultaneous analysis of Figure 1 and the Lindeman coefficient [11] shows that the SF_6 cluster solidifies at lower temperatures than the TeF_6 cluster containing the same number of molecules.

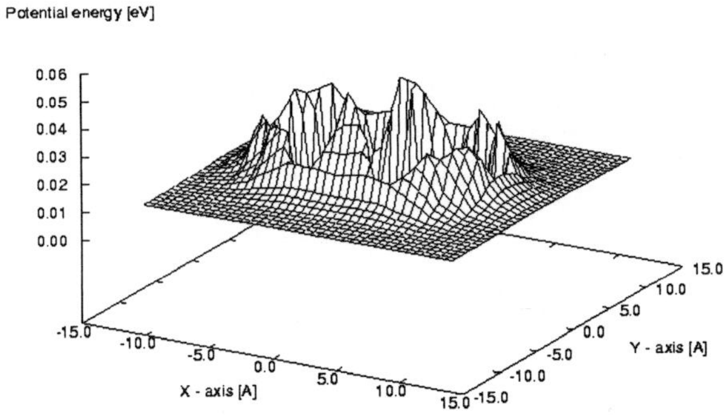

Figure 2. Potential energy surface of SF_6

3. Results for Dynamical Coexistence of Different Phases

The discontinuous change of a small cluster is characterized with appearance of coexisting phases under specific conditions, [11]. First, the cluster must be in equilibrium. Second, the temperature rate of cooling (heating) should be consistent with the landscape of the PES. A rugged, staircase PES requires slow change of the system's temperature in order to avoid trapping in a metastable state. If this trapping happens, for realistic computational times the system cannot escape from a local minimum that acts as if it were an artificial attractor inducing a quasi-periodic orbit. In such a situation, the phase space available to the system is not properly spanned. Figure 2 gives an impression of the PES (maxima and minima are reversed in the plot!) in the simplest case of a three-body system: one SF_6 molecule probes the potential field of other two molecules aligned in the Euler space.

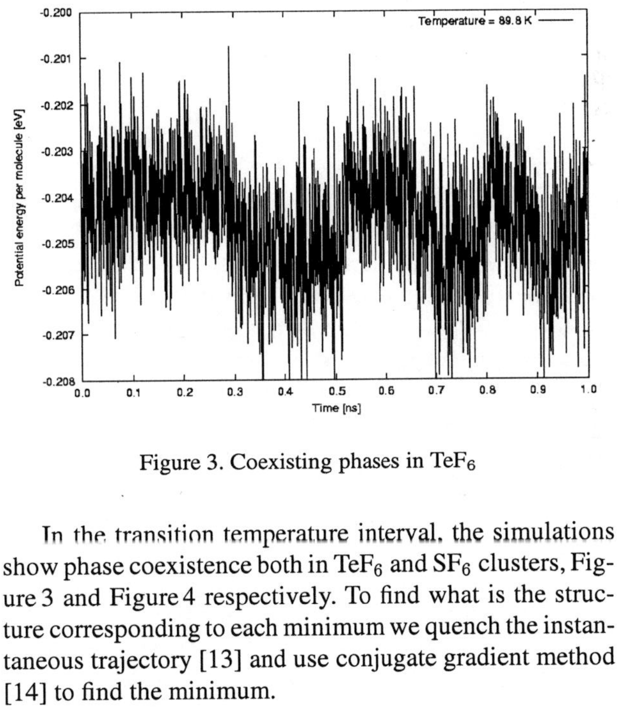

Figure 3. Coexisting phases in TeF_6

In the transition temperature interval, the simulations show phase coexistence both in TeF_6 and SF_6 clusters, Figure 3 and Figure 4 respectively. To find what is the structure corresponding to each minimum we quench the instantaneous trajectory [13] and use conjugate gradient method [14] to find the minimum.

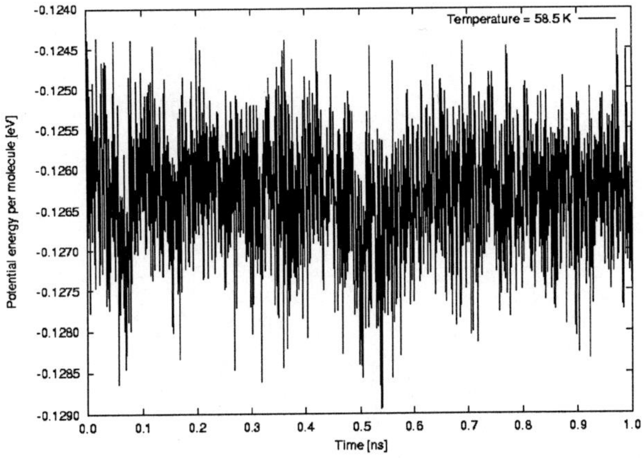

Figure 4. Coexisting phases in SF_6

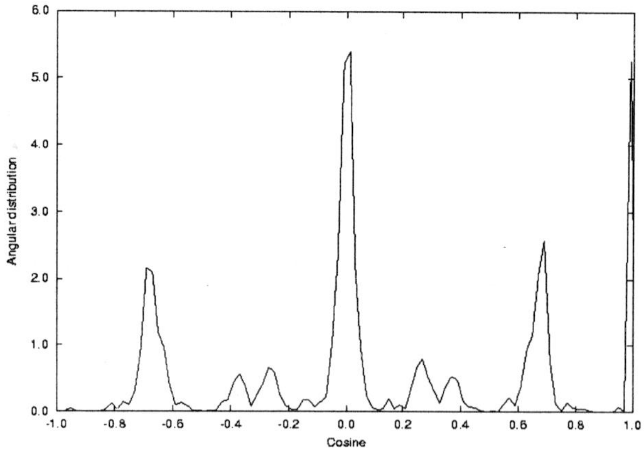

Figure 5. Angular distribution – TeF_6

The orientation order of the molecular axes is recognized in Figure 5, which shows a normalized angular distribution at atemperature of 20 K as a function of $\cos\theta$ where θ is the angle between the Z-axes of any two molecules. Note that for an octahedral molecule the three axes are indistinguishable, so $\cos\theta = 0, \pm 1$ denote the same mutual orientation. At high temperatures this distribution is an isotropic function of θ.

4. Conclusions

For the case of 89 SF_6 cluster the landscape is shallower and the relative energies of most of the linked minima differ only slightly, the barriers between them are low and the competition between metastable states facilitates the system to escape fast from each minimum. Although the SF_6 and TeF_6 molecules have the same symmetry and we use the same form of interatomic interaction potentials to describe their clusters, the topographies of their PES differ significantly in the values of minimum-saddle-minimum ratios that determine the system dynamics. The latter readily exhibits coexistence of the ordered and disordered forms. This results in better separation of the higher-energy local minima of the TeF_6 cluster, and longer residence times in the partially-ordered form. This makes the TeF_6 substance more promising for technological purposes.

Acknowledgement

The work has been supported by the NATO Collaborative Linkage Grant Award Number: SA (PST.CLG.976363) 5437.

References

1. Stanley HE. Introduction to Phase Transitions and Critical Phenomena, Clarendon Press, Oxford (1971).
2. Creswick RJ, Farach HA, Poole CP, Jr. Introduction to Renormalization Group Methods in Physics, (Wiley, New York 1992).
3. Bartell LS, Valente EJ, Caillat J. J. Phys. Chem. (1987); 91:2498.
4. Beniere FM, Fuchs AH, de Feraudy MF, Torchet G. Mol. Phys. (1992); 76: 1071.
5. Ball K,Berry RS, Kunz R, Li FY, Proykova A, Wales D. Science. (1996); 271:963
6. Pisov S, Proykova A. Proceedings of the Second Meeting of Young Researchers in Physics, Heron Press (2001).
7. Proykova A, Berry RS. Eur. J. Phys.D (1999); 9: 445.
8. Nose S. Mol. Phys. (1984); 52:255.
9. Proykova A, Berry RS. Z. Phys D. (1997); 40:215
10. Daykov I, Proykova A. Proceedings of the Second Meeting of Young Researchers in Physics, Heron Press (2001).
11. Lindeman FA. Phys. Z. (1910); 11:609.
12. Kunz RE, Berry RS. Phys. Rev. E (1994); 49:1895.
13. Stillinger FH, Weber TA.,Kinam (1981); 3: 159.
14. Press WH, Flennery BP, Tuekolsky S, Vetterling WT. Numerical Recipes (Cambridge University, Cambridge, 1986).

Nanoscience & Nanotechnology
eds. E. Balabanova, I. Dragieva, Heron Press, Sofia, 2001

SHOCK-WAVE PROPAGATION IN NANODISPERSE MEDIA

Valentin Gospodinov

Space Research Institute – BAS, Space Materials Division, 1000 Sofia P.O. Box 799, BULGARIA

Abstract. In this paper a simple model is presented for the shock Hugoniot of a two-component monodisperse elemental powder mixture. It is expressed in terms of the thermodynamic attributes of the single components. Together with the conservation laws of mass and momentum it defines in implicit form a surface in the space of the specific volumes of the two components and the pressure. This surface is the two-component analogue of the well-known Hugoniot curve of a single component system. Also, it has been established that in order the assumptions for thermodynamic equilibrium of the compressed material behind the shock front to hold it is necessary the average diameter of the particles to be smaller than a quantity determined by the shock rise time and the thermal diffusivity of the material.

Keywords: shock Hugoniot, powder mixtures, thermodynamic equilibrium

1. Introduction

Shock-wave methods play an important role in basic and applied sciences. They are used to investigate states of matter at high pressures and temperatures and also form a viable technology for materials processing. In the case of powder processing the main advantages include: (i) focusing energy on particle surfaces so that the whole systems do not need to be heated for compaction, (ii) consolidating nanocrystalline powders so quickly that grains do not have time to grow, (iii) bonding powders of composite materials.

Fully understanding the mechanism of shock-wave processing of powders and powder mixtures is dependent upon knowing the shock Hugoniot of the mixture of interest.

The shock Hugoniots of most metals and of many nonmetal elements and materials are already established experimentally within high accuracy. However, it is not possible to determine experimentally the shock Hugoniots of all materials of interest to science and technology. Since shock Hugoniots depend on the initial density of the material, this holds to even a greater extent for powder materials, whose initial density can vary broadly. That is why it is reasonable to introduce a complex approach which integrates theoretical models for multicomponent powder mixtures, with experimentally determined shock Hugoniots of the components.

The objective of the present communication is following this approach to derive the shock Hugoniot of a two-component elemental powder mixture. This is done in Sec. 3. At first, in Sec. 2 the applicability of the basic assumptions of shock physics to powder mixtures is analyzed. In Sec. 4 some conclusions are drawn and a possible direction for continuation of the research is outlined.

2. Theoretical Background

In the course of analyzing the results of shock experiments the following basic assumptions are employed [1]: (i) stresses behind the shock wave are isotropic, (ii) the compressed materials behind the shock front are in thermodynamic equilibrium, and, (iii) the form of the relation between the shock velocity u_s and the particle velocity u_p is linear in a wide range of particle velocities in the absence of phase transitions and chemical reactions:

$$u_s = c_0 + s u_p \tag{2.1}$$

The Hugoniot intercept, c_0, and the slope, s, are determined from the data by the method of least squares. If rigidity effects and possible low pressure phase changes are neglected, the intercept should correspond to the velocity of an infinitesimal pressure pulse, or the bulk sound speed, $c_0 = [(\partial P/\partial \rho)_S]^{1/2}$ (at $P = 0$). Since the slope is linearly related to the pressure derivative of the adiabatic bulk modulus, $(\partial B/\partial P)_S$, a linear u_s–u_p Hugoniot then reflects a nearly linear dependence of B_S on the pressure. Equation (2.1) is the shock Hugoniot in kinematic form.

The additional assumption which we make in the present work, specific to powder mixtures, is that the components have equal particle velocities. This is based on experimental evidence [2].

The assumption for thermodynamic equilibrium behind the shock front applied to a two-component elemental powder mixture has the form:

$$P_A(V_A, E_A) = P_B(V_B, E_B), \tag{2.2a}$$
$$T_A(V_A, E_A) = T_B(V_B, E_B). \tag{2.2b}$$

If the components are not at equal pressures in the beginning of the shock rise, they will come to equilibrium in a few multiples of a time $t \sim d/c_0$, where $c_0 \sim 10^3 m/s$ is some average bulk sound speed typical for a solid, and $t \sim 100\,ns$ is a typical average shock rise time for ductile materials. Then $d \leq 100 \mu m$. This is a typical value for commercial metal powders. The equal pressures requirement is readily met.

Since the two components have different thermal properties they absorb different amounts of thermal energy and, hence, are at different temperatures. Significant heat conduction will occur over a distance $d = \sqrt{at}$, where $a = \lambda/\rho C_p$. For Cu $a \sim 10^{-4} m^2/s$ and, hence $t \sim 10^{-4} s$. This value of t is three orders of magnitude larger than the shock rise time and the equal temperatures requirement is not met. We see that it is necessary $d \leq 1\mu m$ for thermal equlibration to occur within the shock rise time.

3. Derivation of the shock Hugoniot of a two-component elemental powder mixture

Having in mind the restrictions, imposed on the components particles size as a result of the considerations in the previous section, the laws of conservation of mass, momentum and energy for a mixture will be no different from those for a single-component system, provided there is no interaction between the components (no chemical reactions occur):

$$\rho_{00AB} u_s = \rho_{1AB} \left(u_s - u_p\right), \quad (3.1)$$

$$P_1 - P_0 = \rho_{00AB} u_s u_p = \rho_{1AB} \left(u_s - u_p\right) u_p, \quad (3.2)$$

$$\rho_{00AB} u_s \left(E_1 - E_0 + \frac{1}{2} u_p^2\right) = P_1 u_p, \quad (3.3)$$

where ρ_{00AB} is the initial density of the mixture, ρ_{1AB} is the shocked density, P_0 is the initial pressure, P_1 — the pressure in the shock front, and E_0 and E_1 — the specific internal energies in front of the shock front and behind it, respectively.

Substituting Eqs.(3.1) and (3.2) into Eq.(3.3) and dividing the specific internal energies between the two components gives

$$x \Delta E_A + (1 - x) \Delta E_B = \frac{1}{2} P_1 \left(V_{00AB} - V_{1AB}\right), \quad (3.4)$$

where x is the mass fraction of material A, and V_{00AB} and V_{1AB} are the initial and shocked specific volumes of the mixture. E_A and E_B are the specific internal energies of the components, and the initial pressure is assumed to be zero. Equation (3.4) is a simple expansion of the Hugoniot energy equation $\Delta E = \frac{1}{2} P (V_0 - V)$ for the case of a two-component mixture.

The energy and pressure of each constituent may be separated into thermal and cold components [3, 4],

$$E_A = E_{cA} + E_{TA}, \; E_B = E_{cB} + E_{TB}, \quad (3.5)$$

$$P_{cA} + P_{TA} = P_1 = P_{cB} + P_{TB}, \quad (3.6)$$

where the subscripts c and T refer to the cold ($0\ K$) and thermal components, respectively. Using the definition of the Grüneisen parameter $\gamma \equiv V(\partial P/\partial E)_V$ we assume each component's Grüneisen parameter-to-volume ratio is constant and temperature independent: $\gamma/V = \gamma_0/V_0 = const$. Experimental work on a number of materials [5], as well as theoretical considerations [6], indicate this to be an adequate approximation. Now we may relate the thermal pressures of the two components to their thermal energies

$$P_{TA} = \frac{\gamma_{0A}}{V_{0A}} E_{TA}, \; P_{TB} = \frac{\gamma_{0B}}{V_{0B}} E_{TB}. \quad (3.7)$$

Here γ_0 is the thermodynamic value at standard conditions given by $\gamma_0 = 3\alpha c_0^2/C_p$, α is the thermal expansion coefficient, c_0 the sound speed, and C_p the specific heat at constant pressure.

If the ambient energy is used as a reference state, the Δ-s in Eq.(3.4) may be removed and after substituting in it with Eq.(3.5) it takes the form

$$x \left(E_{cA} + E_{TA}\right) + (1 - x) \left(E_{cB} + E_{TB}\right) = \frac{1}{2} P_1 \left(V_{00AB} - V_{1AB}\right). \quad (3.8)$$

If we substitute with Eqs.(3.6), (3.7) into Eq.(3.8) and solve for P_1 we obtain:

$$P_1 = \frac{\sigma_A P_{cA}(\lambda_A) + \sigma_B P_{cB}(\lambda_B) - \dfrac{x E_{cA}(\lambda_A)}{V_{0A} V_{0B}} - \dfrac{(1-x) E_{cB}(\lambda_B)}{V_{0A} V_{0B}}}{\sigma_A + \sigma_B - \dfrac{1}{2}\left(\varphi - \eta_A \lambda_A - \eta_B \lambda_B\right)}, \quad (3.9)$$

where the following quantities are introduced:

$$\sigma_A = \frac{x}{\gamma_{0A} V_{0B}}, \quad \sigma_B = \frac{(1-x)}{\gamma_{0B} V_{0A}},$$

$$\eta_A = \frac{x}{V_{0B}}, \quad \eta_B = \frac{(1-x)}{V_{0A}},$$

$$\lambda_A = \frac{V_{1A}}{V_{0A}}, \quad \lambda_B = \frac{V_{1B}}{V_{0B}}, \quad \varphi = \frac{V_{00AB}}{V_{0A} V_{0B}}.$$

Equation (3.9) is the shock Hugoniot of a two-component elemental powder mixture. It gives the relation between the shock pressure and the specific volumes of the two components. The parameter φ accounts for the initial porosity of the mixture. Let us introduce the usual porosity $\alpha_{0AB} = V_{00AB}/V_{0AB}$. We shall assume it known, as well as the quantities u_s and u_p — the shock and the particle velocities of the mixture.

Then from Eq.(3.1) we obtain

$$V_{1AB} = V_{00AB}\left(1 - \frac{u_p}{u_s}\right) = \alpha_{0AB} V_{0AB}\left(1 - \frac{u_p}{u_s}\right),$$

$$V_{0AB} = x V_{0A} + (1 - x) V_{0B},$$

$$V_{1AB} = x V_{1A} + (1 - x) V_{1B},$$

$$x V_{1A} + (1 - x) V_{1B} = \alpha_{0AB}\left[x V_{0A} + (1 - x) V_{0B}\right]\left(1 - \frac{u_p}{u_s}\right).$$

Dividing both sides of the last equation by $V_{0A}V_{0B}$ gives

$$\eta_A\lambda_A + \eta_B\lambda_B = \alpha_{0AB}\left(\eta_A + \eta_B\right)\left(1 - \frac{u_p}{u_s}\right), \quad (3.10)$$

After some work on Eq.(3.2) and assuming $P_0 = 0$ it takes the form

$$P_1 = \rho_{00AB}u_s u_p = u_s u_p / V_{00AB}$$

or

$$P_1 = u_s u_p \left[\alpha_{0AB}V_{0A}V_{0B}\left(\eta_A + \eta_B\right)\right]^{-1}. \quad (3.11)$$

Solving Eqs.(3.10), (3.11) together with Eq.(3.9) we obtain a set of values $\{\lambda_A, \lambda_B, P_1\}$. They define a surface which is the two-component analogue of the well-known Hugoniot curve of a single-component system. To apply this scheme we must obtain expressions for the cold pressures and energies of the two components. The derivation of these quantities from experimental shock Hugoniots of dense materials is given elsewhere [3, 4].

4. Conclusions

A model is presented in the form $P_1 = P_1(\lambda_A, \lambda_B)$ to determine the Hugoniot of a two-component elemental powder mixture. Only the thermodynamic attributes of the single components, the initial porosity of the mixture, and its experimental shock Hugoniot in kinematic form — $u_s = u_s(u_p)$ are used as input data.

In developing the model, the applicability of the basic assumptions of shock-wave physics to powder mixtures has been extensively discussed. It has been established that in order the assumptions for thermodynamic equilibrium of the compressed material behind the shock front to hold it is necessary the average diameter of the particles to be smaller than a quantity determined by the shock rise time and the thermal diffusivity of the material. For Cu which is ductile and a good heat conductor this average diameter is $\sim 1\ \mu m$.

It has been pointed out that due to the different thermal properties of the two components of the mixture they absorb different amounts of thermal energy. A possible line of continuing this research is to elucidate the explicit form of this thermal energy partitioning ratio.

References

[1] Zharkov V.N. and Kalinin V.A. *Equations of State for Solids at High Pressures and Temperatures*, (New York: Consultants Bureau, 1971) Ch.5.

[2] Krueger B.R., Mutz A.H., and Vreeland Th. *J. Appl. Phys.*, 1991, v. **70**, No 10, pp. 5362–5368.

[3] Gospodinov V.V. *LANL e-print archive*, 1999, cond-mat/9911407.

[4] Gospodinov V.V. *4^{-th} general conference of the Balkan Physical Union*, 22–25 Aug. 2000, Veliko Turnovo, Bulgaria.

[5] Carter W.J., Marsh S.P., Fritz J.N. and McQueen R.G. *Accurate Characterization of the High-Pressure Environment* ed. E.C. Lloyd NBS Special Publication 326, (1971).

[6] Walsh J.M., Rice M.H., McQueen R.G. and Yarger F.L. *Phys.Rev.* **108**, 196, (1957).

Nanoscience & Nanotechnology
eds. E. Balabanova, I. Dragieva, Heron Press, Sofia, 2001

DYNAMICAL PHASE COEXISTENCE OF MOLECULAR NANO-CLUSTERS: KINETICS VERSUS THERMODYNAMICS

A. Proykova[1], R. Radev[1], S. Pisov[1], I. Daykov[2], and R.S. Berry[3]

[1]University of Sofia, Faculty of Physics, 5 J. Bourchier Blvd., Sofia-1126, Bulgaria
[2]Cornell University, Department of Physics, 117 Clark Hall, Ithaca, NY 14850, USA
[3]The University of Chicago, Department of Chemistry, 5735 S. Ellis Ave., Chicago IL, 60637, USA

Abstract. We report results from Molecular Dynamics and Monte Carlo simulations, which show that free molecular clusters can coexist in two different solid structures: an orientation-disordered bcc phase and a strained monoclinic orientation-ordered phase. The coexistence is observed in a temperature interval, which becomes narrower as the cluster grows larger. Appearance of a finite temperature interval of phase transformation is a phenomenon that distinguishes the finite systems from their bulk counterpart. In this interval the phases are stabilized by kinetics.

Keywords: nano-sized clusters, phase transformations, theoretical analysis, Monte Carlo, Molecular Dynamics

1. Introduction

Transitions between different phases play a central role in many technological processes in industry, and are also of fundamental interest for theoretical physics in understanding the behavior of large number of interacting particles.

Nanosized clusters exhibit some of the same solid-like behavior of bulk solids and liquids, [1]. They have been used for a while as tools for understanding the phase changes of bulk matter. Recently, their peculiar properties have attracted the scientific interest and have become the major research objective.

Molecular clusters offer elements of interest that are not found in atomic clusters. Experimental studies of phase transitions have been performed on molecular clusters, [2]. The results, obtained in these experiments, demonstrate that most of the phase transformations have hardly been understood. It seems worthwhile to initiate computer experiments (simulations), which would shed light on the nature of phase transitions in small systems.

The phase transitions are associated with singularities of measurable quantities like heat capacity. These singularities are caused by fluctuations, which cover all length scales up to macroscopic wavelengths. None of these scales can be neglected. In finite systems, however, the range of fluctuations is limited by the system size. As a result, the phase changes occur in a temperature interval rather than at a unique temperature [5]. This finite-size effect hampers elucidating the nature of phase transitions in both real and computer experiments, [6].

The very first question to be answered in a computer study of the cooperative behavior of a interacting many body system is the existence of a stable ordered phase at finite temperatures. Theoretically, it is known to be a very difficult problem. Having established the existence and structure of the ordered phases, we turn to the question of transition nature between the phases. The transition is triggered when a thermodynamic parameter, e.g. temperature is varied, [2-4]. Here we study transitions between various structures found in TeF_6 molecular clusters implementing Monte Carlo and Molecular Dynamics simulations of canonical ensemble to answer the following questions. The first one is related to the limitations, if any, of the validity of the ergodicity in small systems. The second question is why variety of structures are observed in small clusters, while in bulk those structures disappear.

2. Clusters with Rugged Potential Energy Surfaces

One such system is a cluster made of tellurium hexafluoride molecules, whose kinetics of freezing has been examined in some detail, [2-4]. These rigid molecules have an octahedral symmetry: the positively charged tellurium atom is in the center of the molecule. The charge (0.6e) has been computed using LCAO with planar basic functions. The six fluorine atoms are located at the corners of an octahedron. Due to the high symmetry of the molecules, only hexadecapole moment exists with a small contribution to the total potential, which contains a Lennard-Jones and a Coulomb term, [3]. The intermolecular potential is orientation dependent. This is the reason to present it as a sum over atom-atom potentials. Variety of structures originates from the orientation dependence of the intermolecular potential, [4]. Due to that anisotropy, the transition between the different structures involves both lattice reconstruction and orientation ordering, achieved by molecular rotation. At very low temperatures (below 20 K), we observe a strained monoclinic structure with ordered molecular axes, Figure 1. To observe stable structures, we quench the trajectories from the constant temperature MD simulations at different temperatures. Nose's thermostat is implemented to ensure a constant temperature of the cluster. Quenching is a means to associate arbitrary points in the phase space with the local minima in whose wells they lie on the potential energy surface, [7]. The quenched trajectory at the transition temperature (cluster-size dependent) reveals two coexisting structures, Figure 2. In fact, the coexistence is not observed at a

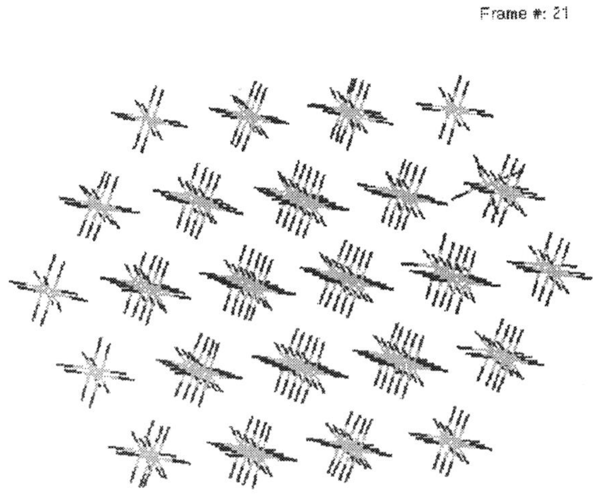

Figure 1. Stable structure 89 TeF_6 molecule at 20 K

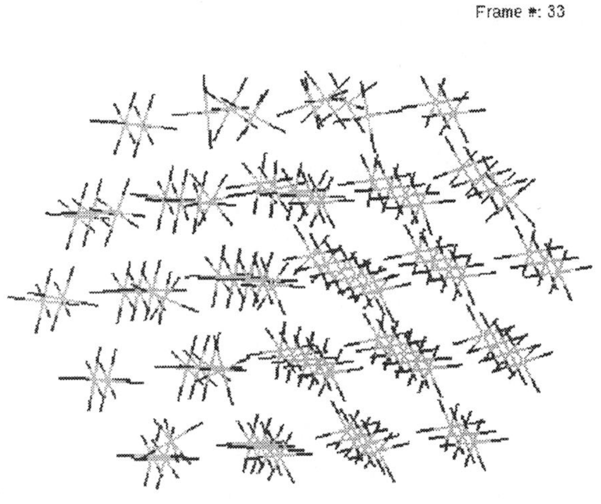

Figure 2. Coexisting structures at 89.4 K

unique temperature. It is observed in a temperature interval, which becomes narrower as the cluster grows larger.

Appearance of a finite temperature interval of the phase transformation is a phenomenon that distinguishes the finite systems from their bulk counterpart. Above the transition temperature the cluster forms a *bcc* orientation disordered structure.

We have investigated why the thermal coexistence of the two solid structures is more difficult to be observed in bigger clusters, than in smaller ones, [6]. The main reason is that small systems exist in metastable states, which are stabilized by kinetics not by the thermodynamics. Bulk systems do not reach these states because they occupy a small fraction of the phase space available to the system. We should wait infinitely long time until the bulk system equilibrates and visits these states. However, a wise Monte Carlo Method [8] can be implemented to investigate systems with rugged potential energy surface. This method is based on the jump-walking technique for spanning the configuration space, [9,10].

The Monte Carlo simulations give complementary information about the studied system by generating a Markov sequence of configurations. These are used to obtain the statistical (ensemble) averages of the measured quantities. In a molecular dynamic simulation we compute a long trajectory in the phase space over which we calculate the time averages of the same quantities. Under the conditions of the ergodic hypothesis these two should be the same,

$$\langle \mathrm{A} \rangle \{\mathrm{time}\} = \langle \mathrm{A} \rangle \{\mathrm{configurations}\}.$$

We have developed a parallel J-Walking code that enables us to carry out a Monte Carlo simulation efficiently in a multiprocessor-computing environment. We apply the code in the study of the thermodynamic properties of a 59-molecule cluster. The results have been compared to the results from the canonical molecular dynamic simulations.

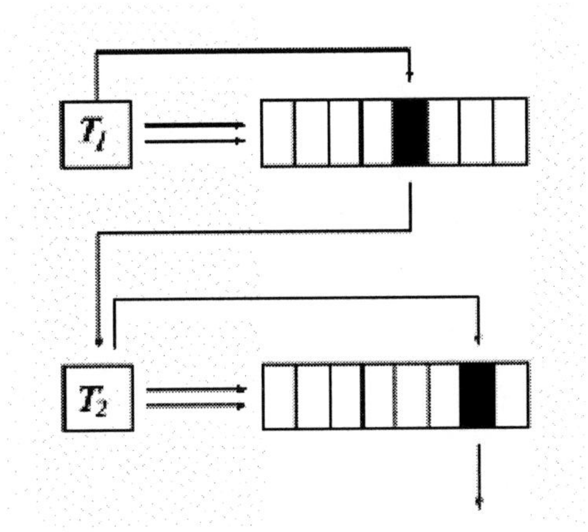

Figure 3. The diagram shows the J-W process between two temperatures

A schematic representation of the Jump-Walking (J-W) procedure is given in Figure 3. Each square represents a Metropolis Monte Carlo simulation at a particular temperature. The set of boxes in the right-hand side denotes the array of previous configurations of the system, which are stored in memory to avoid correlation between the lower and the higher temperature. Such a procedure is equivalent to the following physical situation: a cluster at a given temperature could contain grains, which do not let it transform into the most stable structure. In order to get rid of them, we heat the cluster for a short time, just to let the domains to melt their boundaries. Mathematically, we do a similar procedure by mixing different states into the same simulation. The essence of mixing is that we increase the probability of population of states, which are hardly accessible for the cluster before heating. Such kind of techniques are needed to explore rugged potential surfaces that usually trap the cluster in a specific well, normally corresponding to a glassy structure. In order to obtain reliable results we should perform a lot of sampling – millions or, even better, tens of millions. One way to speed up the calculations that take many hours of computational time and to increase the accuracy of the results, is to use parallel computers. Our computer code is for parallel machines (SUN3500 and CRAY-T3E). In Figure 4 we give an example of simultaneous simulations of 4 clusters. In the example the four systems at temperatures are distributed between 8 processors.

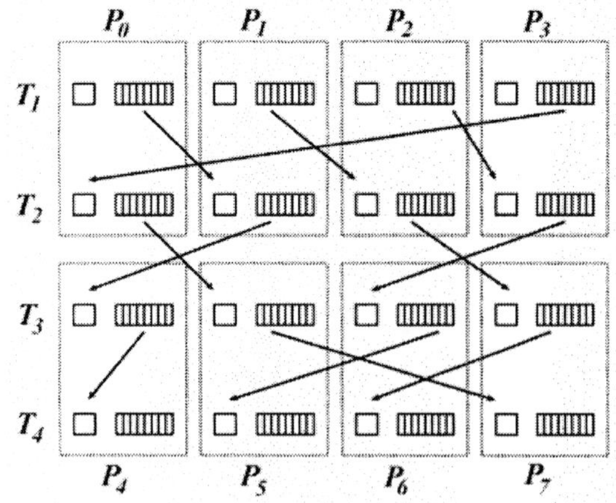

Figure 4. Diagram of a parallel decomposition of data and data-exchange between processors in J-Walking algorithm

In Figure 4 we show the parallel decomposition of computation. Each process computes part of one of 4-multi stage J-walking, and exchange configurations and energy with others. We have used the code to compute the caloric curve of four clusters, containing 59 molecules each; in order to see if the phase transition occurs in the same temperature interval as it has been retrieved from the molecular dynamics computations. We have found that the phase transition appears almost at the same mean temperature – 76 K for this cluster size. The temperature interval is about 4 K as it is seen in Figure 5. The transition is smooth, although we have proved in [11] that it is a first order-like one. This is the size-effect mentioned in the introductory part. The Lindeman criterion suggested in [12] to distinguish between liquid and solid phases can in principle be used for finding solid-solid transformations, too. This criterion is proportional to the ratio of the second to the first moment of the displacement distribution. The diffusion coefficient is another characteristics, which is computed from our data. It shows that the two solid structures are different, as well.

4. Conclusions

Both simulation methods indicate the same transition temperature interval for the cluster size studied ($\sim$ 2.4 nm) if long MD runs are carried out ($\sim$ 1 ns) and a million of configurations are used in MC. We conclude that the ergodic requirements can be satisfied in solid-solid transformations of nanoclusters under those special limitations.

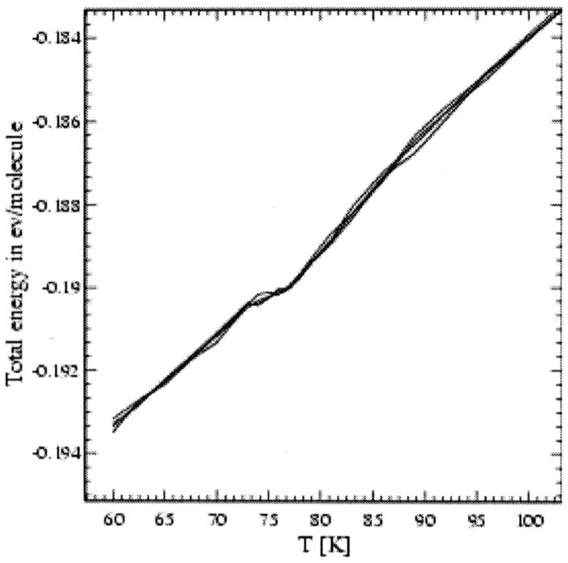

Figure 5 Caloric curves of 4 transforming clusters. In the plateau region about 76 K coexisting phases are observed in molecular dynamic simulations

Acknowledgement. The work has been supported by the NATO Collaborative Linkage Grant Award Number: SA (PST.CLG.976363) 5437.

References

1. Berry RS, Beck, Davis HL, Jellinek J. in Evolution of Size Effects in Chemical Dynamics, Part 2, (New-York, 1988); p.75
2. Bartell LS. J. Phys. Chem. (1992); 96:108
3. Proykova A, Berry RS. Z. Phys D. (1997); 40:215
4. Proykova A, Radev R, Li FY, Berry RS. J. Chem. Phys. (1999); 110:3887
5. Binder K. Monte Carlo Methods in statistical physics. (Springer - Berlin 1979)
6. Daykov I, Proykova A. Proceedings of the Second Meeting of Young Researchers in Physics, Heron Press (2001).
7. Nose S. Mol. Phys. (1984); 52:255
8. Allen MP, Tildesley DJ. Computer Simulation of Liquids Clarendon Press, Oxford, (1994).
9. Radev R, Parallel Monte Carlo simulation of critical behavior of finite systems, Report from the TRACS program, Edinburgh-UK (May-June 2000)
10. Frantz DD, Freeman DL, Doll JD. J. Chem. Phys. (1990); 93:2769
11. Frantz DD, Freeman DL, Doll JD. J. Chem. Phys. (1992); 97:5713
12. Lindeman FA. Phys. Z. (1910); 11:609

Nanoscience & Nanotechnology
eds. E. Balabanova, I. Dragieva, Heron Press, Sofia, 2001

MODELLING OF SILICA NANOPARTICLE PRODUCTION BY ARC PLASMA METHOD - NEED AND CHALLENGES

Emilia Balabanova

Institute of Electronics, Bulgarian Academy of Sciences Blvd. Tzarigradsko chaussee 72, Sofia 1784, Bulgaria

Abstract. The work reviewed the main results from our one dimensional modelling of silica nanoparticle production by arc plasma method. The challenges in the modelling are considered. They related to the presentation of the complex processes carried out in a plasma flow reactor (mixing of the flows with different temperature, chemical generation of particle precursors, particle growth etc.).

Keywords: nanoparticles, thermal plasma, modelling

1. Introduction

Silica nanoparticles are amorphous silica particles having submicron mean sizes (from a few to few-hundred nm). They are building elements of ultrafine silica powders well known under the commercial names – "aerosil", "cab-o-sil", etc. [1]. The ultrafine silica powders have wide application as fillers and tixotropic agents in pharmaceutical, cosmetic, elastic-gum, etc industries.

Silica nanoparticles are also the product, obtained in the optical waveguide preforms, by MCVD method [2].

The industrial methods of ultrafine silica powders production are mostly flame methods. By them easy vaporising liquids as $SiCl_4$, or gases as SiF_4 and SiH_4 are oxidised or hydrolysed to produce the precursors of silica nanoparticles – SiO_2 gas molecules. The precursor conversion to nanoparticles is carried out in an aerosol flame reactor [3]. The arc plasma method is an alternative to the flame methods allowing the use of refractory materials (e.g. silica sand) as initial substances. The SiO_2 vapour is converted to SiO_2 nanoparticles in an aerosol plasma reactor. The reactor conditions provide the main difference between the flame and plasma methods.

While the flame reactor is one-flow isothermal device, the plasma reactor is a complex one, where many phenomena occur simultaneously – mixing of flows with different temperature, chemical and condensation processes, aggregation etc. For investigations of the phenomena taking place in any type of reactor modelling is needed. It allows the finding of the relations between the reactor operating parameters and particle characteristics (size and structure) which are important for silica nanoparticle applications.

Despite the wide industrial production of silica nanoparticles, until now many problems concerning the mechanism of their flame generation still are not studied [4]. The situation concerning the plasma reactors is worse. The development of models presenting the complexity of the processes in a plasma flow reactor faces a lot of challenges.

2. Modelling the processes in our aerosol plasma reactor

In our laboratory we have some experience in the production of ultrafine silica powders from silica sand. The sand is vaporised in the plasma chamber by use of argon arc plasma at atmosphere pressure. The vapour enters the plasma flow reactor (see Figure 1) and mixes with oxygen. During that the vapour is cooled, reacts chemically and condenses, producing nanopaticles. All these processes are realised in a very short time (in order of some milliseconds). The problem is the adequate modelling of the phenomena carried out in the plasma flow reactor.

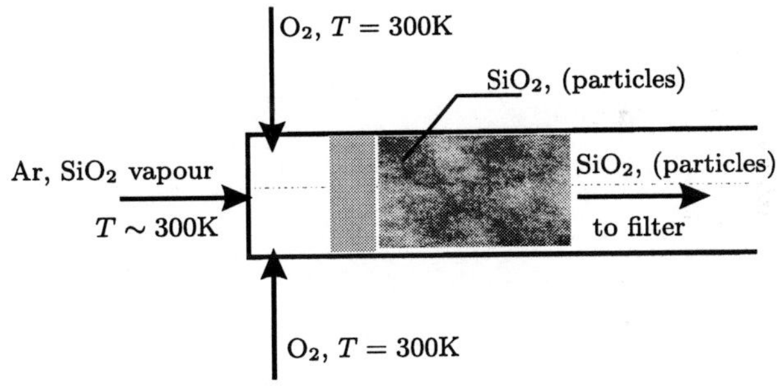

Figure 1. Schema of the plasma flow reactor

Challenges in the modelling:

1. Presentation of the mixing and the cooling processes
2. Presentation of the monomer synthesis
3. Presentation of the nanoparticle growth
3. Presentation of the nanoparticle aggregation

What we have made:

1. Modelling of the mixing and the cooling based on empirical equations [5]

2. Modelling of the nanoparticle formation and growth based on chemical and free-molecular -coagulation kinetic sets of equations [6-8]

3. Estimation of the aggregation ability of the plasma produced nanoparticles, accepting coalescence mechanism [9]

Our model concerning the simultaneous processes in the plasma flow reactor is one-dimensional. It accepts the previously obtained [10] empirical function for mean mass temperature variation in a tube plasma flow reactor. We assume [5] that the temperature variation in our reactor is correlated with the variation of the cooling gas (O_2) concentration (see Eqs. 1, 2). In some calculations [6-8] simpler approximation of the mixing and cooling processes is used, namely – instantaneous mixing ($g(t) = 0$) and constant cooling rate ($\alpha = 10^5$; 10^6 K/s). This rough approximation allows the influence of the cooling rate and the initial monomer concentration on the nanoparticle growth to be studied.

$$\alpha \equiv \frac{dT}{dt} = \frac{(T_E - T_0)\pi}{2t_c} \sin\frac{\pi t}{t_c} \qquad (1)$$

$$g(t) \equiv \frac{d[O_2]}{dt} = \frac{([O_2]_f - [O_2]_0)\pi}{2t_m} \sin\frac{\pi t}{t_m} \qquad (2)$$

Here, $T_0 = 3000\,\mathrm{K}$ and $T_E = 2000\,\mathrm{K}$ are the initial and the final mean mass temperatures of the flow, respectively. $[O_2]_0 = 0\,\mathrm{cm}^{-3}$ is the initial O_2 concentration; $[O_2]_f = 3 \times 10^{18}\,\mathrm{cm}^{-3}$ is the final O_2 concentration reached at the mixing; t_c is the cooling time; t_m is the mixing time.

The chemical reactions (3–5) are considered for the production the monomers – nanoparticle precursors (SiO_2 gas molecules). The consideration corresponds to the fact of silica vapour destruction to SiO and O_2 and the following reoxidation during the mixing with the excess of O_2 (cooling gas). The set of equations (6–9) is applied in the model for presentation of the monomer generation.

$$\mathrm{SiO} + \mathrm{O_2} \xrightarrow{k_1} \mathrm{SiO_2} + \mathrm{O} \qquad (3)$$

$$\mathrm{SiO} + \mathrm{O} + \mathrm{O_2} \xrightarrow{k_2} \mathrm{SiO_2} + \mathrm{O_2} \qquad (4)$$

$$\mathrm{O} + \mathrm{O} + \mathrm{O_2} \xrightarrow{k_3} \mathrm{O_2} + \mathrm{O_2} \qquad (5)$$

$$\frac{d[\mathrm{SiO}]}{dt} = -w_1 - w_2 - \lambda[\mathrm{SiO}] \qquad (6)$$

$$\frac{d[\mathrm{O}]}{dt} = w_1 - w_2 - 2w_3 - \lambda[\mathrm{O}] \qquad (7)$$

$$\frac{d[\mathrm{O_2}]}{dt} = -w_1 + w_3 - \lambda[\mathrm{O_2}] + g(t) \qquad (8)$$

$$\frac{d[\mathrm{SiO_2}]}{dt} = w_1 + w_2 - \lambda[\mathrm{SiO_2}] \qquad (9)$$

Here w_1, w_2, w_3 are the chemical reaction rates; λ is a coefficient that take into account the variation of the concentration with variation of the gas volume during cooling (the pressure is constant; 1 atm).

The set of equations (Eq.10) describes the nanoparticle growth. The free-molecular coagulation is particle growth mechanism since the high initial saturation of the silica vapour is observed in the experiments. (The Knudesen number in the system is higher than 10) We use discrete-sectional method for approximation of the free-molecular growth processes. This leads to the reduction of the computational time and memory.

$$\frac{dz_n}{dt} = \sum_{l<n-1} 2^{l-n} C_{l,n} z_n z_l + \frac{3}{4} C_{n-1,n-2} z_{n-1} z_{n-2} + \frac{1}{2} C_{n-1,n-1} z_{n-1}^2 - \sum_{l=n-1}^{M} C_{l,n} z_n z_l - \lambda z_n \qquad (10)$$

Here z_n, z_l are the number concentrations of particles in groups n and l; M is the number of groups; $C_{n,l}$ is the coagulation coefficient.

In some calculations [7] the influence of interparticle interactions has been studied by use of coefficients based on Hamaker potential for correction of the pure free-molecular coagulation coefficients ($C_{n,l}$).

3. Results and discussion

In the frame of a simple model [6,8] accepting constant cooling rate and instantaneous mixing it is obtained that the faster the cooling rate and/or the lower the initial monomer concentration, the smaller the mean size of nanoparticles. In Figure 2 the particle size distribution functions (PSDFs) are shown. PSDF has log-normal character. The mean particle size is determined at PSDF = 50%. The inclusion of interpaticle attraction (Hamaker potential) in coagulation coefficient leads to an increase of the mean size of about 16%.

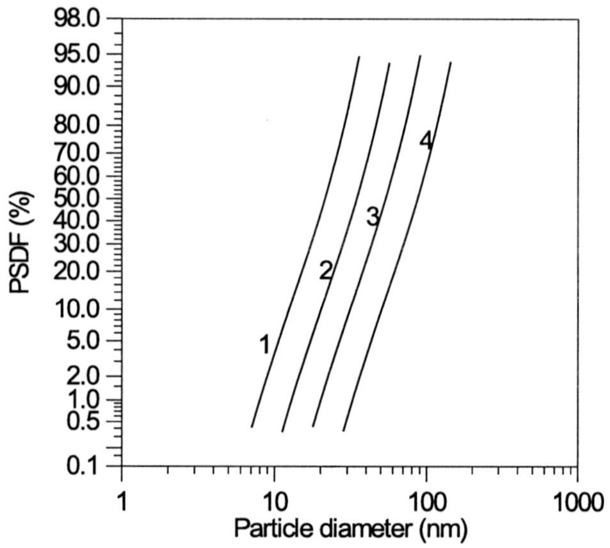

Figure 2. PSDF calculated for initial monomer concentrations $6 \times 10^{16}\,\mathrm{cm}^{-3}$ and $2 \times 10^{17}\,\mathrm{cm}^{-3}$ at cooling rate of 10^6 K/s (1,2), at cooling rate of 10^5 K/s (3, 4), respectively.

In the frame of a more detailed model [5] considering the introduction of the cooling gas and the corresponding temperature variation in the reactor (see Figures 3, 4) it is shown that the mixing process strongly affects nanoparticle formation and growth. A delay in the monomer generation, as well as a delay in the establishing of the self-preserving form of the PSDF is observed as a result of the mixing of the cooling gas (see Figures 5, 6).

Since the aggregation of nanoparticles is an important problem, we have considered the aggregation ability of silica nanoparticles produced in our plasma reactor and those produced by flame method [9]. It is found that plasma produced nanoparticles are less aggregated than the flame that. The reason is the higher temperature and the shorter particle residence time in the plasma flow reactor.

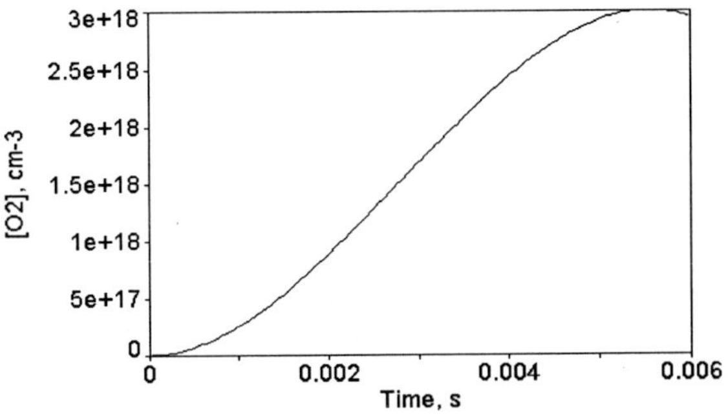

Figure 3. Variation of the cooling gas concentration

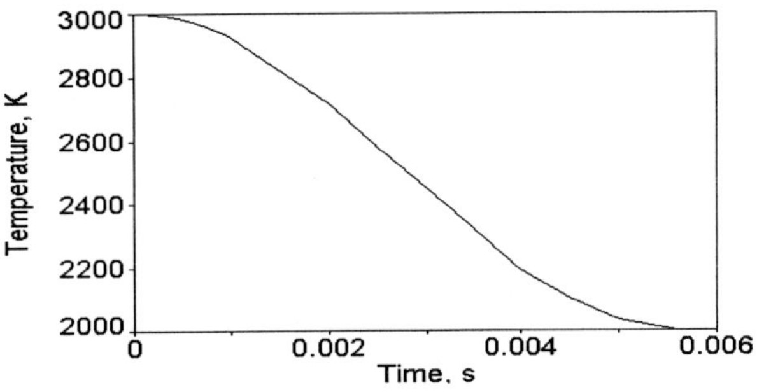

Figure 4. Variation of the mean mass temperature

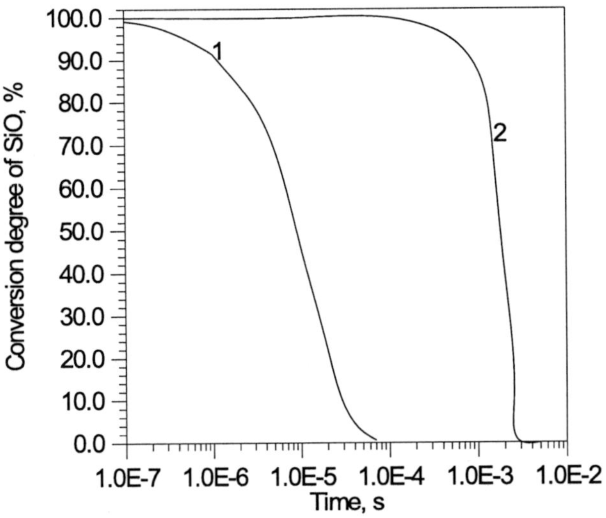

Figure 5. Conversion degree of SiO at instantaneous mixing (1), and at noninstanteous mixing (2)

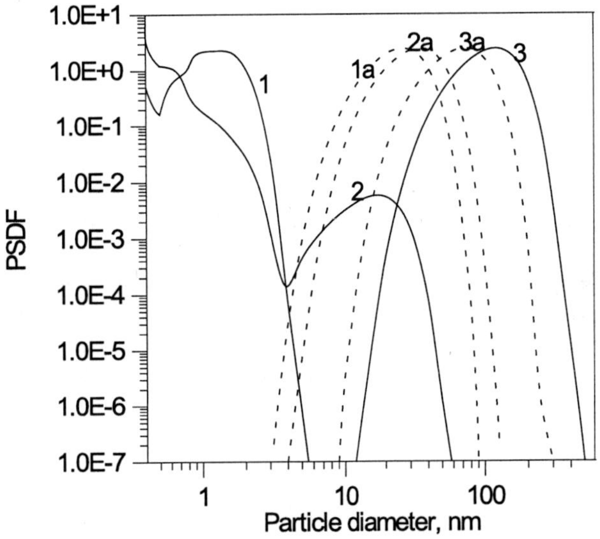

Figure 6. Evolution of PSDF during mixing at: $t = 0.5 \times 10^{-3}$ s (1), $t = 1.0 \times 10^{-3}$ s (2), $t = 5.6 \times 10^{-3}$ s (3); during instantaneous mixing: 1a, 2a, 3a, respectively

4. Conclusions

On the basis of the one-dimensional model for the processes of silica nanoparticle production by arc plasma method, the relations between process parameters (cooling rate, initial monomer concentrations, etc.) and particle characteristics (mean size, structure) are found. It is obtained that the mixing of the flows in the reactor plays an important role in the nanoparticle growth mechanism. Further, more precise description of the fluid dynamic processes in the reactor is needed. The incorporation of the fluid dynamic equations in the model of particle formation and growth is a serious challenge. Also special attention should be paid on the examination of the nanoparticle aggregation mechanism in relation to the conditions in the reactor. The developed more sophisticated models can be used for studying the synthesis of any other nanoparticles taking place in any other aerosol reactor.

References

1. Iler R., The chemistry of silica, Wiley-Intercience Publication, 1979
2. Kim Kyo-Seon, Pratsinis S.,1988 AIChE J.34, 912
3. Ulrich G., Riehl J., 1982, Colloid Interface Sci, 87, 257
4. Koch W. Fridlander S. K., 1990, J Coll. Interface Sci. 140, 419
5. Balabanova E., 1998 High Temp. High Press. 30, 493
6. Balabanova E., Levitsky A., Oliver D., 1994 Czech. J. Phys., 44, 139
7. Balabanova E., Oliver D., 1996 Czech. J. Phys. 46, 881
8. Balabanova E, Oliver D., 1997 High Temp. High Press. 29, 711
9. Balabanova E., 2000, Vacuum 58, 174
10. Polak L. S. (Ed.) Teoretitcheskaia i prikladnaia plasmochimia (in Russian,) Moskwa , Nauka, 1975, 205

Nanoscience & Nanotechnology
eds. E. Balabanova, I. Dragieva, Heron Press, Sofia, 2001

FAST-REACTION STAGE IN THE GROWTH OF SEMICONDUCTOR NANOPARTICLES

T. Petrova[1], C. Dushkin[1,2], S. Saita[2], K. Yoshie[2], Y. Yamaguchi[2,3]

[1]University of Sofia, Department of Inorganic Chemistry, Faculty of Chemistry, 1 James Boucher Blvd., 1126 Sofia, Bulgaria;
[2]Mitsubishi Chemical Corporation, Yokohama Research Center, Non-Equilibrium Laboratory, 1000 Kamoshida-cho, Aoba-ku, Yokohama 227-8502, Japan;
[3]University of Tokyo, Department of Chemical System Engineering, Faculty of Engineering, Hongo 7-3-1, Bunkyo-ku, Tokyo 113-8656, Japan

Abstract. The present paper deals with the kinetics of growth of two kinds of semiconductor nanoparticles, CdSe and InAs, as deduced from their optical properties such as absorption and photoluminescence. The data for the average nanocrystal radius, dispersion and growth rate are processed with two-stage model. This allows revealing initial fast-reaction stage where a nanoparticle is growing due to the adsorption of atoms from the vicinal suspension and their incorporation in the crystalline lattice. The process is looking as a pseudo-reaction rather different from the usual diffusion-controlled kinetics operating at long times. Since the reaction stage is observed with both II-VI and III-V compound semiconductors, we report it as important process for the quick and energy-safe manufacturing of various nanomaterials from liquid suspension.

Keywords: nanoparticles, nanocrystals, CdSe, InAs, growth kinetics

1. Introduction

A semiconductor nanoparticle consists of an inorganic core, for example CdSe, and of organic shell coming from the matrix of synthesis - normally TOPO (trioctylphosphine oxide). Due to the quantum confinement effect, the wavelengths of absorption and emission increase from green to red with increasing of the nanoparticle size. The peculiar optical properties make the semiconductor nanoparticles of interest in a variety of practical applications [1]. In fact, they can replace the conventional semiconductor materials whenever possible, even in the lasers at the latest [2]. To make them applicable in the industrial scale, however, one needs the synthesis of vast amount of nanoparticles, which opens the questions about the way of their growth and the possible control of this process.

The nanoparticle synthesis is carried out in a three-necked round bottom flask, containing molten TOPO connected with a cooler; the precursor is injected with a syringe [3–6]. The temperature is hold with a mantle heater at about 300°C and is measured with a thermocouple. A good mixing is ensured with a magnetic stirrer in inert atmosphere achieved by passing of an inert gas (argon).

It is found [5,6] that the growth kinetics shows an initial process of fast increase in the average particle size accompanied with narrowing of the size distribution. This process is theoretically attributed to a sort of reaction where the precursor adsorbs on the particle surface [6]. At long times the growth is dominated by the usual diffusion. In this paper, the proposed model of reaction-driven kinetics is confirmed with data for CdSe and InAs nanoparticles, the latter taken from the literature. The growth parameters of different materials are extracted by processing the dimensionless radius, dispersion and growth rate as a function of time.

2. Theoretical Model

The examination of experimental data for CdSe and InAs nanoparticles [5,6] reveals that initially the particle radius R is increasing rather quickly as a function of time compared to the rest of growth process. The time scale of this specific stage is [6]

$$\tau_r = \delta / k\Omega c_\infty .$$

Here δ is the thickness of the subsurface layer around a nanoparticle in the suspension, k is the rate constant of pseudo-reaction, Ω is the molar volume of the semiconductor compound and c_∞ is the precursor concentration. As seen from Figure 1, the biggest increase of the dimensionless radius, $\rho = R/\delta$, happens within one-two time intervals, $T = t/\tau_r$. It can be tens to hundreds of times smaller than the complete duration of synthesis process normally described in the literature. This duration is comparable with the characteristic time of diffusion-limited process τ_d listed also in Table 1.

The nature of two processes can be understood as follows: At the beginning, there are a lot of precursor atoms, metal and metalloid, in the vicinity of a nucleus comprising a crystalline lattice of few periods. The atoms are bound to the respective organic carriers (TOPO, other alkane phosphines or another surfactant) in the form of metalorganic compounds. In this state they have a strong affinity to adsorb onto the nucleus surface in a way that the organic tail will stay outside thus forming the capping layer, whereas the metal and metalloid will be incorporated in the crystalline lattice of the core. As the reaction stage we consider the sequence of sub-stages such as crossing of the surface barrier (adsorption on the nanoparticle surface), reorientation of the metalorganic complex and bonding of atoms to their lattice neighbors. The driving force of its progressing is the supersaturation maintained in the bulk medium of

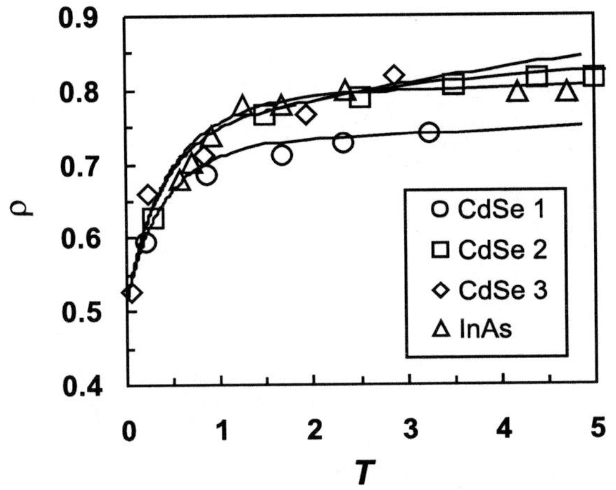

Figure 1. Dependence of the dimensionless radius of nanoparticles, ρ, on the dimensionless time, T. Our data are denoted with CdSe 1 and 2 [6]; the literature data are CdSe 3 and InAs [5]. The solid lines are theoretical fits with the scales seen in Table 1.

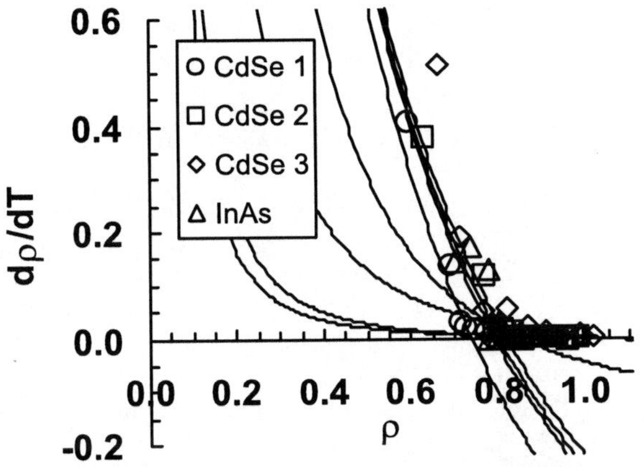

Figure 2. The growth rate $d\rho/dT$ versus the nanoparticle radius ρ for the same data as in Figure 1. The thick lines are drawn by Eq. (1) for the fast reaction mode. For comparison are given also the predictions for diffusion driven growth (thin lines).

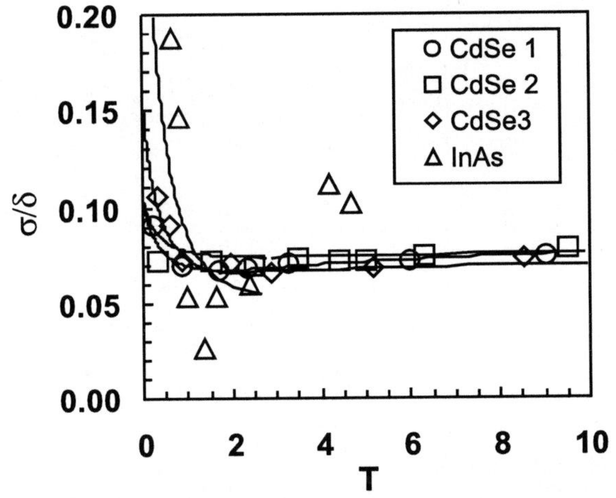

Figure 3. The scaled dispersion of nanoparticles σ/δ as a function of time T for the experiments depicted in Figures 1 and 2. The lines are theoretical fits of the data.

thickness δ surrounding the nucleus. In the course of nanocrystal growth, however, this domain is depleted of precursor, which leads to decrease in the growth rate dR/dt plotted in dimensionless form in Figure 2. Zero rate will correspond to drop of the subsurface concentration to less than the threshold for reaction-driven growth, i.e., below the value in equilibrium with a nanocrystal of core radius ρ_r – cf. Eq. 1.

Fortunately, this does not happen, because the created concentration gradient brings new material from the bulk of the suspension to the subsurface layer. It serves as the basis of the diffusion-limited growth of nanocrystals, which will stop at $\rho = \rho_d$.

Here we put emphasis on the reaction-driven kinetics, which is described by the following equations:

$$\frac{d\rho}{dT} = \frac{1}{\rho} - \frac{1}{\rho_r} \tag{1}$$

for the growth rate,

$$T(\rho) = \rho_r\left(\rho_0 - \rho + \ln\frac{\rho_r - \rho_0}{\rho_r - \rho}\right) \tag{2}$$

for the radius, and

$$\sigma(\rho) = \sigma_r + (\sigma_0 - \sigma_r)\frac{\rho_0(\rho_r - \rho)}{\rho(\rho_r - \rho_0)} \tag{3}$$

for the dispersion [6]. Equations (1) and (2) are valid for the radius of any nanoparticle from the ensemble, although they are applied in Figures 1 and 2 to calculate the mean particle radius and growth rate. Gaussian size distribution of the nanoparticles is assumed in deriving of Eq. (2); therefore it applies only for the mean radius.

Table 1. Growth parameters of different semiconductor nanomaterials

Constants	CdSe 1	CdSe 2	CdSe 3	InAs
τ_r (s)	20	10	250	2000
τ_d (s)	2000	1500	6000	16000
δ (nm)	1.9	1.9	1.9	1.9
R_0 (nm)	0.92	0.92	0.92	0.92
R_r (nm)	1.4	1.52	1.52	1.5
R_d (nm)	1.77	1.9	1.85	1.6
σ_0 (nm)	0.2	0.2	0.3	0.7
σ_r (nm)	0.13	0.16	0.12	0.1
σ_d (nm)	0.14	0.17	0.16	0.14

The calculations use also the counterparts of Eqs. (1)–(3) derived for the diffusion growth; they can be found in Ref. [6] together with the matching procedure. For example, the thin lines in Figure 2 and also the parts of theoretical curves at $T \geq 2$ in Figures 1 and 3 are plotted using the diffusion expressions.

3. Results and Discussion

The comparison between our data for CdSe and those from Ref. [5] shows a similarity in the growth conditions. It is not surprising bearing in mind that we are using the same materials as precursor and matrix. This reflects the system-oriented parameters such as the initial radius of nuclei R_0 and dispersion σ_0, which should be determined by the size of reverse micelles pre-existing or forming in the course of injection. Also, this is the thickness of subsurface layer δ, of the order of the length of a TOPO molecule. The system type introduces also constrain on the average size and dispersion which could not vary so much in the course of synthesis. For this reason the respective constants $R_{r,d}$ and $\sigma_{r,d}$ are of very close values.

However, there are also process-dependent parameters whose values can vary in fairly large limits. These are the time constants τ_r and τ_d. Looking on the expression for τ_r, one can reveal the factors affecting the fast-reaction process: the temperature (via the rate constant k) and the precursor concentration c_∞. Increasing the temperature should generally increase k and decrease, therefore, τ_r. The increase of metal atoms in the precursor will obviously make the reaction faster (decreasing τ_r). From both compounds present, Cd and Se, the cadmium atoms are known to have strong precipitation effect, for which reason they can be the rate determining reagent put in excess. Subsequently, the much slower growth of CdSe 3 can be due to lower content of Cd (smaller c_∞) and/or to lower temperature of synthesis.

Comparing the growth of various kind of semiconductor nanoparticles, CdSe and InAs, shows similar trends. The system-dependent parameters for InAs are close to that for CdSe except the much larger initial dispersion. However, the characteristic times are much longer. Assuming nearly the same other conditions, such as temperature and precursor concentration, the tremendously large τ_r can be explained with much smaller reactivity of In and As.

4. Conclusions

The growth kinetics shows initial process of fast increase of the average particle size accompanied with narrowing of the size distribution, irrespective to the semiconductor material ($A^{II}B^{VI}$ or $A^{III}B^{V}$). This process is attributed to a sort of reaction where the precursor adsorbs on the particle surface. Due to lower reactivity, the reaction is going much slower for $A^{III}B^{V}$ compounds.

At long times, the growth is dominated by the usual diffusion leading to vanishing increase in the size at appreciable broadening of the size distribution within substantial time interval.

Practically important is the fast reaction driven growth where the nanoparticles can reach the desirable size within a minute depending on the conditions. This saves energy and time for the synthesis of large amount of uniform nanoparticles in a continuous regime.

The growth parameters for controlling the process are calculated using two-stage theoretical model. The model predictions are confirmed with data for both CdSe and InAs nanoparticles and used to design the reaction pathway for nanocrystal synthesis in the industrial scale.

References

1. S. Maenosono, C. D. Dushkin, S. Saita, Y. Yamaguchi, Jpn. J. Appl. Phys. Part 1, 39, 4006 (2000).
2. V. I. Klimov, A. A. Mikhailovsky, D. W. McBranch, C. A. Leatherdale, M. G. Bawendi, Science 287, 1011 (2000).
3. C. B. Murray, D. J. Norris, M. G. Bawendi, J. Am. Chem. Soc. 115, 8706 (1993).
4. J. E. B. Katari, V. L. Colvin, A. P. Alivisatos, J. Phys. Chem. 98, 4109 (1994).
5. X. Peng, J. Wickham, A. P. Alivisatos, J. Am. Chem. Soc. 120, 5343 (1998).
6. C. D. Dushkin, S. Saita, K. Yoshie, Y. Yamaguchi, Adv. Colloid Interface Sci. 88, 37 (2000).

Nanoscience & Nanotechnology
eds. E. Balabanova, I. Dragieva, Heron Press, Sofia, 2001

MODES OF NUCLEATION OF NANOPARTICLES PREPARED BY BOROHYDRIDE REDUCTION (IN THE CASE OF Co-B-H-N PARTICLES)

P. Kashchiev[1], B. Rashkova[1], S. Stoeva[2], P. Stoimenov[2] and I. Dragieva[1]

[1]Bulgarian Academy of Sciences, Central Laboratory of Electrochemical Power Sources, 1113 Sofia, Bulgaria;
[2]Kansas State Univ., Depart. Chemistry, KS 66506, USA (since September 1999);

Abstract. The thermodynamic criteria and the validity of 2D and 3D modes of nucleation process for the case of nanoparticles synthesis are discussed, as well as the most general and universal solution - nucleation theorem of Kashchiev. For the cases of heavy and light atoms participation into the nucleus, the dependence of its critical size on the density and volume are explained. The electron microscope investigations by TEM, by high resolution and selective area electron diffraction of samples with Co-B-H and Co-N-B-H compositions, prepared by borohydride reduction with $NaBH_4$ using water solutions of cobalt salts, or of cobalt complexes with ethylenediamine at prior constructed bonds between light and heavy atoms for the second case, are shown.

Keywords: modes of nucleation, nanoparticles, borohydride reduction

1. Introduction

The thermodynamic criteria and the validity of 2D and 3D modes of nucleation process for the case of nanoparticles' synthesis are discussed, as well as the nucleation theorem of Kashchiev [1]. The dependence of the critical size on the density and volume of the nucleus are explained for the cases of heavy and light atom participation by means of the above theorem. The nucleation work W^* and nucleus size n^* are different functions of $\Delta\mu$ (supersaturation):

Classical theory [2,3]:
$W^* = (1/2)n^*\Delta\mu$, 3D Modes – Volmer – Weber
$W^* = n^*\Delta\mu$, 2D Modes – Frank – van der Merwe

Nucleation theorem of Kashchhiev [1]:
$dW^*/d\Delta\mu = -n^*$, Universal relationship between the nucleation work and the nucleus size.

2. Experimental

2.1. Synthesis of the precursors (*trans*- and *cis*-$[Co(en)_2Cl_2]Cl$). The above mentioned compounds are geometrical isomers. The presence of chelate ethylenediamine rings in them provides an enhanced stability in these precursors and strongly reduces the rate of the reduction process. *Trans*-$[Co(en)_2Cl_2]Cl$ was prepared using $CoCl_2.6H_2O$ and fuming ethylenediamine. The green crystals of *trans*-$[Co(en)_2Cl_2]Cl$ were filtered, washed with alcohol and dried at 110°C. The violet *cis*- form was prepared by heating aqueous green solution of the *trans*- complex on a steam bath [4].

2.2. Synthesis of CoNBH particles. The nanocrystalline and amorphous samples were prepared by means of a reduction of aqueous solution of *trans*- or *cis*-$[Co(en)_2Cl_2]Cl$ (4.86 g) with $NaBH_4$ (2.84 g) using the "ideal mixing method" by application of intensive mechanical stirring of the reacting solutions (200 ml) (Table 1). The particles were filtered, washed with acetone, and dried in a vacuum furnace at an ambient temperature [4].

Table 1. Experimental conditions for synthesis and composition of the prepared nanoparticles

Type of the precursor	pH range of the initial solution of $Co(en)_2$ complex	Number of samples	Analysis of the elemental composition (atomic %) B	H	C	N	Fig. #	Ref.
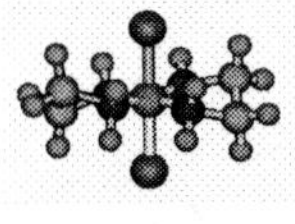	1.7–5.6	1374	—	—	—	—	1	[7]
		1395	9.24	48.42	5.18	2.90	3	—
		1513	9.41	43.84	3.59	1.58	2, 3, 4	—
Trans-$[Co(en)_2Cl_2]Cl$	5.6	1514*	7.07	53.17	4.28	1.69	—	—
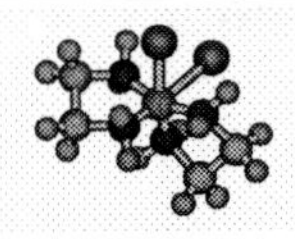	3.0–5.4	1386	—	—	—	—	5	[4]
		1515	9.48	42.00	4.33	2.35	6, 7, 8	—
Cis-$[Co(en)_2Cl_2]Cl$								

*during the application of a d.c. magnetic field (~ 700 Oe)

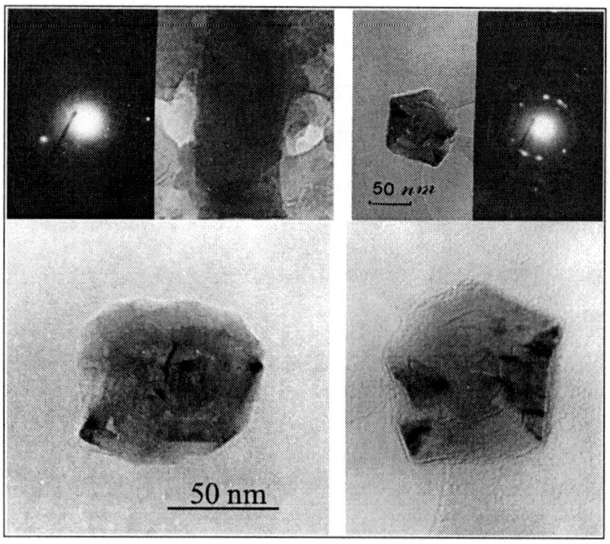

Figure 1. TEM and SAED of crystalline CoNBH nanoparticles [7]

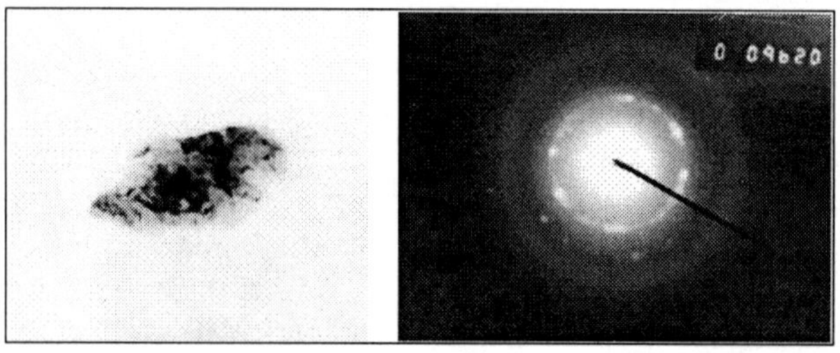

Figure 2. TEM and SAED of crystalline CoNBH nanoparticles

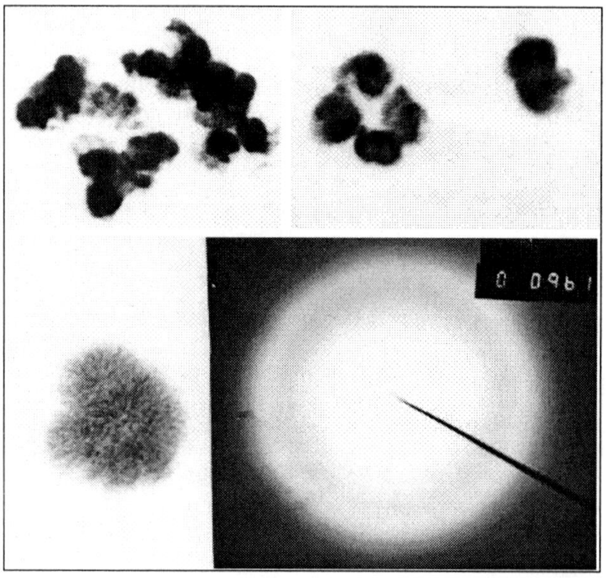

Figure 3. TEM and SAED of amorphous CoNBH nanoparticles

3. Investigations

The electron microscope investigations by TEM, by high resolution and selective area electron diffraction (JEM-200CX, Jeol, Japan) of samples with Co-B-H [5, 6] and Co-N-B-H [7] compositions, prepared by reduction with $NaBH_4$, using water solutions of cobalt salts, or of cobalt

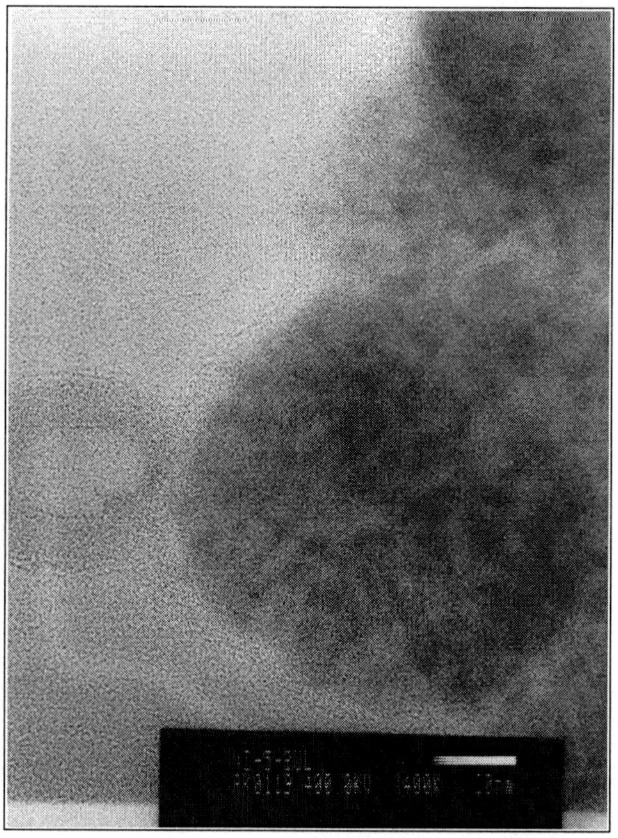

Figure 4. High resolution of CoNBH nanoparticles

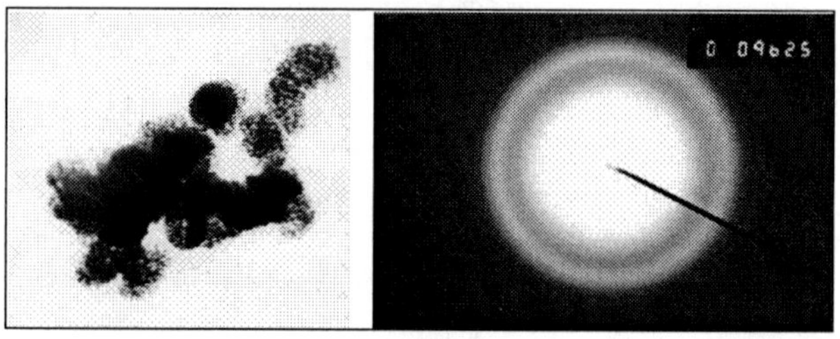

Figure 5. TEM and SAED of CoNBH nanoparticles

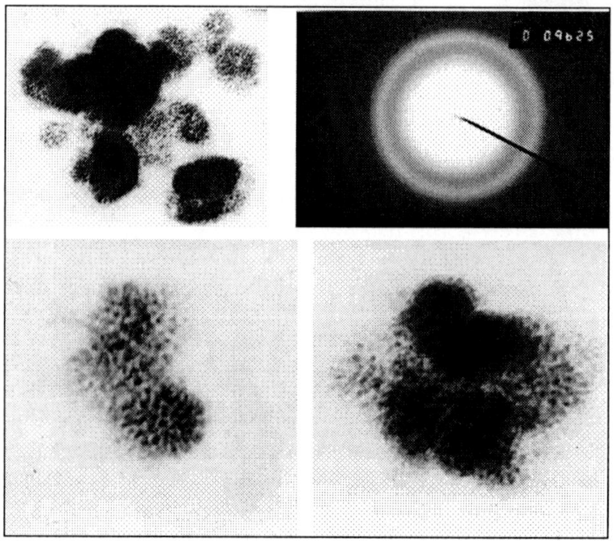

Figure 6. TEM and SAED of CoNBH nanoparticles

ethylenediamine complexes at prior constructed bonds between light and heavy atoms for the second case, are shown. The 2D and 3D modes of nucleation and growth of crystalline or amorphous nanoparticles respectively are assumed for the cases of kinetically controlled or diffusionally limited processes of synthesised nanoparticles using the chemical method of reduction with $NaBH_4$.

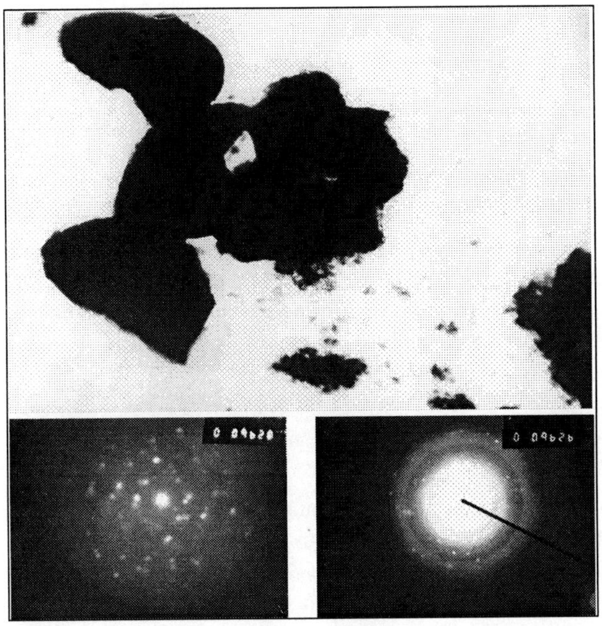

Figure 7. TEM and SAED of CoNBH (see [8])

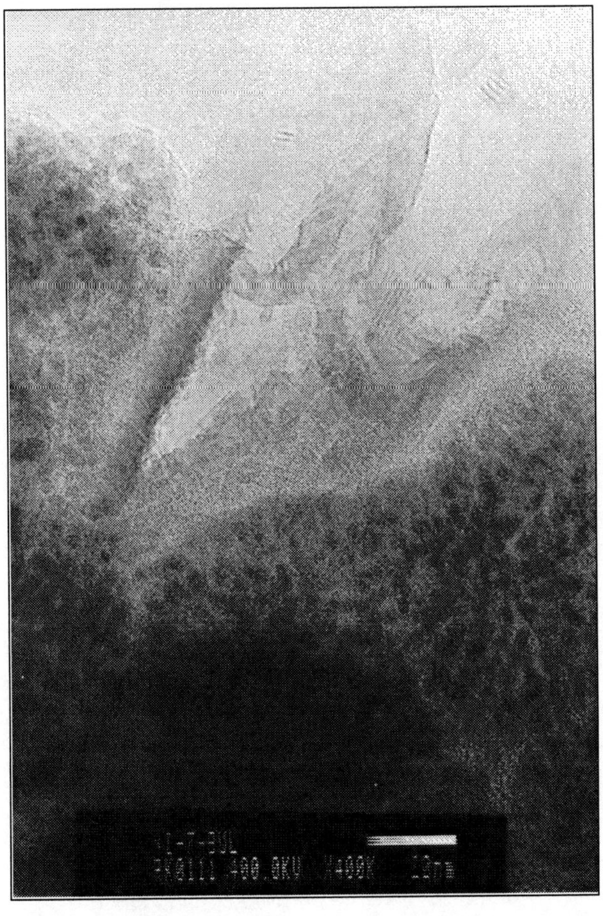

Figure 8. High resolution view of CoNBH nanoparticles

4. Thermodynamic Criterion for Modes of Growth

4.1. Frank-van der Merwe growth, when $d\mu/dn > 0$.

$\varphi_{Me-B,H} > \varphi_{Me-Me}$ – the more densely packed atoms, the greater the tendency toward 2D layerlike growth. Higher nucleation and growth rates favour layerlike growth. Result – 2D growth, crystalline particles (see Figures 1, 2 and 7) [2]. All photos from TEM have the same magnification as shown in Figure 1.

4.2. Volmer-Weber growth, when $d\mu/dn < 0$.

$\varphi_{Me-B,H} < \varphi_{Me-Me}$ – when interatomic bonding $\varphi_{Me-B,H}$ is weaker than the Me-Me atomic bonding φ_{Me-Me}: result is 3D-island growth, spherical amorphous particles (see Figures 3, 4, 5, 6 and 8) [2].

4. Conclusions

The modes of nucleation influence on the structure, size and properties of produced by borohydride reduction CoNBH nanoparticles and their chemical composition. The increase of the hydrogen content follows the decrease of the boron – dependence well known for the techniques of ideal mixing solutions [7].

References

1. D. Kashchiev, Nucleation. Basic Theory with Applications, Oxford: Butterworth-Heinemann, 2000
2. I. Markov, Crystal Growth for Beginners. Fundamentals of Nucleation, Crystal Growth and Epittaxy, Singapore; New Jersey; London; Hong Kong: World Scientific 1995
3. E. Budevski, G. Staikov, W. J. Lorenz, Electrochemical Phase Formaton and Growth. An Introduction to the Initial Stages of Metal Deposition, Weinheim; New York; Basel; Cambridge; Tokyo: VCH, 1996
4. V. Krastev, P. Stoimenov, S. Stoeva, I. Dragieva, K. Klabunde, XPS Investigation of Oxidation State of CoNBH Nanoparticles, Nano-crystalline and Thin Films Magnetic Oxides, Kluwer Academic Publishers, Proceedings of NATO Serial ARW,(1999) 301-308.
5. V. Krastev, M. Stoycheva, E. Lefterova, I. Dragieva, Z. Stoynov, XPS, TEM and SAD investigations of nanosized CoxByHz particles obtained by two different borohydride methods Journal of Alloys and Compounds, 240, 1-2 (1996) 186.
6. I. Dragieva, Z. Stoynov and K. Klabunde, Boron-Containing Nanoparticles and Their Contribution to the Theory and Production of New Materials, Nanostructured Materials, 12 (1999) 613.
7. I. Dragieva, S. Stoeva, P. Stoimenov, K. Klabunde, Complex Formation in Solutions for Chemical Synthesis of Nanoscaled Particles Prepared by Borohydride Reduction Process, Nanostructured Materials, 12 (1999) 267.
8. I. Dragieva, K. Klabunde, Z. Stoynov, P. Stoimenov, S. Stoeva, G. Ivanova and S. Bliznakov, Nucleation, Growth and Properties of Nanoparticles Produced by Borohydride Reduction, International Symposium on Cluster and Nanostructured Interfaces, October 25–28, 1999, Richmond, Virginia, USA, M-11 in Cluster and Nanostructure Interfaces, Eds. P. Jena, S. N. Khanna, B. K. Rao, World Scientific Publ. Co., Singapore, New Jersey, London, HongKong, 2000, 55 - 59.

Nanoscience & Nanotechnology
eds. E. Balabanova, I. Dragieva, Heron Press, Sofia, 2001

ELECTRON MICROSCOPIC STUDY OF LEAD-BORATE COMPOSITES CONTAINING $PbMoO_4$ NANOCRYSTALS

V. D. Ivanova[1], E. P. Kashchieva[1], B. T. Jivov[2], Y. B. Dimitriev[2]

[1]Department of Physics, University of Chemical Technology and Metallurgy, 8 Kl.Ohridsky Blvd, 1756-Sofia, Bulgaria
[2]Department of Silicate Technology, University of Chemical Technology and Metallurgy, 8 Kl.Ohridsky Blvd, 1756-Sofia, Bulgaria

Abstract. Composite materials based on $2PbO.B_2O_3$ (g-PbB) glass and crystalline $PbMoO_4$ (c-PbMo) are prepared by incorporation method combined with melting of mixtures from both components at different heat treatment times and cooling rates of the melts. The density of the samples is measured. The phase formation is studied by X-ray diffraction analysis. Using TEM structural changes are investigated in dependence on the composition, cooling rate and heat treatment time. It is established that the applied method reduces the gas inclusions and formation of micropores in the final materials. Fast cooled glassy samples with 5 and 10 wt% c-PbMo possess a background structure about 100nm. Aggregation trend is found in the samples containing over 20 wt% c-PbMo and crystallization, mixture structure of immiscibility droplets and microcrystals and extracted c-PbMo are observed.

Keywords: nanocomposite, incorporation method, lead-borate glass, $PbMoO_4$ crystal

1. Introduction

Recently the properties and methods for fabrication of various crystalline or amorphous nanostructured materials and nanodevices strongly attract the attention of scientists and technologists and the nano science takes its merited place in the modern word. The investigations concern generally an appropriate selection of precursors and combinations of non-traditional synthesis methods for nanopowders, nanoparticulate dispersions and coatings, bulk nanocomposite materials, nanophase alloys, ultrathin wires and others [1–5]. All studies are obligatory combined with information on the structural evolution in the materials allowing to control and to guide the choice of the composition, methods of preparation and specific behavior.

Borate and lead-borate nanostructured composites were subject of our previous works [6,7]. The incorporation techniques [8–10] was applied and nano- and micro-scale sized $PbMoO_4$ crystalline phase was included in pure borate or lead-borate glassy matrix. The crystalline phase was chosen as a material with promising acousto-optics applications [11]. It was established that the glasses used, possessing low melting temperature and reduced reaction ability, are suitable to form kinetically stable amorphous matrix with slightly pronounced tendency towards crystallisation at heat treatment. Correlation was found between composition, mode of preparation, crystal sizes, density and specific electrical conductivity of the nanocomposites.

The present study continues these investigations with modifications of the compositions and mode of preparation. The aim is to follow their influence on the nano- and microstructure of the composite materials using transmission electron microscopy (TEM).

2. Experimental Procedure

Composite materials based on previously obtained lead-borate glass with composition $2PbO.B_2O_3$ (g-PbB) and crystalline $PbMoO_4$ (c-$PbMoO_4$) are prepared by incorporation method [8–10]. The choice of the matrix composition ($2PbO.B_2O_3$) is made according to the phase diagram of the system PbO-B_2O_3 [12]. The precursors used for g-PbB and c-$PbMoO_4$ are chemically pure reagents: B_2O_3, Pb_3O_4, PbO and MoO_3. The glass was melted in platinum crucibles at 900°C for 10 min and c-$PbMoO_4$ – in porcelain crucibles at 800°C for 10 min. The c-$PbMoO_4$ content varies from 5 to 40 wt% in relation to the matrix. The scheme of sample preparation includes melting of the mixtures from both components (g-PbB and c-$PbMoO_4$) at 850°C for 5 min, 10 min, 20 min, and 2 hours. Two different cooling rates of the melts are applied: fast cooling (f.c.) by pouring the melts onto a copper plate and slow cooling (s.c.) by cooling the melts in the crucibles. The density (ρ) is measured by a pycnometer method using diiodomethane as a medium.

The phase formation was followed by X-ray diffraction analysis (XRD) using DRON UM-1 X-ray diffractometer and Cu-K_α radiation. The structural changes are investigated in dependence on the composition, cooling rate and heat treatment time. The observations were performed by TEM with EM-400 Philips electron microscope using single stage C+Pt replica technique. In order to distinguish nano- and micro-scale crystals selected area electron diffraction (SAED) was applied on small particles of the samples.

3. Results and Discussion

Some characteristics of the obtained samples are presented in Table 1. The formation of c-$PbMoO_4$ is confirmed by X-ray diffraction analysis. The c-PbMo content and different cooling rates lead to their visual changes from transparent yellow glasses (f.c.) to opaque brown materials (s.c.).

Table 1. Characteristics of the samples

No	Composition (wt%)		Heat treatment time	Density, ρ, (g/cm^3)
1.	95% g-PbB,	5% c-PbMo	2 h	5,92
2.	90% g-PbB,	10% c-PbMo	2 h	5,75
3.	80% g-PbB,	20% c-PbMo	2 h	5,94
4.	70% g-PbB,	30% c-PbMo	2 h	6,31
5.	80% g-PbB,	20% c-PbMo	5 min	5,94
6.	80% g-PbB,	20% c-PbMo	10 min	6,09
7.	80% g-PbB,	20% c-PbMo	20 min	6,10

The density of the starting materials are as follows: 1.83 g/cm^3 – for vitreous B_2O_3, 6.61 g/cm^3 – for g-PbB [13] and 6.92 g/cm^3 – for c-PbMo. The sample densities are not strongly influenced by the increasing of c-PbMo content from 5 to 20 wt.% and vary from 5.75 g/cm^3 to 5.94 g/cm^3.

For samples containing 30 wt.% c-PbMo the density increase is higher and $\rho = 6.31\ g/cm^3$. Comparing with the values for the same composites obtained by pressing and sintering [6, 7], (ρ is from 4.85 g/cm^3 to 5.93 g/cm^3) conclusion can be drown, that when the present melting method is applied the gas inclusions and formation of micropores in the final materials are reduced.

The XRD analysis show that all samples containing 5, 10 and 20 wt% c-PbMo are amorphous, while at 30 and 40 wt% c-PbMo well pronounced diffraction patterns of c-PbMo are identified. By SAED nanocrystals ranging in size from 50 to 100 nm are found in compositions containing over than 10 wt% c-PbMo. The results of TEM analysis show that the fast cooled glassy samples with 5 and 10 wt% c-PbMo possess a background structure with size about 100 nm (Figure 1). This value is much higher than that one in pure g-PbB [14] and the presence of c-PbMo provokes some structural changes. Aggregation trend is stimulated by increasing of the c-PbMo content. For samples containing 20 wt% c-PbMo the slow cooling leads to development of crystallization process (Figure 2a), while mixture structure of immiscibility droplets and microcrystals are typical at fast cooling of the melt, when also an extraction of the separate crystals on the replicas begins (Figure 2b). This extraction tendency increases at 40 wt% c-PbMo and it is clearer at fast cooling (Figure 3). The crystals are ranging in size from 100 nm to 300 nm. A comparison with our previous investigations where sintering and powdered monocrystals c-PbMo as precursor were used [6, 7], the density and crystal size increase, but a large variety of nano- and microstructural formations is observed. These structures can be appropriate basis for further development of the incorporation method for obtaining of composite based on g-PbB and c-PbMo.

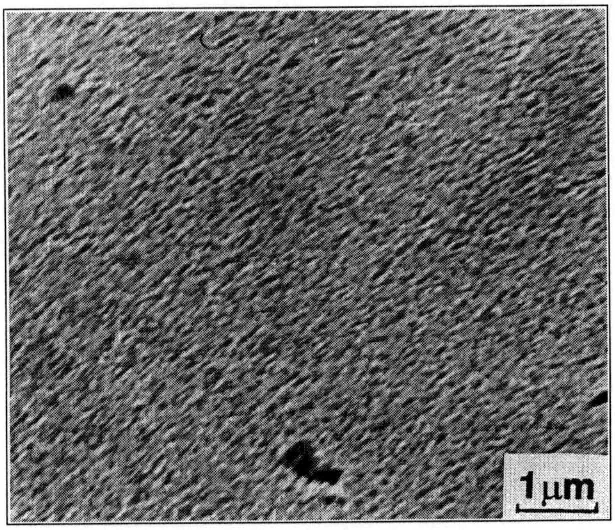

Figure 1. TEM micrographs of 90% g-PbB, 10% c-PbMo, f.c. sample head treated 2 h at 850°C

4. Conclusions

Composite materials based on previously obtained lead-borate glass $2PbO.B_2O_3$ and crystalline $PbMoO_4$ are prepared by incorporation method. The application of the method leads to reduction of gas inclusions and formation of micropores in the final materials. Some fast cooled glassy samples possess a background structure with size about 100 nm showing some rearrangements of the structure of pure g-PbB.

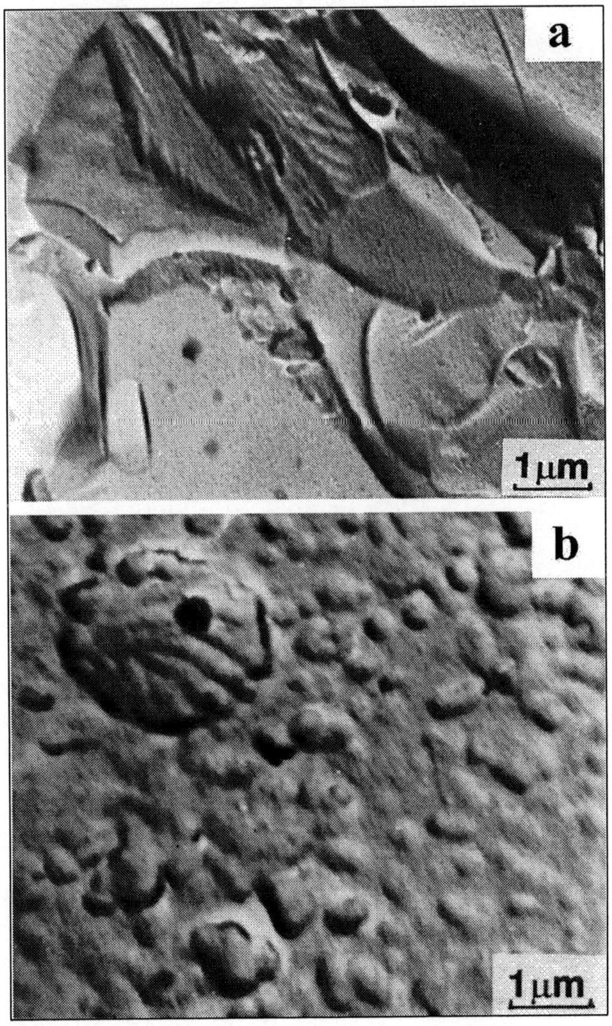

Figure 2. TEM micrographs of 80% g-PbB, 20% c-PbMo, head treated 5 min at 850°C: a) s.c. sample; b) f.c. sample

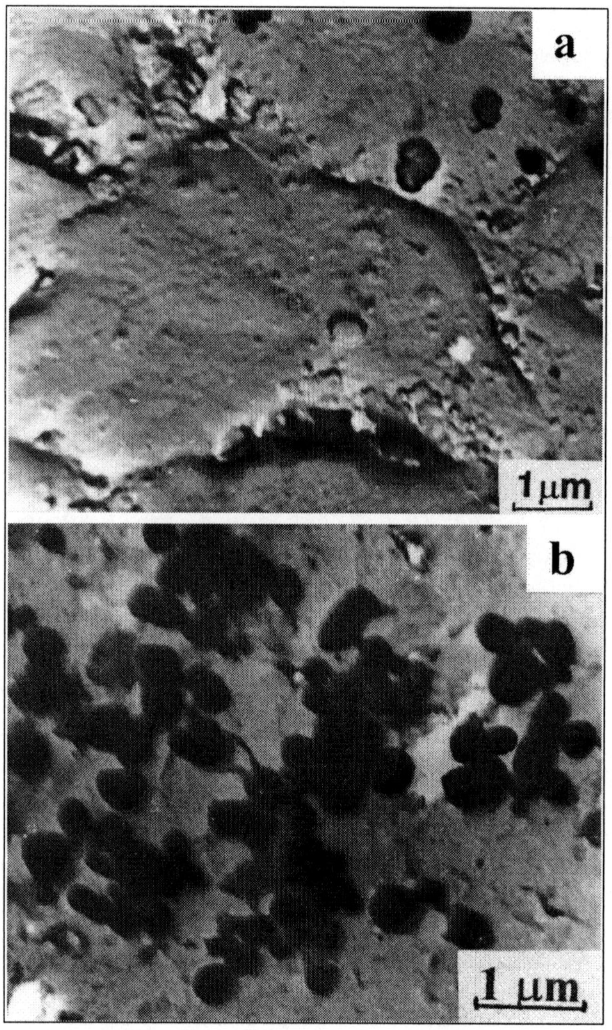

Figure 3. TEM micrographs of 60% g-PbB, 40% c-PbMo: a) s.c. sample; b) f.c. ample.

The observed crystallization, mixture structures of immiscibility droplets and microcrystals and separate extracted crystals can be appropriate basis for further development of the incorporation method for obtaining of composite based on g-PbB and c-PbMo.

Acknowledgements. This work is financially supported by the Bulgarian National Foundation of Science, under Grant X-376 / 2001.

References

1. S. Davis, K. Klabunde, J. Amer. Chem. Soc. 82 (1982), 153.
2. U. Koster, Mater. Sci. Forum 235 (1997), 377.
3. H. Gleiter, Nanostructured Mater., 1 (1992)1.
4. A. J. Burggraaf, A. J. A.Winnubst, H. Verweij, Third Euro-Ceramics, 3, 1993, Eds. P. Duran, J. F. Fernandez. Faenza Editrice Iberica S.L., Spain, p.561.
5. S. Kanzaki et al. , Bull. Ceram. Soc. Japan, 29 (1994) 124.
6. E. Kashchieva, V. Ivanova, B. Jivov, P. Petkov, Y. Dimitriev, Proc. XIII-th Conf. on Glass and Ceramics, Varna, 29 Sept. - 1 Oct. 1999, Eds. B.Samuneva, S.Bachvarov, I.Gutzov, Y.Dimitriev, "Science Invest ", Sofia, 1999, Vol.1 Glass p.159.
7. E. Kashchieva, V. Ivanova, B. Jivov, Y. Dimitriev, Phys. Chem. Glasses 41, 6 (2000) 355.
8. T. Komatsu, K. Shioya, K. Matusita, J. Amer. Ceram. Soc. 76 (1993) 2923.
9. H. Kim, T. Komatsu, R. Sato, K. Matusita, J.Non-Cryst. Solids 162 (1993) 201.
10. L. Hu, Z. Jiang, Phys.Chem. Glasses 37,(1996) 19.
11. D.D. Nihtianova, Current Topics in Crystal Growth Res. 4 (1998) 115.
12. R. F. Geller , E. N. Bunting, J. Res. Nat. Bur. Stand. 18 , 5 (1937) 585.
13. H. Doweidar, A.H.Oraby, Phys. Chem. Glasses 38 (1997) 69.
14. E. Kashchieva, Y. Dimitriev, Proc. of the 5-th ESG Conf. Glass Sci. Technol. for the 21-st Century, June 21 - 24, 1999, Prague, Eds. A. Helebrant, M. Maryska, S. Kasa, Czech Glass Soc., Prague 1999, ISBN 80-238-3861-X, CD-ROM, pp. B4-65 – B4-68.

Nanoscience & Nanotechnology
eds. E. Balabanova, I. Dragieva, Heron Press, Sofia, 2001

ION IMPLANTATION AND LITHOGRAPHY WITH SUBMICRON DIMENSIONS

P. Bankov, N. Drandarov, D. Karpuzov

Institute of Electronics – BAS, Sofia

1. Introduction

Finely focused beams are required for the fabrication of various quantum devices. At present, beams with diameters less than 100 nm are commonly achieved. For example, the minimum pattern size which can be delineated in PMMA resist is 12 nm to 15 nm.

The development of focused ion beam (FIB) systems in early 90's boosted numerous ion-beam applications to sub-micron technology such as narrow metal strip deposition, trench etching, patterning, lithography of submicron devices, resist modification etc.

2. Experimental: Focused Ion Beam System IBL-100

The focused ion beams (FIB) system IBL-100 at the Institute of Electronics, BAS is based on liquid metal ion sources (LMIS) with extremely high brightness compared to the conventional ion sources. LMIS is a point source since the ions are extracted from an apex of a sharp emitter needle in a high electric field. The ion beam is accelerated up to $\sim$ 100 keV and focused by condenser and objective lenses onto the target. An $E \times B$ mass analyser can be used to select the desired ions.

3. Ion Species

The operation of LMIS is limited to materials that have low melting point and are inert to the material of the emitter. In many cases this limitation can be removed by using alloys with low melting points. Various ions including Be, B, Si, As, Ga and Au have been produced by LMIS. Stable ion sources of Be, Si and Ga ions have operating life of several hundred hours or longer and a relatively high emission current intensity.

4. Beam Current

Beam current depends on the beam energy, probe size and source brightness. For our FIB system, the beam current is typically several ten picoamps at a probe size of about 100 nm, which corresponds to a current density of > 0.1 A/cm^2 (i.e. higher than conventional ion beam systems). The beam current is generally limited by the chromatic aberration of the electrostatic lenses. Although the low beam current intensity limits the application of FIB for large-scale fabrication, the high current density makes it a useful tool for local modification of the materials. The beam can be well positioned using stepmotors (controlled by a laser ellipsometer) and scanned over the designed area.

5. Beam energy

Damageless etching, deposition and shallow doping can be achieved by low energy beams (below 1 keV). At these energies the lens aberration becomes large, and it becomes extremely difficult to obtain submicrometer beam dimensions unless a retarding field optical configuration is used.

6. Application of fine Focused Ion Beams

FIB lithography. The focused ion beam lithography is most attractive as nanofabrication technique. A few 10 nm sized patterns, for example, can be fabricated using focused ion beam.

Advantages:

— high sensitivity (resulting from the high energy-deposition rate to the resist material);

— very low proximity effect (easier to resolve patterns in close proximity due to reduced backscattering of the beam)

— high resolution (due to the low proximity and the fine pattern delineation capability);

— maskless (no mask is needed);

— easy automation (patterning control by using computer).

Disadvantages:

— low throughput.

FIB etching. Physical sputter etching using FIB is simple and useful for various "in situ" patterning and characterization by SIMS composition mapping [1,2]. This, however, has disadvantages. Deep groves with high aspect ratio, for example, are difficult to be formed due to redeposition of sputtered particles. Ion beam assisted etching is expected to overcome these disadvantages.

Ion beam assisted etching is performed by irradiating FIB in reactive gas ambient. The irradiation activates the reaction between adsorbed etchant gas and a target, and enhances the desorption of the reaction products.

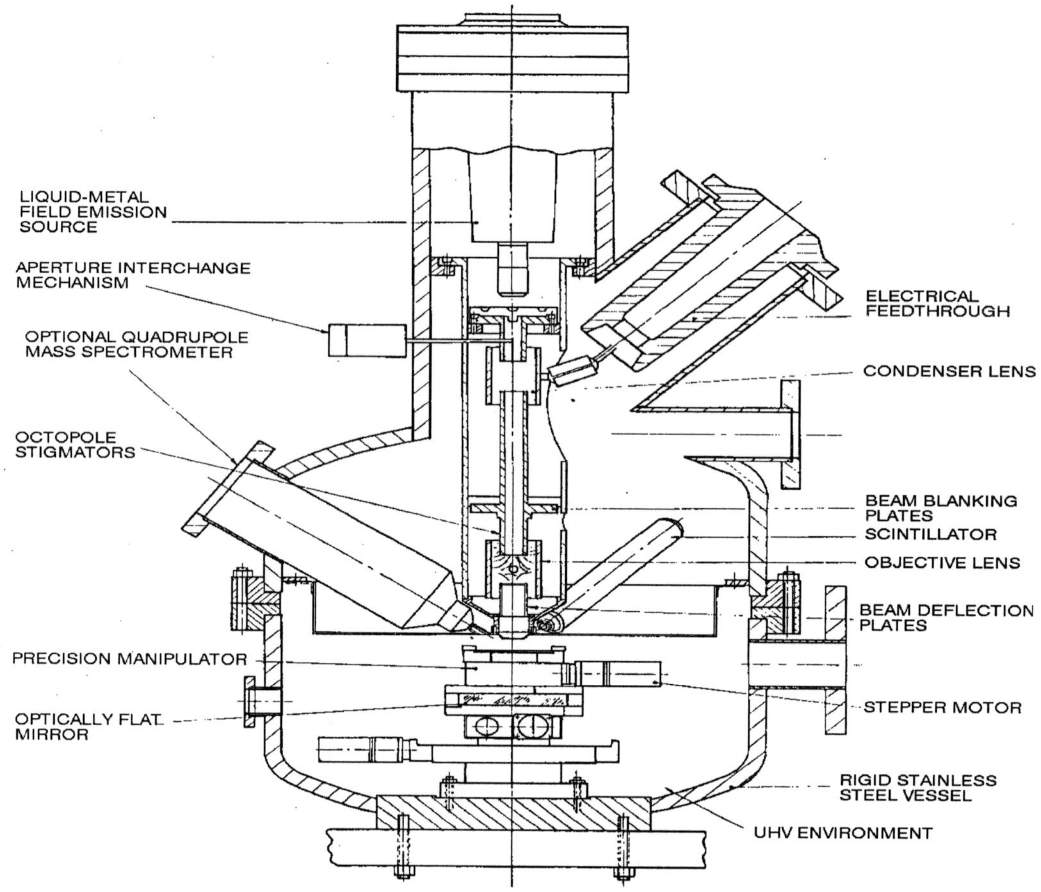

FIB implantation for synthesis/doping. FIB can be efficiently used for selective formation of a two-dimensional electron gas system (2DEG) or quantum wire and box structures. Recently, Gamo et al implanted Si^+ FIB into GaAlAs through a thin (1,5 nm) layer of GaAs [1]. By layers of GaAs, GaAlAs and GaAs overgrown by molecular beam epitaxy (MBE), inverted modulation-doped layers were formed, see the figure below for illustration.

Focused Ion Beam System IBL-100. One of the most promising material for device metallization and Schottky barrier contacts, the cobalt disilicide $CoSi_2$, is routinely synthesized by high-fluence ion implantation. By FIB implantation of cobalt in silicon (e.g. Co^{2+} in Si (111) at 70 keV, 400°C) [3] and subsequent annealing, a direct patterning without a mask can be achieved, and the implantation parameters like dose, energy or target temperature can be changed on any single area of the wafer.

References

1. Kenji Gamo, Annual Research Report - Gamo Laboratory, (1994), Research Center for Extreme Materials, Osaka University
2. Kenji Gamo, Research Center for Extreme Materials, Osaka University, Nucl. Instr. and Meth. B65 (1992) 40-49
3. St. Hausmann, L. Bischoff, M. Voelskow, J. Teichert, W. Moller, H. Fuhrmann, Nucl. Instr. and Meth. B148 (1999) 610-614

Nanoscience & Nanotechnology
eds. E. Balabanova, I. Dragieva, Heron Press, Sofia, 2001

IR SPECTROSCOPY STUDY OF AMORPHOUS, CRYSTALLINE NANOPARTICLES AND SINTERED COMPOSITES

I. Markova-Deneva[1], K. Alexandrova[2], G. Ivanova[2], M. Pankova[1], I. Dragieva[2]

[1]Univ. Chem. Tech. Met., 1756 Sofia, 8 Kl. Ohridski Blvd., Bulgaria
[2]Central Lab. Electrochem. Power Sources, Bulgarian Acad. of Sciences, 1113 Sofia, Bulgaria

Abstract. Investigations by means of infrared spectroscopy (IR-spectroscopy) of amorphous and crystalline $Co_xB_yH_z$ nanoparticles and their composites with AlN in the 4000–400 cm^{-1} range are carried out. Proofs for creation of chemical bonds between boron and oxygen atoms and a formation of BO_3 and BO_4 type of groups incorporated in boroxol rings $(B_3O_{4,5})_x$ and structured as diborate, triborate and tetraborate units are given. Composites with AlN, which are heat-treated at high temperatures show the existence of chemical bonds between aluminum and oxygen atoms forming coordination groups of the types AlO_4 and AlO_6.

Keywords: infrared spectra, $Co_xB_yH_z$ nanoparticles, composites, AlN additive

1. Introduction

Infrared spectroscopy is a powerful method for nanosized materials investigation, including for their application in a lot of new products, such as nanowires and nanoparticles by templating fabrication technique [1], as well as in the preparation of composites with them.

The aim of the present publication is to document investigations of composite materials, done by infrared spectroscopy, produced with the use of isotropic in shape and various in structure nanosized particles obtained by the borohydride reduction process[2, 3]. Thus, we try to achieve the transition towards producing anisotropic composite materials, for which one has to know, to determine and to estimate the kind and the presence of –B–O–B– bonds, such as in the groups BO_3 and BO_4 [4]. The isotropic and micronized AlN [5], applied by us in some specific applications, and in this one, in the role of an additive to nanoparticles, produced by borohydride reduction, is also a subject of this study.

2. Experimental

Two types of reactors working under different hydrodynamic conditions of mixing of both water solutions – this of the aqua salts solution of $CoSO_4$ with that of $NaBH_4$ solution are used [2–4]. Nanoparticles with the composition $Co_xB_yH_z$ are synthesized at room temperature with amorphous structure using a reactor with ideal mixing conditions and particles with nanocrystalline structure by means of "antigravity" reactor, working in ideal displacement regime [4]. Thus prepared amorphous or crystalline nanoparticles are pressed under press loading of 100 MPa and sintered at lower (200, 300, 400°C) or higher temperatures (500, 550, 600, 700, 750, 800, 850°C) in a flow of pure N_2. Pressed and sintered samples of nanosized particles are investigated by means of infrared spectroscopy IR Spectrophotometer with Furriet transformation of Brukner EKINOX 55 using KBr matrix. AlN powder (® Aldrich) is introduced to the nanoparticles mixture amounting to 10, 20 and 30 mass%, pressed and sintered at more than 550°C temperature. Years ago comparable chemical, electron microscopic, XRD, BET and infrared analyses of this type of AlN to other types of AlN have been carried out [6].

3. Results and Discussion

In Figure 1 is presented an infrared spectrum of pressed and sintered at 200°C amorphous nanoparticles. The same dependence has been observed earlier [3].

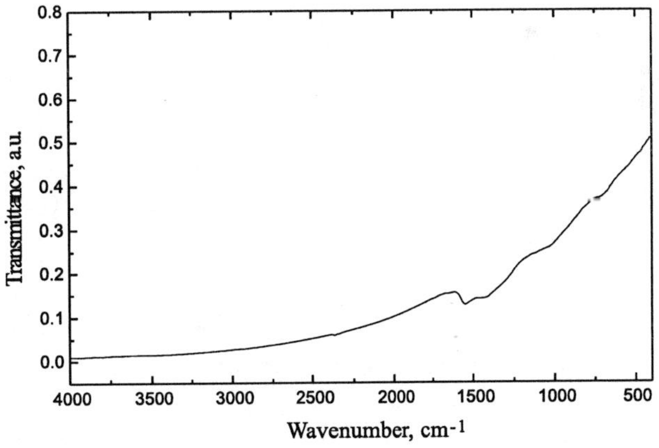

Figure 1. IR spectra of amorphous nanoparticles pressed and heat-treated at 200°C

The observed absorption bands are broad with not clearly expressed absorption peaks in the whole range from 4000 to 400 cm^{-1} typical for the nanoparticles in an amorphous state.

In Figure 2 the IR spectrum of the same nanoparticles, but obtained in a crystalline state is presented, after pressing and sintering of the probe at the same (200°C) temperature. The difference between the IR spectra shown in Figure 1 and 2 is evident, irrespective of the same chemical composition and the same conditions for pressing and sintering.

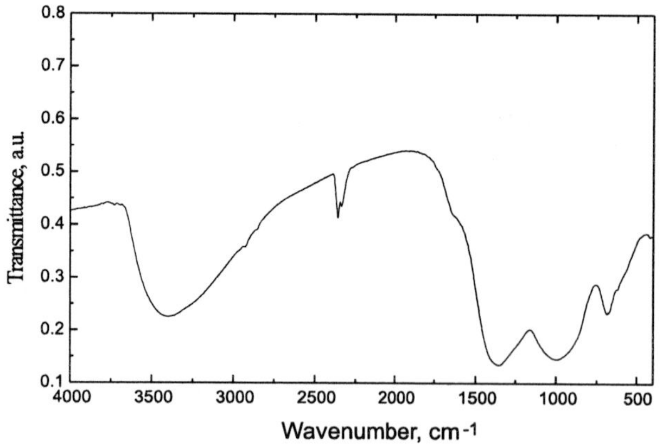

Figure 2. IR spectrum of crystalline nanoparticles pressed and heat-treated at 200°C

It is known from previous IR spectroscopy study [5] of borate glasses that in the frequency region of 4000 to 400 cm^{-1} these glasses have revealed three modes of vibrations: symmetric stretching ν_1 (BO_4) at 750 cm^{-1}, asymmetric stretching ν_3 (BO_3) at 1020 cm^{-1} and deformation mode ν_4 (BO_4) at 520 cm^{-1}. These vibrations are typical for the B–O–B bonds in the BO_3 and BO_4 groups forming different structural units. The BO_3 groups form the so-called boroxol ring $(B_3O_{4,5})_x$ while the BO_4 group can be incorporated in several ways into the boroxol ring. This is believed to occur due to the generation of BO_4 groups in a system with diborate, triborate and tetraborate units.

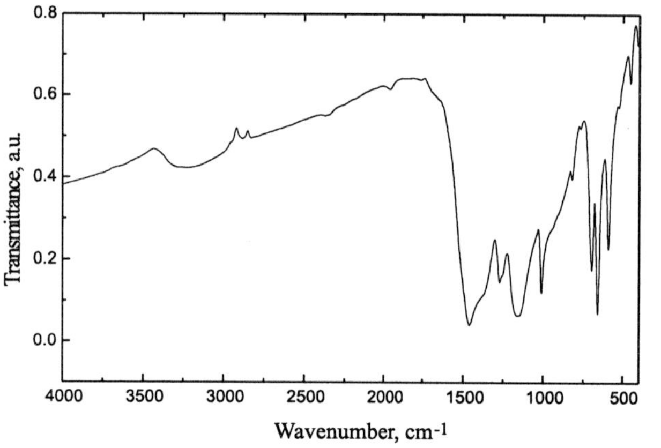

Figure 3. IR spectrum of crystalline nanoparticles pressed and heat-treated at 750°C

In Figure 3 the infrared spectrum of crystalline nanoparticles, pressed and sintered at 750°C, is shown. The IR spectra of crystalline nanoparticles heat-treated in the interval between 200 and 850°C are characterized by intensive sharp bands observed at 1400–1100 cm^{-1}, which are very sensitive to the B–O–B bond angle and to the B–O bond length in the BO_3 groups, forming the boroxol rings with three non-bridging oxygen. The absorption between 1400–1100 cm^{-1} region accounts for the ν_3 (BO_3) mode of vibration.

The bands of absorption at the lower frequencies between 1000–500 cm^{-1} have been referred to as the stretching vibration of B–O bonds in BO_4 groups involving the oxygen connecting the groups. The absorption at 750 cm^{-1} characterized by symmetric stretching vibration ν_1 (BO_4) mode, while the absorption at 500 cm^{-1} is a result of deformation mode ν_4 (BO_4) vibrations.

The bands of absorption grow in intensity and shift to lower frequencies with increasing of the heat-treated temperatures from 200 to 850°C. The quantity of BO_4 groups increases relatively, at the expense of BO_3 groups from the boroxol ring with three non-bridging oxygen. The higher temperatures of heat-treatment stimulate the forming of non-bridging oxygen, the generation of triborate units, respectively. At lower temperatures (200–500°C) the bands are broad and weak, while at higher temperatures (550–850°C) they are intensive and sharp.

The IR studies of the composites with AlN have characterized stretching vibrations typical for the Al–O bonds in the AlO_4 tetrahedra (900–650 cm^{-1}) and in the AlO_6 octahedra (650–500 cm^{-1}) [7].

In Figures 4 and 5 the IR spectra of composites prepared on the base of nanoparticles with AlN as an additive (30 mass%), pressed and sintered at the same conditions are shown.

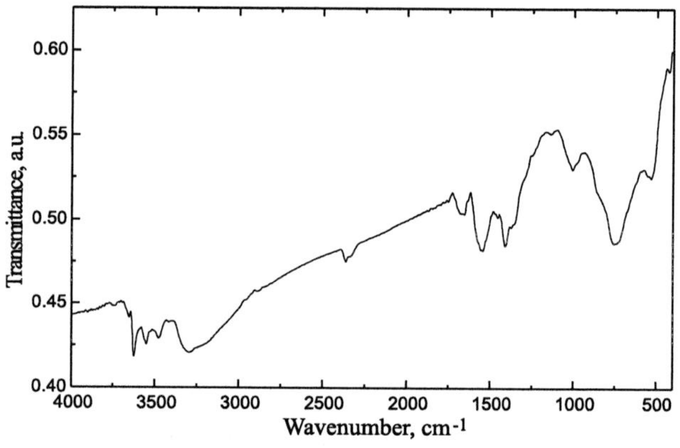

Figure 4. IR spectrum of composite with 30 mass% AlN (on the base of amorphous nanoparticles) pressed and heat-treated at 700°C

In the cases of composites sintered at 550°C we observed bands of absorption characterized by stetching vibration of Al–O bonds in AlO_4 tetrahedra between 900–650 cm^{-1} and in AlO_6 octahedra between 650–500 cm^{-1}.

With incerase of the AlN content the quantity of evaluated AlO_4 and AlO_6 coordinated groups is also increasing. Their bands of absorption at lower frequencies double the bands due to BO_4 groups. The supplement of AlN reveals a change of boron coordination from three (BO_3) to four BO_4. The latter group would have an inherently higher local symmetry. The presence of AlN decreases relatively to the intensity of the absorption bands of BO_4 groups at the lower frequencies.

Upon IR spectra of composites from nanocrystalline particles and AlN, sintered at the heat-treatment temperatures above 550°C, an interested fact is observed: with the increase of the percentage content of AlN the intensity of the bands of absorption in the frequency range 1400–750 cm^{-1} decreases, and at 30% AlN those bands disappear and a new intensive band of absorption at 740 cm^{-1}

with clearly determined peak appears. That behavior is logical. The temperature of 550°C is not enough for the Al to transform into the oxidation form as AlO_4-tetrahedra, or AlO_6-octahedra, respectively. The aluminum atoms remain connected to the nitrogen. The hydrogen included in AlN (0.20–0.26 mass% hydrogen [6]), delivers and reduces (BO_4) groups to (BO_3) groups. Besides, AlN shoes bands of absorption in the range 800–600 cm^{-1} with maximum at 740 cm^{-1}, corresponding to the valence vibrations of the Al $\equiv$ N bonds [6]. At lower content of AlN the observed bands of absorption are characterized by valence symmetric vibrations of B–O–B bonds in diborate and triborate units with different number of (BO_3) (BO_4) groups.

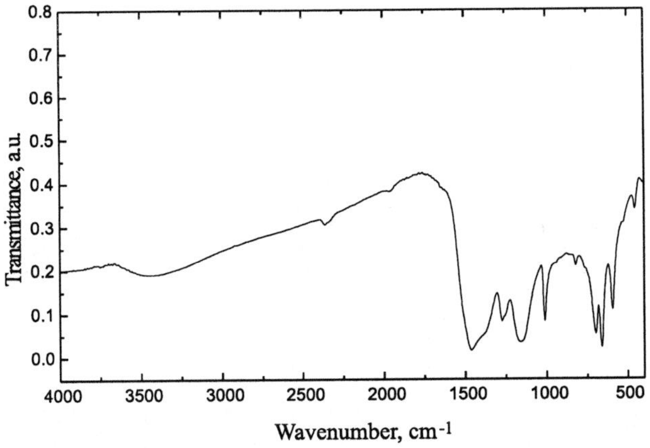

Figure 5. IR spectrum of composite with 30 mass% AlN (on the base of crystalline nanoparticles) pressed and heat-treated at 700°C

The IR spectra of the composites of crystalline nanoparticles with AlN, heat-treated at 700°C, are characterized by small bands with clearly determined absorption bands, with clearly expressed peaks, corresponding to the valency vibration of the B–O–B bonds (BO_3) groups (1400–1000 cm^{-1}) and in the (BO_4) groups (900–500 cm^{-1}), at which valence vibration are doubled on the Al– bonds in AlO_4 tetrahedra (800–650 cm^{-1}) and AlO_6 octahedra (650–500 cm^{-1}). The heat-treatment temperature of 700°C is sufficient for transition of the Al in its oxidation form. With increase of the percentage content of AlN, the absorption bands in the low-frequency range (750–700 cm^{-1}) become broader, with more weakly expressed intensity and even splitting (Figure 5).

Thus, the increase of the quantity of the AlN in the prepared isotropic composites has a reverse to the heat-treatment temperature effect.

4. Conclusions

The investigations by means of infrared (IR) spectroscopy in (4000–400 cm^{-1}) range of amorphous and crystalline $Co_xB_yH_z$ nanoparticles, obtained by chemical reduction, and also pressed and heat-treated in 200 to 850°C temperature interval, have proved the creation of chemical bonds between boron and oxygen atoms, as well as the formation of groups of the types BO_3 and BO_4 incorporated in boroxol rings $(B_3O_{4,5})_x$ and in structures of diborate, triborate and tetraborate units.

The applying of micronized AlN powder as an additive to the thus prepared composites exhibits an activity, characterized by the desorption of hydrogen and determined upon IR spectra as a (740 cm^{-1}) valence vibration typical for Al $\equiv$ N bond. At high temperatures of heat-treatment procedure (and respectively) low content of AlN in the composite, the presence of chemical bonds between Al atoms and oxygen ones forming the coordination groups of type AlO_4 and AlO_6 are determined.

References

1. M. Sasaki, M. Osada, N. Sugimoto, S. Inagaki, Y. Fukushima, A. Fukuoka, M. Ichikawa, "Novel templating fabrication of nano-structured Pt clusters and wires in the ordered cylindrical mesopores of FSM-16 and their unique properties in catalysis and magnetism", Microporous and Mesoporous Materials, **21**(1998) 597.
2. I. Markova, M. Pankova, I.Dragieva, "Infrared spectroscopy study of nanoparticles produced by borohydride reduction process", Bulgarian Chemistry and Industry , **70** (1999),No2, 55.
3. . I.Dragieva, G. Ivanova, S. Bliznakov, E. Lefterova, Z. Stoynov, I. Markova, M. Pankova, "Synthesis, properties and applications of nanoparticles produced by borohydride reduction process, Phys. Chem. Glasses, **41** (2000) 264
4. I. Dragieva, Z. Stoynov, K. Klabunde, "Boron-containing nanoparticles and their contribution to the theory and production of new materials", NanoStructural Materials, **12** (1999) 613.
5. S. Ram, K. Ram, "IR and Raman studies and effect of g radiation on crystallization of some lead borate glasses containing Al2O3", J. Mat. Sci., **23** (1988) 4541.
6. I. Markova, I. Dragieva, V. Vassilev, D. Lepkova, A. Vutzova, "AlN Powders", Chem. Industry (Bulg.), **LXIII** (1992) 1, 35.
7. P. Tarte,"The determination of cation co-ordination in glasses by infrared spectroscopy", Spectra Chem. Acta, **18** (1962) 549.

Nanoscience & Nanotechnology
eds. E. Balabanova, I. Dragieva, Heron Press, Sofia, 2001

FREE VOLUME MODEL CONSIDERATIONS OF DENSITY CHANGES OF AMORPHOUS SOLIDS DUE TO RELAXATION ANNEALING OF STRUCTURAL NANOVOIDS AND THEIR EXPERIMENTAL PROOF

K. Russew[1,2], L. Stojanova[1] and F. Sommer[2]

[1] 1Bulgarian Academy of Sciences, Institute for Metal Science, 67 Shipchenski prohod Boul., 1574 Sofia, Bulgaria
[2]Max-Planck-Institut fur Metallforschung, Seestrabe 92, 70174 Stuttgart, Germany

Abstract. A theoretical consideration of length and density changes caused by isothermal and non-isothermal relaxation heat treatment of amorphous solids is carried out in the framework of the Free Volume Model (FVM). The model predictions have been proved experimentally by using bulk $Pd_{40}Cu_{30}Ni_{10}P_{20}$ amorphous metallic alloy as an example. For a first time it has been shown that the density changes as a function of the isothermal heat treatment conditions below and in the close proximity to the glass transition temperature T_g can be well described with the derived FVM model equations.

Keywords: amorphous solids, density, nanovoids, free volume model, relaxation

1. Introduction

The Free Volume Model (VFM) of Turnbull and Cohen [1,2] describes the atomic mobility of amorphous solids in terms of the volume available for the atoms. Atoms possessing a critical free volume v^*, allowing atomic movements in their closed proximity are called structural defects. The probability for an atom to possess a free volume v^* [1,2] is called a structural defect concentration c_f. The relaxation changes of the properties of amorphous solid are due to atomic rearrangements and thus depend on the concentration of structural defects. The as-quenched defect concentration $c_{f,o}$ tries to reach the quasi-equilibrium defect concentration $c_{f,e}$ which depends on the temperature of annealing only. The aim of this study is to present model considerations about the length and density changes of amorphous solids in the framework of FVM, and to prove them experimentally by using bulk $Pd_{40}Cu_{30}Ni_{10}P_{20}$ amorphous metallic alloy as an example.

2. Theoretical Considerations

It has been shown [1,2,3] that by relaxation annealing of metallic glasses annihilation and/or production of structural defects takes place. The annihilation follows a bi-molecular reaction kinetics, whereas the production is a mono-molecular reaction [3]. The proper differential equation describing the change of the concentration c_f of structural defects at a constant temperature T along with the time t of relaxation annealing can be presented as [3]

$$\frac{dc_f}{dt} = -k_r c_f (c_f - c_{f.e}) \quad (1)$$

where $k_r = \nu_r \exp(-Q_r/RT)$ is the rate constant of relaxation, Q_r is the activation energy for relaxation, and ν_r is the attempt frequency. Its solution is

$$c_f(t)^{-1} = c_{f,e}^{-1} + (c_{f,o}^{-1} - c_{f,e}^{-1}) \exp(-k_r c_{f,e} t)\,. \quad (2)$$

The temperature dependence of $c_{f,e}$ is given by [1,2]

$$c_{f,e} = \exp\Big(-\frac{B}{T - T_0}\Big) \quad (3)$$

where B and T_0 are two model parameters corresponding to the empirical constant B and to the ideal glass transition temperature T_0 of the empirical Vogel-Fulcher-Tamman equation, respectively.

Under nonisothermal relaxation condititions at a constant heating rate q, eq. (2) transforms into [4]:

$$\frac{1}{c_f(T)} = \Big(-\int_{T_0}^{T} Q(x) e^{-\int_{T_0}^{T} P(x)dx} dx + \frac{1}{c_{f,o}}\Big) e^{\int_{T_0}^{T} P(x)dx} \quad (4)$$

where

$$P(T) = \frac{\nu_r}{q} \exp\Big(-\frac{Q_r}{kT} - \frac{B}{T - T_0}\Big)$$

and

$$Q(T) = -\frac{\nu_r}{q} \exp\Big(-\frac{Q_r}{kT}\Big).$$

It is obvious that the annihilation or production of structural defects (nanovoids) should be accompanied by negative or positive length, volume and density changes of the amorphous solid, respectively. In a previous paper [5] we have shown, that the relaxation caused deviation $\Delta L_f(T)$ of the sample length from the extrapolated to higher temperatures simple linear temperature dependence of the thermal expansion under heating at a constant heating rate q, can be presented as follows:

$$\Delta L_f(T) = \frac{\gamma v^* N_A L_0}{3V_{mol}[1 + \alpha_l^0 (T - T_B)]^2} \Big(\frac{1}{\ln c_{f,o}} - \frac{1}{\ln c_f(T)}\Big) \quad (5)$$

where γ is a geometrical overlapping factor of atomic free volumes, N_A is the Avogadro number, L_0 is the initial sample length at the starting temperature of heating T_B, and α_l^0 is the coefficient of linear thermal expansion.

Under isothermal relaxation conditions the corresponding length change is presented as:

$$\Delta L_f(t) = \frac{\gamma v^* N_A L_0}{3V_{mol}}\Big(\frac{1}{\ln c_{f,o}} - \frac{1}{\ln c_f(t)}\Big). \qquad (6)$$

The initial sample density ρ_0 is obviously equal to $\rho_0 = W/V_0 = W/L_0^3$ where W and V_0 are the sample weight, and initial volume, respectively. The density change of a cube of initial edge length L_0, caused by nonisothermal or isothermal relaxation annealing can be presented as:

$$\begin{aligned}\Delta\rho(T) = \rho(T) - \rho_0 &= \frac{W}{(L_0 + \Delta L_f(T))^3} - \frac{W}{L_0^3} \\ &= -\frac{3W(\Delta L_f(T)/L_0)}{L_0^3(1 + 3(\Delta L_f(T)/L_0)} \qquad (7)\end{aligned}$$

and

$$\frac{\Delta\rho(t)}{\rho_0} = -\frac{3(\Delta L_f(t)/L_0)}{1 + 3(\Delta L_f(t)/L_0)}, \qquad (8)$$

respectively. Equations (5) and (8) were used for FVM evaluation of the experimental data obtained by studying the non-isothermal thermal expansion anomalies and isothermal density changes of bulk amorphous $Pd_{40}Cu_{30}Ni_{10}P_{20}$ samples.

3. Experimental results

The bulk amorphous $Pd_{40}Cu_{30}Ni_{10}P_{20}$ metallic alloy was obtained in the form of rods of 6 mm diameter by rapid water quenching of its melt, sealed in silica glass tubes under vacuum of 10^{-5} Torr. The fully amorphous state of the samples was proved by X-ray. The length change of the samples at a constant heating rate of 10 K/min was monitored up to the temperature of crystallization with the aid of a Perkin Elmer TMS-2 silica glass dilatometer.

Figure 1 shows the temperature dependence of the anomalous deviation $\Delta L_f(T)$ of the sample length from the extrapolated to higher temperatures simple linear temperature dependence of thermal expansion. The experimental data are presented with graphic symbols (o), whereas the solid line presents the fitting curve obtained from the nonlinear multi-parameter regression analysis of the experimental data according to eq. (5). The FVM fitting parameters are given in Table 1.

Table 1. Fitting FVM parameters

Parameter	Value	Parameter	Value
ν_r, [1/s]	3.12×10^{19}	α_l^0, [1/K]	1.86×10^{-5}
Q_r, [kJ/mol]	130.45	V_{mol}, [m^3]	7.96×10^{-6}
$c_{f,o}$	1.42×10^{-9}	v^*, [m^3]	1.2×10^{-29}
T_0, [K]	307	γ	0.8
B, [K]	6324	N_A	6.02×10^{23}
L_0, [m]	5.05×10^{-3}	T_B, [K]	323

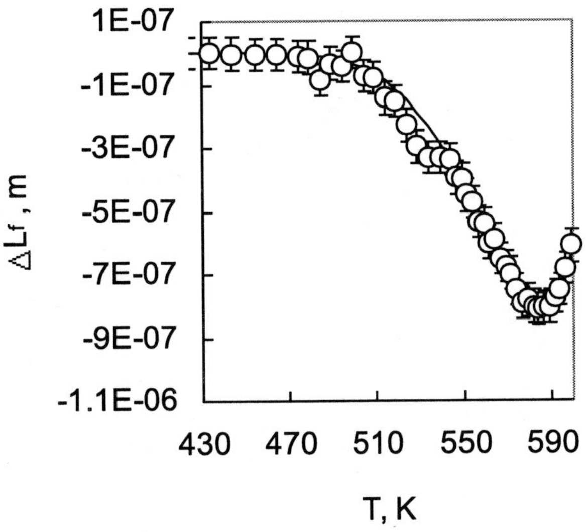

Figure 1. Temperature dependence of ΔL_f of a bulk amorphous $Pd_{40}Cu_{30}Ni_{10}P_{20}$ sample at a heating rate of 10 K/min

The density changes of $Pd_{40}Cu_{30}Ni_{10}P_{20}$ samples of approximately 3–4 g weight under isothermal annealing conditions were studied at 500, 520, 550, and 573 K. The buoyancy method of density determination was used. The density ρ_0 of as-quenched amorphous samples varied between 9.2472 ± 0.0007 and $9.2886 \pm 0.0007\,g/cm^3$. The value of 9.275 g/cm^3 reported by Inoue in [6] for the same amorphous alloy lies within these limits. The density scattering is due to the existence of small casting voids of varying volume fraction in the different samples. Such casting voids were observed metallographically in the cross section of the samples and some of them could be seen even without any optical means. However, as far as the relative density change $\Delta\rho/\rho_0$ is of importance only, the scattering of the initial density does not affect unfavorably the density change measurements. Figure 2 shows the $\Delta\rho/\rho_0$ experimental data (graphic symbols) together with the fitting curves (solid lines) calculated with the aid of eq. (8) inserting the FVM model parameters, given in Table 1. The coincidence of the experimental data with the calculated model curves is very good.

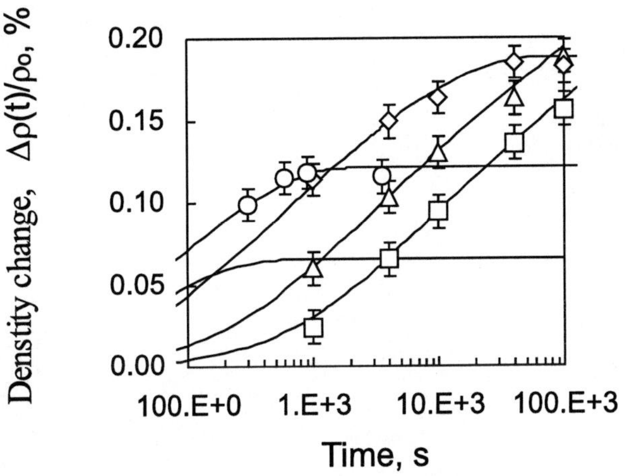

Figure 2. Density change $\Delta\rho/\rho_0$ of a bulk amorphous $Pd_{40}Cu_{30}Ni_{10}P_{20}$ sample as a function of annealing time at: □ – 500 K; △ – 520 K; ◇ – 550 K; ○ – 573 K

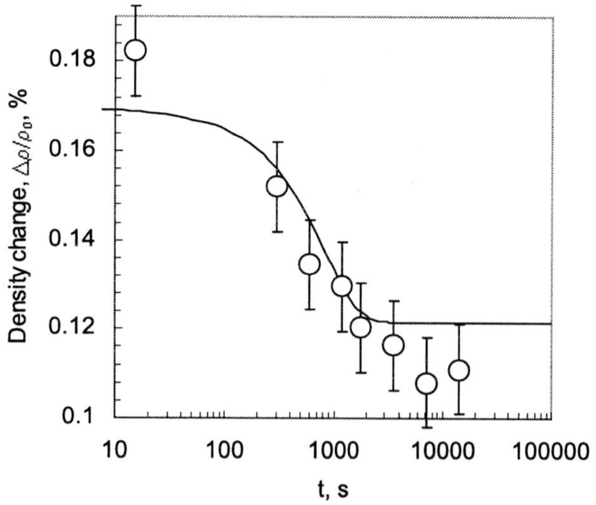

Figure 3. The reversible relative density changes $\Delta\rho/\rho_0$ (in %) of a bulk amorphous $Pd_{40}Cu_{30}Ni_{10}P_{20}$ sample, firstly annealed at 550 K and than at temperature 573 K

The reversible nature of density change in the temperature range below the glass transition temperature was experimentally proved by changing the annealing temperature of the sample used for density measurements at 550 K, from 550 to 573 K. The resulting density changes as a function of the annealing time at 573 K is shown in Figure 3. The solid line represents the FVM model curve, calculated for the new annealing conditions by replacing the $c_{f,o}$ value of 1.42×10^{-9}, see Table 1, with the final value of the defect concentration reached after the last anneal of the sample at 550 K, namely 1.58×10^{-12}. The coincidence between the experimentally measured and the expected density changes under these conditions is relatively good. The quasi-equilibrium defect concentration at 573 K has been reached after annealing time of approximately 1×10^4 s.

4. Conclusions

4.1. The determination of length- and density changes, respectively, due to relaxation annealing under non-isothermal and isothermal conditions, respectively, is a useful method for studying the structural relaxation of amorphous solids.

4.2. The Free Volume Model provides an outstanding theoretical basis for model description of structural relaxation of amorphous solids.

Acknowledgements. The financial support of the Bulgarian Fund for Scientific Research (Contract TN 807/98) is gratefully acknowledged.

References

1. M.H. Cohen and D. Turnbull, J. Chem. Phys. 31, 1164, 1959.
2. D. Turnbull and M.H. Cohen, J. Chem. Phys. 52, 3038, 1970.
3. P. A. Duine, J. Sietsma and A. vd. Beukel, Acta Met. Mat., 40, 743, 1992.
4. K. Russew, B, J. Zappel and F. Sommer, Scripta Met. Mater, 32, 271, 1995.
5. K. Russew, L. Stojanova and F. Sommer, Int. J. of Rapid Sol., 8, 267, 1995.
6. A. Inoue, Bulk Amorphous Alloys, Preparation and Fundamental Characteristics, Vol. 4 of Materials Science Foundations, Trans Tech Publ. Ltd, Switzerland, 1998.

Nanoscience & Nanotechnology
eds. E. Balabanova, I. Dragieva, Heron Press, Sofia, 2001

APPLICATION OF ELECTROCHEMICAL NOISE ANALYSIS FOR DETECTION OF HETEROGENEITY ON METAL SURFACES

E. Slavcheva[1], P. Plagemann[2], G. Schmitt[2]

[1]Central Laboratory of Electrochemical Power Sources, Bulg. Acad. of Sci., Akad. G. Bonchev Str., Bl 10, 1113 Sofia, Bulgaria
[2]Lab. for Corrosion Protection, University of Applied Science, Frauenstuhlweg – 58644 Iserlhon

Abstract. ECN has been applied to study propensity of Cu to pitting corrosion in synthetic water. The influence of anion composition was identified in terms of standard deviation SD(I) and PSD(I) data, as well as interaction coefficients. The high sensitivity of ECN diagnosis permits fast detection of heterogeneity on metal surface, which at later stages can cause initiation of localised corrosion.

Keywords: electrochemical noise, Cu, pitting corrosion

1. Introduction

Electrochemical Noise Analysis (ECN) is a modern experimental technique, which requires no external signal in order to receive a response from system under study. Although many ECN studies are available for crevice and/or pitting of different metals [1–5], there are only few reports on the application of ECN to investigate localised corrosion of copper [6,7]. Earlier own investigations on Cu pitting showed that ECN analysis allows to correlate the current noise intensities with corrosion and pitting intensities [8]. The ECN pattern reacts quickly on changes in the electrolyte composition and the surface condition of the metal and therefore, should indicate pit initiation at very early stage.

Here we report an investigation on the effect of different anions in synthetic tap water on the development of the electrochemical noise intensity on copper within the first 20 hours of contact. The aim was to investigate effects of single ions, as well as combination effects on the likelihood of copper pitting corrosion.

2. Evaluation of ECN-Data

The investigation was performed with a self-developed ECN measuring equipment consisting of a signal conditioning box and a PC with an analogue data acquisition card and a corresponding ECN-software. The signal conditioning box is needed to amplify and filter the measured raw signals with a low pass filter to avoid aliasing effects. After collecting the raw data in defined time intervals the DC part of the signal was removed to yield only the signal fluctuations around zero. Due to the stochastic origin of these fluctuations the standard deviations of all measured intervals were calculated. Additionally the data were transformed via FFT into the frequency domain and the slopes of the power spectral density PSD(I) were calculated. High noise intensities correlate with high standard deviations and high negative slopes of PSD(I).

3. Experimental

The parameter study was carried out using a pair of two identical copper tube sections (length of 2 cm) from commercial tubes (Cu-DHP, DIN EN 1057, 22×1) in a flow loop at room temperature. Current noise data were collected for about 20 h at a flow rate of 0.05 m/s. Tests were performed in synthetic tap waters containing only one single anion (chloride, sulphate or bicarbonate as Na-salts) or with all anions in one solution but with different molar ratios in concentrations within the tap water limits. The screening was designed according to a statistical 2^3 factorial plan. The cell was stored in a Faraday cage to avoid external electromagnetic influences.

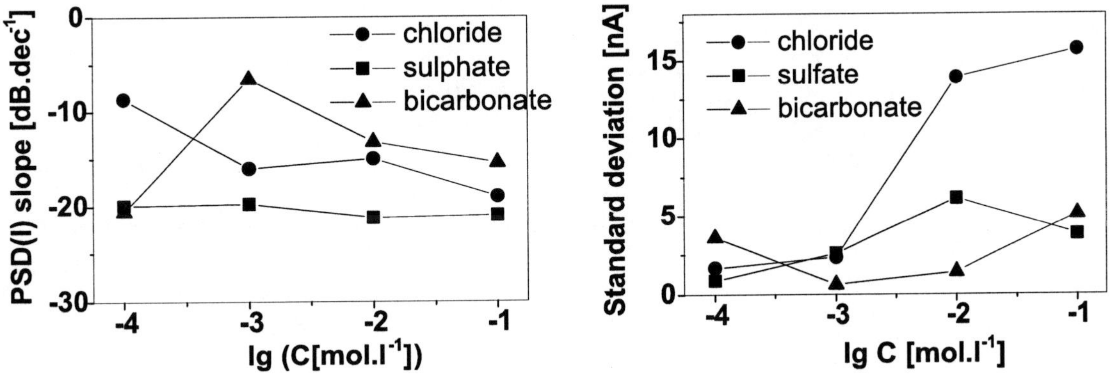

Figure 1. Noise data after 20 h test duration in single anion solutions

4. Results

Single Anions. After 20 h test duration all systems reached quasi steady-state condition, therefore the reproducibility of the noise intensities in the different experiments was very high. The standard deviations indicated that the current noise increases with the chloride and sulphate concentration. The effect of bicarbonate is very low and nearly independent on the concentration. Chloride-containing solutions yielded significantly higher noise activities than the rest of the anions. A clear trend of increasing standard deviation and negative PSD(I) slope with the increase in chloride concentration was obtained (Figure 1).

Interaction of Anions. In the experiments with chloride, sulphate and bicarbonate selected according to the 2^3 factorial plan, the corrosion system also reached quasi steady-state condition after 20 h-exposure time. The time development of both standard deviation and slope of PSD(I) indicated that at the low bicarbonate concentration the influence of the other anions was higher than at high bicarbonate concentration. The lowest noise activities were encountered when the concentration of all anions was on the low level. The highest noise intensity was found in case of lowest bicarbonate and highest chloride and sulphate levels (Figure 2).

The noise intensity increased with increasing chloride and sulphate concentration, specifically at low bicarbonate level.

Interactions between anions can be indentified from interaction coefficients calculated from full factorial analysis (Figure 3). Single ion solutions exhibit the highest effects but the coefficients of interaction are of the same order. Therefore, the noise data cannot be interpreted from single ion experiments. The interaction coefficients indicated that it is the combination of the chloride, sulphate and bicarbonate concentrations, which determines the corrosion activity at the copper surface.

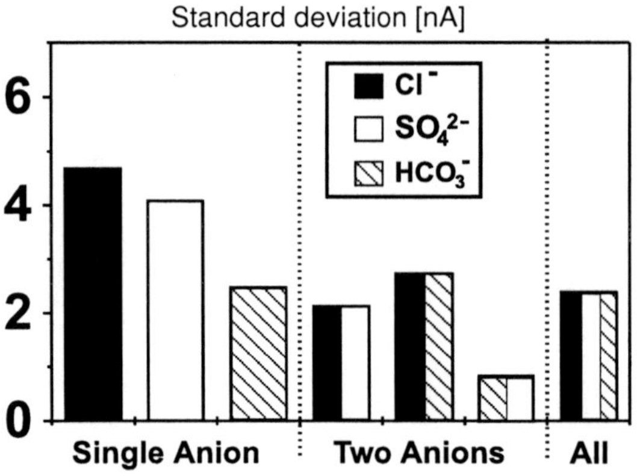

Figure 3. Coefficient of interaction calculated for standard deviations from full factorial analysis

The consistency of the information obtained from the results of the 2^3 factorial plan was proved in an experiment with intermediate concentrations of the three anions (Cl^-: 50 ppm, SO_4^{2-}: 200 ppm, HCO_3^-: 30.5 ppm) (Figure 4).

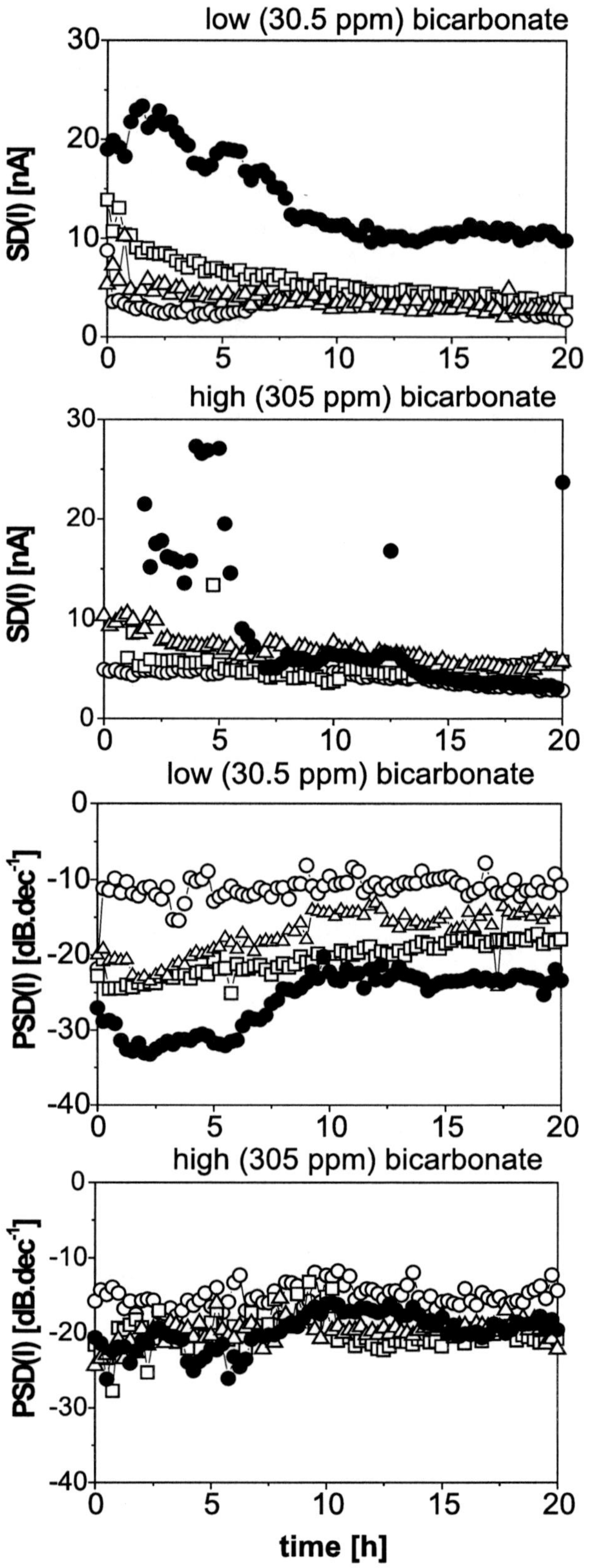

Figure 2. Time records of current noise data in solutions containing different anions: –○– 10 ppm Cl^-, –■– 10 ppm SO_4^{2-}; –□– 250 ppm Cl^-, 10 ppm SO_4^{2-}; –△– 10 ppm Cl^-, 250 ppm SO_4^{2-}; –●– 250 ppm Cl^-, 250 ppm SO_4^{2-}

The noise data obtained fit well into the general trend found from the previous set of experiments.

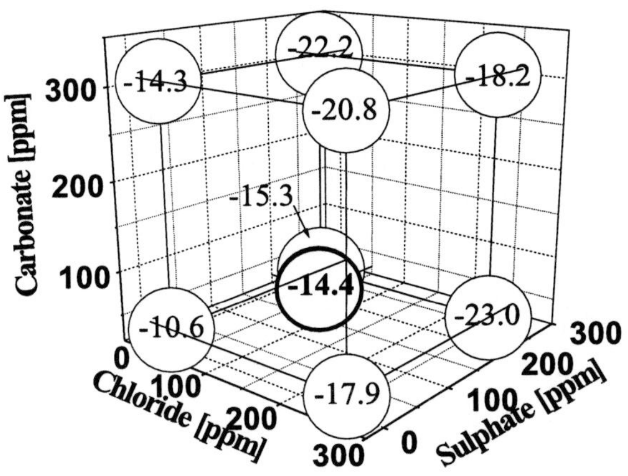

Figure 4. 3D-plot of PSD(I) slopes after 20 h test duration, effect of anion composition

5. Conclusions

ECN diagnosis proved to be a valuable tool for online monitoring of corrosion intensities even on a low level of activity. The effect of system changes on the corrosivity of copper in tap water can be obtained instantanously. The influence of chloride, sulphate and bicarbonate as single ions and the interaction between them in three anion solutions were identified in terms of standard deviation and PSD(I) data as well as interaction coefficients. The results let conclude that sulphate has a passivating and chloride an activating effect. Bicarbonate seems to acts as a buffering agent. It is obviously the combination of the presence of all three anions in a narrow range of concentration ratios, which finally leads to pitting corrosion. The relevance of these short term exposure results for the long term performance of copper in tap water and the likelihood of pitting has been currently investigated in long term exposure tests.

References

1. T. Schauer, H. Greisiger and L. Dulog *Electrochemical Acta* **43** (1998) 2423.
2. W. Oelsner, F. Berthold and H. Kaden *Materials and Corrosion* **49** (1998) 70.
3. F. Mansfeld and H. Xiao *J. Electrochem. Soc.* **140** (1993) 22057.
4. R. A. Cottis, C. A. *Lotto Corrosion Science* **46** (1990) 12.
5. A. Legat, C. Zevnik *Corrosion Science* **35** (1993) 1661.
6. S. Smith, R. Francis *Br. Corr. J.* **25** (1990) 285
7. R. F. Sandenberg, P. C Pistorius, A. Peterkin *13th Int. Corrosion Congress, 25-29 November* Australia, paper 131
8. G. Schmitt and P. Plagemann *Materials and Corrosion* **49** (1998) 677

Nanoscience & Nanotechnology
eds. E. Balabanova, I. Dragieva, Heron Press, Sofia, 2001

ON THE PROBLEM OF REINFORCEMENT OF POLYMER NANOCOMPOSITES

Ya. Ivanov[1], Yu. Yanovsky[2], V. Zgaevsky[2], Yu. Karnet[2], V. Stoyanov[1], N. Balabaev[3]

[1]Central Laboratory of Physico-Chemical Mechanics, Bulgarian Academy of Sciences, Acad. G. Bonchev St., Block 1, IV km, 1113 Sofia, Bulgaria
[2]Institute of Applied Mechanics, Russian Academy of Sciences, 32A, Leninsky Prospect, 117334, Moscow, Russia
[3]Institute of Mathematical Problems in Biology, Russian Academy of Sciences, 142292 Pushchino, Moscow region, Russia

Abstract. Presented study is based on the proposed structural model of interphase layer formed from polymer macromolecules on the surface of the filler. The segments of the macromolecule are connected consecutive with hard junctions: some junctions are 'bend' to the filler surface and the free ends of the macromolecules are fixed to the other junction. Those form the interphase layer. Under these circumstances the properties of the interphase layer are determined from the theory of hyperelasticity.

On the basis of this modal, using Hookean law the modulus of elasticity of interphase layer in normal direction to the filler surface, the modulus of elasticity of interphase layer in tangential direction to the filler surface, and the shear moduli are evaluated. These moduli are calculated depending on both of the number of segments in the polymer chain and the 'bend' of the macromolecules to the filler surface.

Keywords: polymer nanocomposite, structural model, interphase layer, modulus of elasticity, shear modulus

1. Introduction

The most polymer composites are inhomogeneous media with high developed contact surface between polymer matrix and fillers, i.e. with relatively high area of the interface. The reinforced plastics, filled thermoplastics and thermosets, the matrix polymer composites and the polymer nanocomposites appertain to this category.

The polymer matrix consisted of polymer chains with different nature play an important role in formation of the properties of polymer composite. Their unique physico-chemical and elastic properties are result of the conformation changes of macromolecules caused by the forces with different nature (for example various types of loading).

The individual macromolecules in contact with nanoparticles are attached to the surface of filler, which influence on the their conformation. This condition is determined by the physical and mechanical potentials of interactions of macromolecules with active centres of the filler surface. The macromolecules are 'bend' to the solid surface when bonds are formed. The description of the behaviour of these polymer macromolecules is one of the fundamental problem both of the physics and the mechanics of the polymers. This behaviour is determining largely of the properties of composite.

This study deals with some results obtained from a model analysis of the formation of these nanoscale macromolecular (interphase) layers.

2. Structural Model of Interphase Layer

The accepting structural model of the interphase layer is presented in Figure 1. It is assumed that this layer is formed from identical macromolecules with certain number of segments ($N = 15$ for this study) with weight m. These segments are connected consecutive with hard junctions of length l. Thus, some junction of the macromolecules is fixed to the filler surface (i.e. these macromolecules are 'bend' to the surface). The free ends of the macromolecules are fixed to the other junction and one interphase layer of thickness h surrounding each nanoparticle is formed.

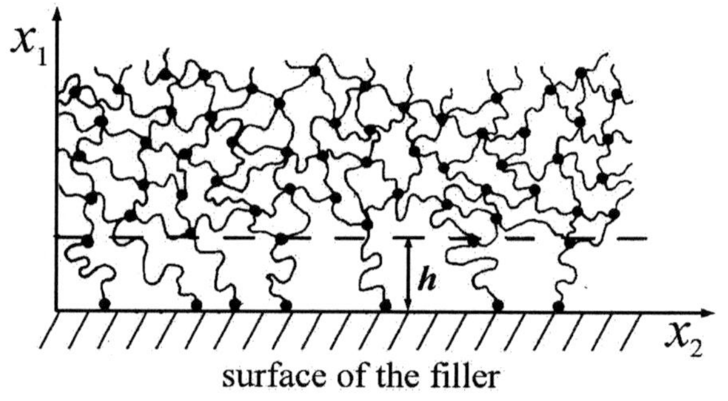

Figure 1. Structural model of the interphase layer

3. Elastic Properties of the Interphase Layer

The influence of the surface is considered as a force acted on each fixed macromolecule and changed its conformation (The macromolecule is deformed.). Under these circumstances the properties of the interphase layer are determined from the theory of hyperelasticity. To take the influence of surface into consideration [1] the phenomenological coefficient, named front-factor [2], was added. Thus the stress tensor of the interphase layer is as follows:

$$\sigma_{is} = 2C\lambda_{ln}\lambda_{sn}a^2_{(n)} - p\delta_{ln} - \sigma^0_{ls} \qquad (1)$$

where $C = \dfrac{T\chi R^{0^2}}{6Na^2h}$.

Here T is the absolute temperature expressed in absolute units, R^{0^2} – mean quadratic distance between the ends of the macromolecules far off the filler surface, N – number of segments of linear dimension a, $\lambda_{ln} = \partial x_l / \partial x_n$ (x_l and x_n are the co-ordinates of the point of interphase layer before and after deformation respectively), $a_{(n)}$ – relative change of the point co-ordinates, connected to the surface influence (Figure 1), p – hydrostatic pressure, δ_{ls} – Kronecker symbol, σ_{ls} – components of the initial stress field in the interphase layer, created from the influence of the surface:

$$\chi = a^2 \varpi \tag{2}$$

where ϖ is the number of the macromolecules joined to the unit area.

The components of the stress tensor σ^0_{ls} according to the Figure 1 are:

$$\begin{aligned} \sigma^0_{11} &= 2C(a_1^2 - a_2^2) \\ \sigma^0_{22} &= \sigma^0_{33} = 0 \\ \sigma^0_{ik} &= 0 \text{ for } i \neq k \end{aligned} \tag{3}$$

Let us consider the small deformations of the interphase layer. The relative elongation λ_{ik} can be presented as follows:

$$\lambda_{ik} = \delta_{ik} + u_{ik} \tag{4}$$

where u_{ik} is the derivative of the displacement vector of the x_i co-ordinate with respect to x_k. We can write u_{ik} as follows:

$$u_{ik} = \varepsilon_{ik} + \omega_{ik} \tag{5}$$

where $\varepsilon_{ik} = 1/2(u_{ik} + u_{ki})$ is the symmetrical part and $\omega_{ik} = 1/2(u_{ik} - u_{ki})$ is the asymmetrical part. From the equilibrium equation for the forces we obtain that $\omega_{ik} = 0$. The combination of Eqs (4) and (5) gives:

$$\lambda_{ik} = \delta_{ik} + \varepsilon_{ik} \tag{6}$$

where ε_{ik} is the component of the tensor of small deformations. Substituting Eq (6) into Eq (1) we obtain:

$$\sigma_{is} = 2C(\delta_{in}\delta_{sn}a^2_{(n)} + \delta_{sn}a^2_{(n)}\varepsilon_{in} + \delta_{in}a^2_{(n)}\varepsilon_{sn}) - p\delta_{is} - \sigma^0_{is}. \tag{7}$$

In the case of uniaxial tension parallel to the x axes we have:

$$\sigma_{11} = 2C(a_1^2 + 2a_1^2\varepsilon_{11}). \tag{8}$$

Thus $\sigma_{22} = \sigma_{33}$ and p can be estimated from:

$$C(a_1^2 + a_3^2 + 2a_2^2\varepsilon_{22} + 2a_3^2\varepsilon_{33}) = 0. \tag{9}$$

Finally, for the incompressible layer (when $\varepsilon_{11} + \varepsilon_{22} + \varepsilon_{33} = 0$), the stress σ_{11} can be obtained as follow:

$$\sigma_{11} = 2C(a_1^2 + 2a_2^2)\varepsilon_{11}. \tag{10}$$

Using Hookean law it is easy to find that modulus of elasticity of interphase layer in normal direction to the filler surface is:

$$E_1 = 2C(2a_1^2 + a_2^2). \tag{11}$$

In a similar way we obtain law for the modulus of elasticity of interphase layer in tangential direction to the filler surface:

$$E_2 = 4C\frac{a_2^2(2a_1^2 + 2a_2^2)}{a_1^2 + a_2^2}. \tag{12}$$

Based on Eq (1) in pure shear condition we get:

$$\begin{aligned} \sigma^0_{12} &= 2C(a_1^2 + a_2^2)\varepsilon_{12}, \\ \sigma^0_{13} &= 2C(a_1^2 + a_2^2)\varepsilon_{13}, \\ \sigma^0_{23} &= 2Ca_2^2\varepsilon_{23}. \end{aligned} \tag{13}$$

Then for the shear moduli we have:

$$G_{12} = G_{13} = C(a_1^2 + a_2^2), \qquad G_{23} = 2Ca_2^2. \tag{14}$$

The analysis of the elastic properties of the interphase layer dependent both of the molecular and the structural characteristics of the layer [3] which enables to determine the value of C as follow:

$$C = \frac{T\chi^{2/3}}{2aN}. \tag{15}$$

In this case the relationship $h(N, \chi)$ can be estimated as:

$$h = aN\chi^{2/3}. \tag{16}$$

Based on the detailed analysis the moduli E_1, E_2, $G_{12}(= G_{13})$, and G_{23} are calculated depending on the number N of segments in the chain (in $2CN$ units) (Figure 2).

The data represented in Figire 2 show that the increase of the number of segments of the polymer chain causes one decrease of the elastic characteristics of the interphase layer. These data confirm established relations in [5,6] for changes of the properties of polymer composites depending on thickness of polymer layer. When the number of segments is above 65 the elastic moduli E_1 and E_2 and respectively shear moduli G_{12} and G_{23} are practically equal.

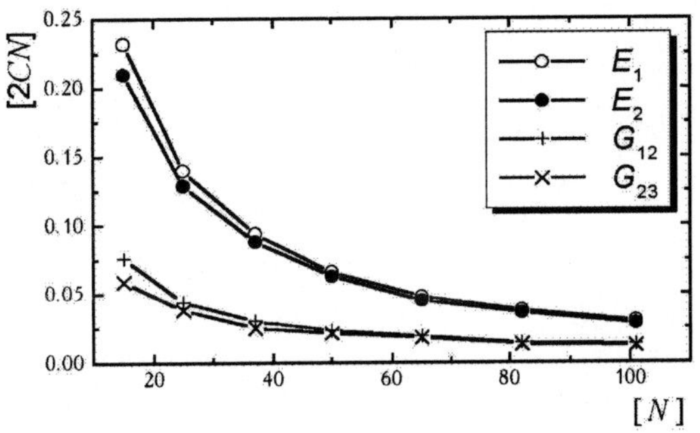

Figure 2. The dependence of E_i and G_{ij} moduli of of the interphase layer on the number of segments in the chain

On the basis of the analysis, used in [3], and Eq (16) the moduli E_1, E_2, G_{12}, and G_{23} are calculated depending on the 'bend' of the macromolecules to the filler surface (in $T/2a^2N$ units) (Figure 3).

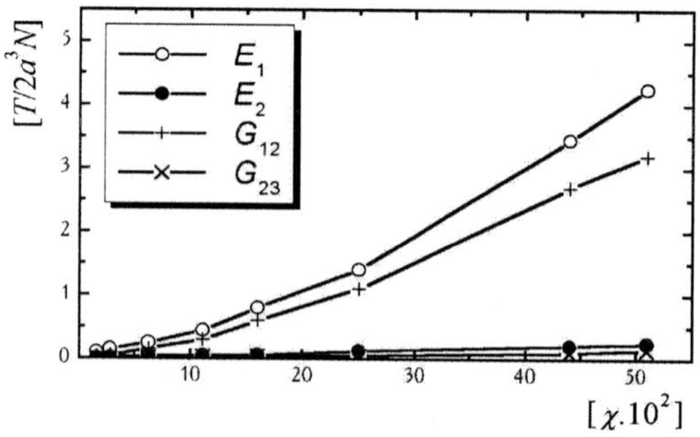

Figure 3. Dependence of the moduli of elasticity and the shear moduli of the interphase layer on the 'bend' of the macromolecules to the filler surface

The data represented in Figure 3 show that the increase of the density of 'bend' macromolecules to the surface of filler cause an increase both of the elastic and the shear moduli of the interphase layer. This shows that under sponge-like to brush-like transition of the structure of polymer layer the all moduli increased.

4. Conclusions

The results of the current study and the analysis of the data represented in Figures 2 and 3 allow to make next conclusions:

— the elastic moduli (E_1 and E_2) are more sensitive than shear (G_{12} and G_{23}) ones to the changes of the number of segments of polymer macromolecules placed in the interphase layer;

— the E_1 and G_{12} moduli are more sensitive than E_2 and G_{23} ones to the increase of the density of fixed polymer macromolecules to the surface of filler.

References

1. Zgaevsky V. E., Yu. Yanovsky *Mech. Comp. Mat. and Str.* **3** (1997) 105.
2. Grosberg A., A. Hohlov *Statistical Physics Macromolecule* (Science, Moscow, 1989) 343 (in Russian).
3. Zgaevsky V. E., et al. *Mech. Comp. Mat. and Str.* **5** (1999) 109.
4. Khalatur P. *Macromol. Chem. Macromol. Symp.* **44** (1991) 23.
5. Ivanov Ya. Microrheology of liquid silicate and polymer composites (D. Sc. Thesis, 1988) (in Bulgarian)
6. Voznesensky V., T. Lyashenko, Ya. Ivanov, I. Nikolov, Computer aided optimisation of composite materials, Budivelnik, Kiev, 1989 (in Russian)

SECTION 2:
Synthesis and Processing (Optical Properties)

Nanoscience & Nanotechnology
eds. E. Balabanova, I. Dragieva, Heron Press, Sofia, 2001

NANO-STRUCTURED MATERIALS FOR SOLAR ENERGY CONVERSION AND STORAGE

J. Schoonman

Delft University of Technology, Delft Institute for Sustainable Energy, Laboratory for Inorganic Chemistry, Julianalaan 136, 2628 BL Delft, The Netherlands

Abstract. Functional ceramics for application in solid-state electrochemical devices for the conversion of solar energy in green electricity and the storage of this electricity are described in this paper. In particular, nanostructured materials for Grätzel-type solar cells are presented. Materials for rechargeable ceramic lithium-ion batteries and dynamic compaction of battery components and wound batteries are described. It is shown that Magnetic-Pulse Dynamic Compaction is an excellent densification technique for obtaining nanostructured micro-structures and improved interfacial contacts between solid battery components. Some recent trends in the storage of hydrogen are presented.

Keywords: Nanostructured materials, green electricity, Li-ion batteries, sustainable hydrogen.

1. Introduction

Environmental issues such as urban air pollution, global warming, acid rain, and other impacts of tailpipe emissions have stimulated research and development of new or improved materials for conversion and storage systems for sustainable energy. Besides, the global transformation of our society into an information society requires substantial use of communication and data processing equipment. Such equipment heavily relies on the availability of high-energy density storage devices for electrical energy.

The trends in the field of Solid State Ionics clearly suggest that rechargeable Li-ion batteries and fuel cells, especially the intermediate-temperature Solid Oxide Fuel Cell (IT-SOFC) and the Polymer-Electrolyte Membrane Fuel Cell (PEM-FC), will dominate the materials research activities in the coming decades.

Solar energy conversion using photovoltaic (PV) solar cells based on crystalline, polycrystalline, or amorphous silicon is currently being introduced into society. New developments in PV focus on the Grätzel cell and Grätzel-type solar cells. These PV cells are based on nanostructured anatase titanium dioxide which is sensitized for visible light using organometallic dyes [1–3].

While PV electricity can be stored in rechargeable batteries, its combination with commercial electrolyzers for the production of hydrogen is attracting increasing attention. So is the combination of wind energy and water electrolysis. The introduction of a hydrogen economy into society requires cheap and safe storage, preferably in hydride forming materials.

In this paper the focus is on materials for Grätzel-type solar cells, Li-ion batteries, and for hydrogen storage.

2. Grätzel-type Solar Cells

The Grätzel solar cell makes use of a nanostructured anatase TiO_2 electrode sensitized by a monolayer of a chemically bound ruthenium(II) metal-to-ligand charge-transfer (MLTC) complex. Nano-structured electrodes provide a very large surface-area enhancement factor of about 1000 with respect to the geometrical surface area, which makes efficient sunlight absorption by a monolayer of the chromophore feasible. A liquid electrolyte, usually acetonitrile, containing a I^-/I^{3-} redox couple provides the intimate electrical contact to the nanostructured electrode and an inert counter-electrode.

For a variety of reasons, the anatase form of TiO_2 appears to be the most effective. Commercial powder, such as Degussa P25 or TiO_2 colloids in a sol-gel process, can be used as precursors in wet-chemical synthesis of porous nanostructured titania photo-electrodes. The morphology of these highly porous electrodes is an agglomeration of spherical particles, 10–30 nm in diameter, and is independent of the type of wet-chemical synthesis process: only the diameters of the particles and their degree of agglomeration can be influenced.

While using the nanostructured approach an operational solar cell with a conversion efficiency of 10% has been realised [1–3], such a liquid-electrolyte based solar cell has disadvantages and research is focussed on a solid-state alternative for the Grätzel cell. An alterative could be the combination of an n-type nanostructured anatase TiO_2 layer with a polymeric p-type conductor, like polythiophene. In order to benefit from the nanostructured approach, the interface between the inorganic and organic semiconductors should also be nanostructured.

It has recently been shown that Chemical Vapor Deposition can be used to deposit fractal type surfaces of anatase TiO_2 [4]. In order to deposit TiO_2 from the gas phase several precursors can be used. These are titanium tetrachloride ($TiCl_4$) and titanium tetra-isopropoxide ($Ti(OCH(CH_3)_2)_4$ (TTIP) in combination with oxygen or water vapor. It has been found that with $TiCl_4$ and O_2 an amorphous deposit is obtained. When $TiCl_4$ and H_2O are used as precursors crystalline TiO_2 is obtained, but X-ray analysis reveals that this deposit does not exhibit the normal anatase structure, because a marked expansion of the lattice is observed. This expansion is related to the incorporation of chloride ions on oxide ion lattice sites as the ionic radius of the chloride ion (181 pm) is larger than that of the oxide ion (132 pm). In addition, the chloride ion on an oxide ion lattice site represents a point defect with an effective positive charge which is electrically compensated for by electrons. This defect structure can account for the deviation from the anatase structure and enhanced electronic conductivity.

In TTIP both titanium and oxygen are present. This precursor could in principle be employed without an oxygen containing precursor. However, the addition of O_2 or H_2O acts to enhance the reaction rate. With TTIP smooth and dense anatase TiO_2 films are usually obtained. These findings reveal that gas-phase synthesis using either $TiCl_4$ or TTIP does not lead to nanostructured TiO_2. Further investigations show that if the CVD conditions were modified to a new "mixed regime", that is the combination of $TiCl_4$ and TTIP together with a mixture of O_2 and H_20, surprising results can be obtained [4]. The substrate temperature turns out to be critical, and values between 300 and 350°C were selected. With these specific combinations of precursors, the mixed regime (diffusion-limited CVD regime for $TiCl_4$ and reaction-limited CVD regime for TTIP) is obtained. In this mixed regime fractal anatase TiO_2 surfaces are obtained. The fractals are nanostructured.

Photovoltaic cells, based on dye-sensitization of the TiO_2 layers with this fractal morphology and using *cis*-$(NCS)_2$-bis(4,4-dicarboxy-2,2-bipyridine)-ruthenium(II) dye give reasonable efficiencies of about 3% provided that the film thickness is about 50 mm, which is approximately five times the optimal thickness in case of the particulate electrodes. Typical values obtained with AM 1.5 irradiation (1000 W/m^2) are: open-circuit photovoltage $V_{OC} = 0.6$ V, short-circuit current $I_{SC} = 9$ mA/cm^2 (geometrical area), and the fill factor $FF = 0.5$. Reducing the light intensity to 500 W/m^2 has no significant influence on the cell efficiency. We realize that the present PV cells are less efficient than the ones reported by Grätzel [1–3] and further optimization is therefore necessary. However, the present CVD morphologies clearly indicate that it is feasible to design an all-solid-state Grätzel-type solar cell using the combination of this fractal morphology of anatase TiO_2 and a p-type conducting polymer. In our laboratory research is focusing on this laminar combination of inorganic n-type and organic p-type materials based on fractal interfaces between the two components.

3. Lithium-ion Batteries

The advantages of rechargeable Li-ion batteries, such as high-energy density, high cell voltage, and cyclability outclass conventional battery systems. Li-ion batteries are rechargeable batteries and often referred to as swing-type or rocking- chair type batteries. Li-ion batteries use an intercalation anode and an intercalation cathode, separated by an electrolyte. This electrolyte is usually a liquid or a polymer-based material, but also ceramic or glassy electrolytes are possible. The liquid electrolytes consist of Li-salts in a stable solvent, normally absorbed in an inert porous polymer separator. The polymer-based electrolytes can be a free-standing foil or a mixture of a polymer and a liquid. The latter can be a gel or also be absorbed in a porous polymer separator. Different cell geometries are possible, depending on, for instance, the type of electrolyte used. A schematic drawing of the discharge process of the Li-ion battery is given in Figure 1 [5].

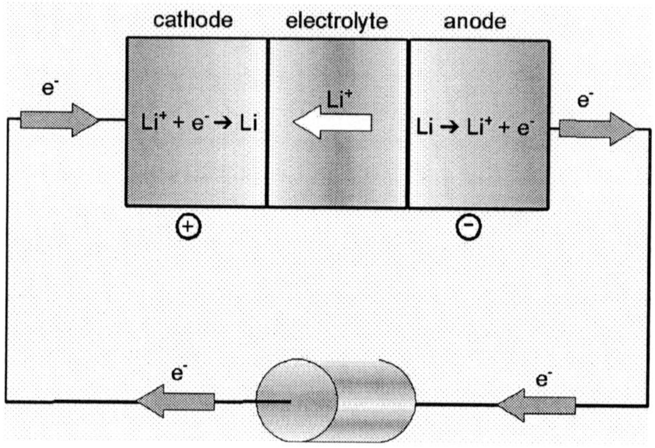

Figure 1. Schematic drawing of a discharging Li-ion battery with simplified overall electrode reactions [5]

A major drawback of the utilization of liquid or polymer-based electrolytes is their inherent instability, caused by high battery voltages and/or temperatures exceeding 50°C. Under these circumstances capacity fade and decrease in cycle life occur, while decomposition of the electrolyte often results in gas evolution leading to irreversible processes and high pressures, which may be detrimental to the battery [6–8].

By using a ceramic electrolyte these problems may be solved, because the stability of a solid electrolyte versus the electrodes is higher, leading to a longer cycle life. Furthermore, the decomposition potential of ceramic electrolytes is higher.

In our recent studies we have found that lithium-doped BPO_4 has excellent lithium-ion conducting properties. This solid electrolyte is synthesized using a very simple synthesis route with readily available precursor compounds, i.e. B_2O_3, P_2O_5, or boric acid and phosphoric acid [5,9–11]. Under appropriate synthesis conditions, nanosized particles of lithium-doped BPO_4 are obtained, i.e. in the range of 30 to 70 nm. Using magnetic-pulse dynamic compaction nanostructured microstructures are obtained which reveal a substantial contribution of grain boundary conduction to the total conductivity. In fact, these results indicate that the dynamical behavior of ions in nanostructured materials is governed mainly by enhanced grain boundary conduction [5,11].

The all-solid-state lithium-ion batteries comprise:

1. a transition metal oxide as cathode, typically $LiCoO_2$, $LiNiO_2$, or $LiMn_2O_4$

2. a solid electrolyte, typically a thin layer of Li-doped BPO_4, or a thin-film Solufill™-Li-doped BPO_4 composite [11,12]

3. an anode material, typically lithiated graphite, coke, or nanostructured SnO_2 or TiO_2.

The anode and cathode materials are mixed with 10 wt.% PVDF (Solef, Solvay) and a solvent (1-methyl-2-pyrolidone, C_5H_9NO, Merck). The ratio of solid material to solvent was about 60:40 by weight. The cathode materials were coated on 15 μm aluminium foil by using a doctor blade. Similarily, the anode materials were coated on 20 μm copper foil. The layer thicknesses of both the anode and the cathode were in the range of 100 to 200 μm.

The electrolyte was a free-standing foil (Solufill™, DSM Solutech) [11,12], in which the nanosized electrolyte powder was homogeneously dispersed in a high-molecular weight polyethylene matrix. After extrusion of the mixture into a base tape, biaxial stretching was applied in order to uniformly reduce the foil thickness down to the desired thickness. While the thickness can be reduced to a few micrometers with the proper starting powder, a foil thickness of about 60 μm was used in our studies.

As an insulation foil, 50 μm Kapton™ (Dupont) was used. The insulation and electrolyte foils were cut into strips of 35 mm wide and about 300 mm long. The electrode foils were cut into strips of 30 mm wide and the same length. This difference in foil width was chosen to prevent short circuiting of the electrodes.

A green battery was made by winding the four separate foils on a copper core (diameter 13 mm). The cathode foil was taped onto a copper core to ensure electrical contact.

The components of this green battery were wound until an external diameter of 16 mm was reached. The copper current collector of the anode foil was wrapped around the outside of the winding, ensuring electronic contact with the copper can. Subsequently, the foils were cut and the cell was taped. Finally, the green battery was placed in a copper tube with an internal diameter of 17 mm that fitted exactly around the green battery ensuring electrical contact of the copper current collector of the anode with the can [11].

The green battery was compacted using Magnetic Pulse Compaction (MPC) [10,11]. With MPC a high-energy pressure pulse can be generated with a typical pressure of 1 to 2 GPa and a pulse duration of 100 to 200 μs. The MPC setup consists of three banks of capacitors which are charged each to a preset voltage of 10 kV to give a maximum energy storage of 90 kJ. The charging voltage determines the delivered pressure. Via a spark gap, the stored energy is discharged through an induction coil with a peak current of 150,000 A. By winding this coil in an appropriate way, a radial magnetic field that acts into the center of the coil is generated.

Any electronically conducting material in the coil will pick up this magnetic field and, as a result, will move into the direction of the center. When the green wound battery is centered in the coil, the magnetic field will compress the copper can of the green battery, resulting in its compaction. After compaction of the battery, the powder particles within the battery components, as well as the separate battery components are pressed together.

Figure 2 shows a cross section of a magnetically pulse-compacted solid-state lithium-ion battery. Here both the anode and the cathode were $LiMn_2O_4$, which after charging exhibits an open-circuit voltage between 1.0 and 1.2 V. Prior to compaction, the internal resistance of the green battery with a foil area of 40 cm^2 was 200 kΩ. After compaction, the internal resistance of the uncharged battery was 150 Ω and after charging 8.3 kΩ. This increase in internal resistance is ascribed to the change in electrical conductivity of the electrode materials, as due to charging the lithium content of the electrodes changes.

Figure 2. Cross-section of an MPC-compacted all solid-state rechargeable lithium-ion battery [11]

The batteries, as compacted, were tested with a MACCOR battery tester. The symmetrical cells exhibit indeed an open-circuit voltage between 1.0 and 1.2 V. The current and voltage graphs during charging at 5 V confirm the stability of the battery system up to at least 5 V. The high internal resistance is due to the fact that, prior to this charging, the battery's internal resistance had been measured and, therefore, the battery was already charged to some extent.

4. Storage of Hydrogen

Hydrogen-based energy systems are attracting increasing attention worldwide. A hydrogen economy can be implemented in industrialised as well as in rural parts of the world.

One of the major problems to be solved is related to the storage of hydrogen, in particular, if hydrogen is to be used in the transport sector. Here weight and volume of the storage system have to satisfy stronger demands than required for more stationary decentralised applications, while safety measures have to be rigorous.

For high-pressure storage, heavy and bulky storage vessels are required. While liquified hydrogen seems an attractive storage option from the viewpoint of weight and volume, the storage system has potential risks and requires energy to liquify and to keep the liquid hydrogen liquified.

A number of metals, metallic glasses, and intermetallic compounds are able to reversibly absorb a considerable amount of hydrogen under ambient conditions of pressure and temperature, while desorption occurs at slightly elevated temperatures. Important research and development has been performed in the past decades [13–17]. A number of hydrogen storage options is gathered in Table I [18].

Table I. Comparison of the hydrogen content of various hydrogen compounds. N_H represents the number of hydrogen atoms per unit volume and C_H gives the weight percentages of hydrogen in the various materials.

material	$N_H \times 10^{-22}$ (cm^{-3})	C_H (wt%)
H_2 gas, 150 atm	0.7	100
liquid H_2, 20 K	4	100
liquid CH_4	6	25
TiH_2	9	4
LaH_3	6.5	2.1
MgH_2	6.5	7.6
$TiFeH_2$	6	1.9
$LaNi_5H_6$	5.5	1.4

Important properties of the hydride systems are the absorption and desorption rates. These rates depend on the bulk diffusion constant, on the surface conditions and on the particle size.

Recently, various forms of carbon have been explored for utilisation as hydrogen storage sorbent. Cryoadsorption, in which an activated carbon is used as a sorbent, reveals advantages in the moderate size and weight of the carbon, but suffers from the severe conditions, i.e. liquid nitrogen temperatures and 20 atm of pressure, required to hold the physically adsorbed hydrogen [19,20]. The adsorption of hydrogen on carbon materials at hydrogen pressures up to 125 bar and 23°C is being studied as is exemplified in [21].

To date, the storage of hydrogen in carbon nanotubes (CNT) or nanofibers at ambient temperatures is attracting much attention. Both single-wall carbon nanotubes [22–24] and multi-wall carbon nanotubes are being studied [25]. The optimization of carbon nanotube arrays for hydrogen absorption indicate that the DOE targets for hydrogen storage for fuel cell vehicles are not yet reached [26].

A high hydrogen uptake by alkali-doped carbon nanotubes under ambient pressure and moderate temperatures has been reported by Chen et al. [27].

Lithium- or potassium-doped carbon nanotubes can absorb about 20 or 14 weight percent of hydrogen at moderate (200–400°C) or room temperature, respectively, under ambient pressure. These values are greater than those of metal hydride and cryoadsorption as is shown in Table II.

The hydrogen stored in the lithium- or potassium-doped carbon nanotubes can be released at higher temperatures, and the sorption-desorption cycle can be repeated with little decrease in the sorption capacity. The problems to be addressed relate to the high hydrogen uptake capacity in relation to a special open-edged, layered structure of the nanotubes, as well to the mechanism of the catalytic effect of the alkali metals.

Unlike lithiated turbostratic and other graphites, which are being used as anode-material in rechargeable lithium-ion batteries [28], the graphitic materials used by Chen et al. [27] and doped with alkali metals cannot be used as electrode materials, because the alkali metals are fixed in the CNT's. In fact, these materials are not CNT's, as the major part of the materials comprise graphene sheets in the shape of circular cones with a hollow center

Table II. Comparison of hydrogen storage properties of various systems. W_i – hydrogen uptake capacity (weight % H_2), V_i – hydrogen density (g/liter) [27].

System	T_{absorb} (K)	P_{absorb} (atm)	H_2 density W_i	V_i	Energy density kWh/Kg	kWh/liter
CNT	298–773	1	0.4	3.2	0.133	0.106
Li-doped						
CNT	473∼673	1	20.0	180	6.66	6.0
Graphite	473∼673	1	14.0	280	4.66	9.32
K-doped						
CNT	< 313	1	14.0	12.6	4.66	4.2
Graphite	< 313	1	5.0	60	1.66	2.0
FeTi-H	> 263	25	< 2	96	0.58	3.18
NiMg-H	> 523	25	< 4	81	1.05	2.69
Cryoad-sorption	~ 77	20	~ 5	~ 20	1.66	0.67

The role of the morphology and the shape of the doped CNT's in relation to the absorption of hydrogen needs to be studied in detail. Besides, the question arises as to whether active carbons with similar high surface areas and metal-doped exhibit similar hydrogen absorption results.

Small lithium and potassium clusters are flat. In contact with graphitic carbon the valence electrons of the metal atoms can be accepted by the carbon substrate and hence binding between the metal and hydrogen can be facilitated, thus forming a dipole structure. Research should be directed towards alkali-metal doping of single-wall and multi-wall carbon nanotubes, and the materials reported by Chen et al. [27] and the mechanism of hydrogen absorption in the presence of alkali-metal dopants.

The alkali-metal doped materials need to be optimised with regard to: cyclability and chemical robustness, while the metallic component may also be optimised. Surface treatments, like fluid-bed plasmas, need to be investigated. In case Chemical Vapour Deposition of carbons is applied, co-doping of the metal needs to be investigated. Alloying the alkali-metal dopant with calcium or a Rare Earth metal may lead to improvements.

A breakthrough in hydride storage technology has been reported by applying nanostructured Mg-based hydrides. The use of nanostructured hydride-forming light-weight metals and intermetallic compounds opens up the possibility to absorb and desorb hydrogen within minutes, which is especially important for the transport sector.

Acknowledgement

The author is grateful to his co-workers Dr. A. Goossens, Drs. B. van der Zanden, Dr. E.M. Kelder, Dr.Ir. M.J.G. Jak, Dr. Hong Huang M.Sc., and Ir. F.G.B. Ooms for important contribu-tions to the research described in this paper. Ir. E-L. Malo-ney of Everest Coating is gratefully acknowledged for his contribution to the growth of the fractal TiO2 surface morpho-logies.

References

1. B.O'Regan and M. Grätzel, Nature 353 (1991) 737
2. Hagfeldt and M. Grätzel, Chem. Rev. 95 (1995) 49
3. M.K. Nazeeruddin, L. Kavan, I. Exnar, and M. Grätzel, Chem. Commun. (1997) 1705
4. Goossens, E-L. Maloney, and J. Schoonman, Chem. Vap. Deposition 4 (1998) 109
5. M.J.G. Jak, Dynamic Compaction of Li-ion Battery Components and Batteries. Ph.D. Thesis. Delft University of Technology. Delft. The Netherlands. (1999). ISBN 90-901-3116-7
6. L.A. Dominey, in Lithium Batteries (Industrial Chemistry Library, Volume 5), G. Pistoia (Ed.), Elsevier Science, Amsterdam (1994) 278
7. Z.X. Shu, R.S. McMillan, and J.J. Murray, J. Electrochem. Soc. 140 (1993) L101
8. Z. Jiang, M. Alamgir, and K.M. Abraham, J. Electrochem. Soc. 142 (1995) 333
9. E.M. Kelder, M.J.G. Jak, F. de Lange, and J. Schoonman, Solid State Ionics 85 (1996) 285
10. M.J.G. Jak, E.M. Kelder, J. Schoonman, N.M. van der Pers, and A. Weisenburger, J. Electroceramics 2 (1998) 127
11. M.J.G. Jak, F.G.B. Ooms, E.M. Kelder, W.J. Legerstee, J. Schoonman, and A. Weisenburger, J. Power Sources 80 (1999) 83
12. J. Schoonman and E.M. Kelder, J. Power Sources 68 (1997) 65
13. F.H.M. Spit, The Thermodynamics and Kinetics of Hydrogen Solution in some Metallic Glasses. Ph.D. Thesis. Utrecht University, 1982.
14. F.A. Kuijpers, RCo5-H and Related Systems. Ph.D. Thesis. Delft University of Technology, 1973. Philips Research Reports Suppl. No.2 (1973)
15. J.J.G. Willems, Metal Hydride Electrodes Stability of LaNi5-related Compounds. Ph.D. Thesis. Eindhoven University of Technology, 1984.
16. P.H.L. Notten and R.E.F. Einerhand, Advanced Materials 3 (1991)343.
17. P.H.L. Notten and P.Hokkeling, J. Electrochem. Soc. 138 (1991) 1877.
18. K.H.J. Buschow, P.C.P. Bouten, and A.R. Miedema, Rep. Prog. Phys. 45 (1982) 937.
19. J. Nitsch, W. Peschka, W. Schnurnberger, M. Fischer, and H. Eichert, in Hydrogen as an Energy Carrier. C. Winter and J. Nitsch eds. Springer Verlag, Berlin, 1988, Part B.
20. S. Hynek, W. Fuller, and J. Bentley, Int. J. Hydrogen Energy 22 (1997) 601.
21. R. Strobel, L. Jorissen, T. Schliermann, V. Trapp, W. Schutz, K. Bohmhammel, G. Wolf, and J. Garche, J. Power Sources 84 (1999) 221.
22. C. Liu, Y.Y. Fan, M. Liu, H.T. Cong, H.M. Cheng, and M.S. Dresselhaus, Science 286 (1999) 1127.
23. P.A. Gordon and P.B. Saeger, Ind.& Eng. Chem. Res. 38 (1999) 4647.
24. Y. Ye, C.C. Ahn, C. Witham, B. Fultz, J. Liu, A.G. Rinzler, D. Colbert, K.A. Smith, and R.E. Smalley, Appl. Phys. Letters, 74 (1999) 2307.
25. X.B. Wu, P. Chen, J. Lin, and K.L. Tan, Int. J. Hydrogen Energy, 25 (2000) 261.
26. Q.Y. Wang and J.K. Johnson, J. Phys. Chem B, 103 (1999) 4809.
27. P. Chen, X. Wu, J. Lin, and K.L. Tan, Science 285 (1999) 91.

Nanoscience & Nanotechnology
eds. E. Balabanova, I. Dragieva, Heron Press, Sofia, 2001

OPTICAL PROPERTIES OF SiO_2 THIN FILMS CONTAINING Ag NANOCLUSTERS

M. Nikolaeva[1], M. Sendova-Vassileva[1], D. Dimova-Malinovska[1], J.C. Pivin[2]

[1]Central Laboratory for Solar Energy and New Energy Sources, Bulgarian Academy of Sciences, 27 Tzarigradsko Chaussee, 1784 Sofia, Bulgaria
[2]CSNSM, IN2P3-CNRS, Batiment 108, 91405 Orsay Campus, France

Abstract. Ag nanoclusters in a SiO_2 matrix are formed in two different kinds of structures. These are co-sputtered SiO_2:Ag thin films and sandwich SiO_2/Ag/SiO_2 structures in which the SiO_2 and Ag layers are electron-gun evaporated. After the deposition these samples are ion irradiated. As a result of this metal clusters are formed. In co-sputtered films their size is 4–5 nm, while in the sandwich films two different types of clusters appear – macro- and nano-clusters. These systems exhibit a surface plasmon resonance in the visible. The plasmon resonance was characterized by optical transmission measurements and interpreted on the basis of TEM observations. In co-sputtered films the intensity of the plasmon peak increases and the peak becomes narrower and shifts toward lower energy with the ion fluence. In the sandwich films the plasmon resonance shift toward higher energy with ion irradiation and its position saturates at 3.15 eV. In this case the plasmon resonance is mainly due to the macroclusters.

Keywords: nanostructures, Ag clusters, SiO_2.

1. Introduction

The formation of metal nanoclusters in a dielectric medium provides it with interesting non-linear optical properties, due to enhanced third-order susceptibility, with potential applications in optoelectronics, e.g. nonlinear waveguide devices. These systems are used for passive optical elements, such as filters, as well [1]. Nanometer-sized clusters of some metal in a dielectric medium exhibit a strong characteristic absorption peak, due to surface plasmon resonance, which is absent in the bulk metal. The shape, intensity and position of the peak depend on the size and shape of the nanoclusters [2]. In this paper we study the optical absorption of SiO_2 thin films containing Ag nanoclusters prepared by two methods – rf magnetron co-sputtering and electron-gun evaporation of sandwich layers followed by ion irradiation.

2. Experimental

The SiO_2:Ag thin films were prepared by magnetron co-sputtering from a complex target in argon atmosphere. Ag chips were put on the surface of the SiO_2 target. The Ag concentration in the films was varied between 0.1 and 0.4 at.% by changing the Ag quantity on the target. The sandwich structures SiO_2/Ag/SiO_2 were prepared by successive electron-gun evaporation of SiO_2 and Ag layers under a residual pressure of 10^{-5} Pa. The thickness of the Ag layer was varied between 4 and 20 nm because it determines the size of the macro-clusters. The samples were irradiated at room temperature with incremented fluences (Φ) of 4.5 MeV Au^+ ions delivered by the ARAMIS accelerator (at CNRS Orsay, France). The ion current was limited to 0.3 μAcm^{-2} and the sample holder cooled with water in order to prevent a target heating over 20°C. The ion range through the presently studied specimen calculated with the TRIM code [3] is 1 μm, greater than the thickness of our samples. So the ions pass through the film and are stopped in the substrate. RBS spectra, TEM experiments and optical transmission were measured to characterize the thin films [4].

3. Results and Discussion

3.1. SiO_2:Ag thin films. The optical absorption spectra of the plasmon resonance of two samples with different Ag concentration are in shown in Figure 1. The as-deposited SiO_2:Ag films exhibit a plasmon resonance, the intensity of which increases with the concentration of Ag. The absorption peaks are broad, flat and not very intense. After irradiation with increasing fluence of heavy ions (Φ) the magnitude of the visible absorption peaks grows and peaks become narrower. Small metal nanoclusters are formed during the deposition, which were observed by TEM. Their size is of the order of magnitude of the resolution ($\sim$ 1 nm) of the TEM instrument so that it could not be measured. They are homogeneously distributed the bulk as demonstrated by RBS. The low intensity of the plasmon resonance in as-deposited films is due to the small size of the clusters.

The size of the particles grows during irradiation with heavy ions by segregation of the dissolved atoms and also probably by Oswald ripening. It reaches 4–5 nm for ion fluence 5×10^{15} ions.cm^{-2}. The observed changes of the shape and position of the plasmon peak can be related to the change of the size of the dispersed metal clusters. The decreasing of the width of a plasmon peak with the radius of nanoclusters is expected theoretically [4]. Experiments to obtain the optimal metal concentration and size distribution of metal clusters for maximum intensity of the plasmon resonance are under way.

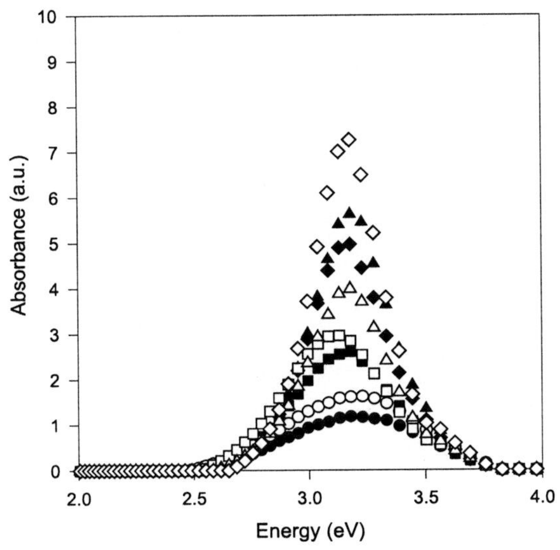

Figure 1. Spectra of optical absorption of co-sputtered SiO_2:Ag films, 750 nm thick, containing 0.1% Ag (full symbols) and 0.4% Ag (open symbols), – as-deposited (circles), irradiated with $\Phi = 1 \times 10^{15}\,cm^{-2}$ (squares), irradiated with $\Phi = 2 \times 10^{15}\,cm^{-2}$ (triangles), irradiated with $\Phi = 5 \times 10^{15}\,cm^{-2}$ (diamonds). The absorption of the SiO_2 matrix was subtracted

3.2. SiO_2/Ag/SiO_2 thin films. After ion mixing of SiO_2/Ag/SiO_2 structures Ag clusters are formed [5]. TEM measurements show (Figure 2) the formation of macro- and nano-clusters. The radii of macro-clusters R_m is roughly proportional to the thickness t_m of the Ag layer and $R_m \approx 1.5t_m$. The radius of nano-clusters decreases with the distance to the macro-clusters from 1.5 to 0.5 nm.

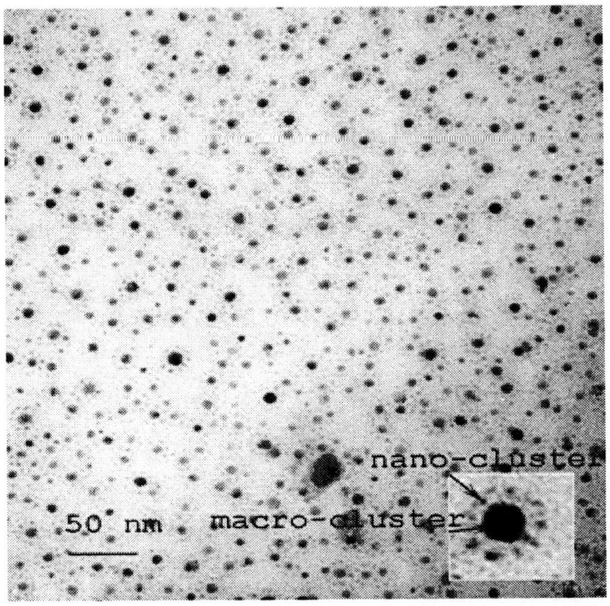

Figure 2. TEM image of Ag clusters in SiO_2/Ag/SiO_2 irradiated with 1×10^{16} ions/cm^2

Figure 3 shows raw optical extinction spectra recorded from an 8 nm Ag layer in SiO_2/Ag/SiO_2. Evaporated layers already exhibit a broad extinction peak in the red region of the visible spectrum, due to their granular nature. The resonance energy in ion beam mixed Ag layers changes until higher fluences are reached and stabilises at a value of 3.15 eV for fluences over 5×10^{15} ions/cm^2, i.e. when macro-clusters observed in TEM are rounded and well separated from each other. These mean resonance energies are very close to those calculated for non-scattering clusters with a diameter in the range of 5 to 30 nm in silica, when quasistatic approximation works [6]. The plasmon resonance of ion beam mixed layers must be ascribed mainly to the macro-clusters formed as a result of their spheroidization. The width (FWHM) of the symmetrical peak after saturation (Figure 3) is 0.28 eV. If the extinction was due to the nano-clusters this width would be in the range 0.6 to 0.8 eV according to the theoretical formula [7] which contains a term proportional to the inverse of the radius of the cluster. The width of the resonance is related to the decrease of the electron free path in small clusters.

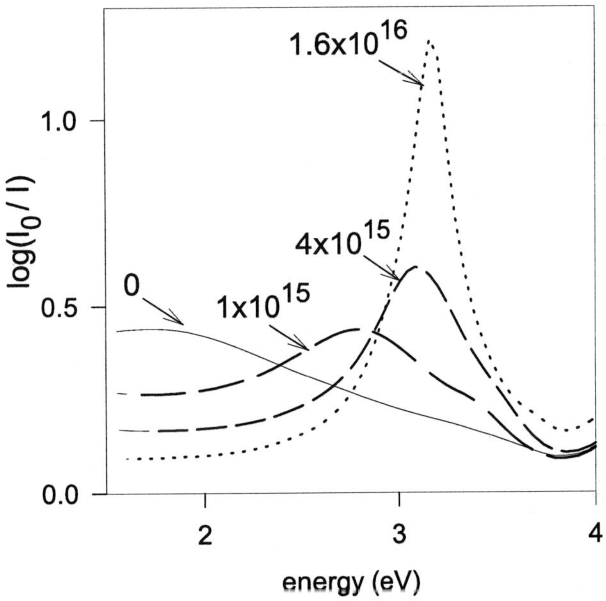

Figure 3. Optical extinction in SiO_2/Ag8nm/SiO_2

4. Conclusion

After irradiation co-sputtered SiO_2:Ag films with high energy ions the size of the Ag metal clusters in them increases. In the case of sandwich structures marco- and nano-clusters of Ag are formed. Both structures exhibit a plasmon resonance at about 3.15 eV. The intensity of this absorption peak grows with the irradiation ion fluence. In ion mixed sandwich structures it is mainly due to the macro-clusters and its position saturates at high ion fluences.

5. Acknowledgement

This study was partially supported by a cooperation project between the Bulgarian Academy of Sciences and CNRS, France.

References

1. R.F. Haglund, Jr., Li Yang, R.H. Magruder III, C.W. White, R.A. Zuhr, Lina Yang, R. Dorsinville and R.R. Alfano, NIM B 91 (1994) 493-504.
2. R.W. Cohen, G.D. Cody, M.D. Coutts, B. Abeles, Phys Rev B 8 (1973) 3689.
3. B.N.J. Persson and A. Liebsch 1982 Solid State Commun. 44 1637
4. M. Nikolaeva, M. Sendova-Vassileva, D. Dimova - Malinovska, J.C. Pivin, In: Proceedings of The meeting of young researchs in physics (1999) p.41
5. J.C. Pivin, D. Dimova-Malinovska, M. Sendova-Vassileva, M. Nikolaeva, A. Martucci, Vacuum (in press)
6. U. Kreibig, L. Genzel, Surf. Sci., 156 (1985) 678.
7. U. Kreibig and Genzel, (1995) "Optical properties of metal clusters", Springer Series in material Science 25 (Berlin:SpringerVerlag)

Nanoscience & Nanotechnology
eds. E. Balabanova, I. Dragieva, Heron Press, Sofia, 2001

INVESTIGATION OF OPTICAL CONSTANTS OF THIN FILMS BASED ON TRANSMISSION SPECTRA MEASUREMENTS

Tamara Pencheva and Milen Nenkov

Rousse University, Physics Department; 8 Studentska Str., 7017 Rousse, Bulgaria

Abstract. Brief description and analysis of colour coordinates method for film optical parameters determination developed by the authors is presented.

Keywords: thin solid films; optical properties

1. Introduction

Thin film optical parameters (film thickness d, refractive index n and extinction k) are often studied after film deposition by various methods based on transmittance and reflectance measurements [1–5]. Optical characteristics of thin dielectric and semiconductor films deposited on transparent substrates may be found from their transmittance spectra $T(\lambda)$.

Classical envelope methods of Manifacier and Swanepoel [2] are based on transmittance spectrum interference extrema and their extrapolation by continuous curves. These methods are appropriate for films of thickness more than 200 nm, because their interference maxima and minima may be easily located and their number is sufficient, see Figure 1a. Films of thickness less than 200 nm can not be analysed by means of envelope methods [2]. The extrema in their spectra become wide and difficult to be located. It makes the application of the envelope methods [2] impossible when the thickness is less than 200 nm (for most of oxide thin films).

Experimental determination of thin film optical characteristics is based on measured transmission or reflection spectra of the film deposited upon transparent substrate. For this aim experimental spectra are fitted with calculated ones from a model of film – substrate optical properties. The models become more complex and developed when refractive index dispersion of the film $n(\lambda)$ and of the substrate $n_S(\lambda)$ together with and the film extinction $k(\lambda)$ are taken into account.

2. Development of curve fitting methods for determination of thin film optical parameters

Development of curve fitting methods in our research group have been directed to determination of the thickness d and refractive index dispersion $n(\lambda)$ of very thin transparent films in the visible spectral region and to investigation of refractive index depth profile (optical inhomogeneity) of weakly absorbing thin films. For this we used many points of transmittance spectra and developed film – substrate optical properties models, fitting procedures and corresponding software.

In our investigation [3] colour coordinates method for determination of optical parameters of transparent very thin films has been proposed. It is based on the numerical minimisation of differences between colour coordinates X, Y, Z corresponding to measured and calculated transmittance spectra. In Figure 1 some typical transmittance spectra of barium titanate thin films with different thickness on glass substrate in the spectral region of the film and the substrate transparence are presented. The spectra of bare substrates are shown as $T_S(\lambda)$, and these of coated ones – as $T(\lambda)$. Spectrum I corresponds to a film of thickness much greater than the critical value of 200 nm. The interference extrema in this spectrum are clearly located and their number is sufficient for application of envelope methods [2]. Spectrum II belongs to a film of thickness near the critical value of 200 nm. The maximum and both minima are wider than those of spectrum I. The total number of extrema in spectrum II is three which makes application of envelopes method still possible. Spectra III and IV correspond to films of thickness less than the critical value of 200 nm. It is difficult to point the exact positions of the minimum in spectrum III. The number of the extrema in spectrum III is only one which makes application of envelopes method impossible. There are no extrema in spectrum IV and this spectrum can not be studied by means of envelopes method too. Figure 1 shows that all maxima are close to bare substrate transmittance and this is a reason to neglect optical absorption in transmittance calculation. Interference minima depth weakly depends on wavelength which indicates very low dispersion.

Description of the approach. The main goal of calculations is to determine the thickness d and optical dispersion $n(\lambda)$ of investigated thin films. For thin films of negligible absorption and weak dispersion (such as barium titanate thin films) the refractive index is often represented by Cauchy expression [2–4]:

$$n(\lambda) = a + b/\lambda^2 \tag{1}$$

where a and b are constants unknown in advance. For every thin film three parameters should be determined: constants a and b and thickness d.

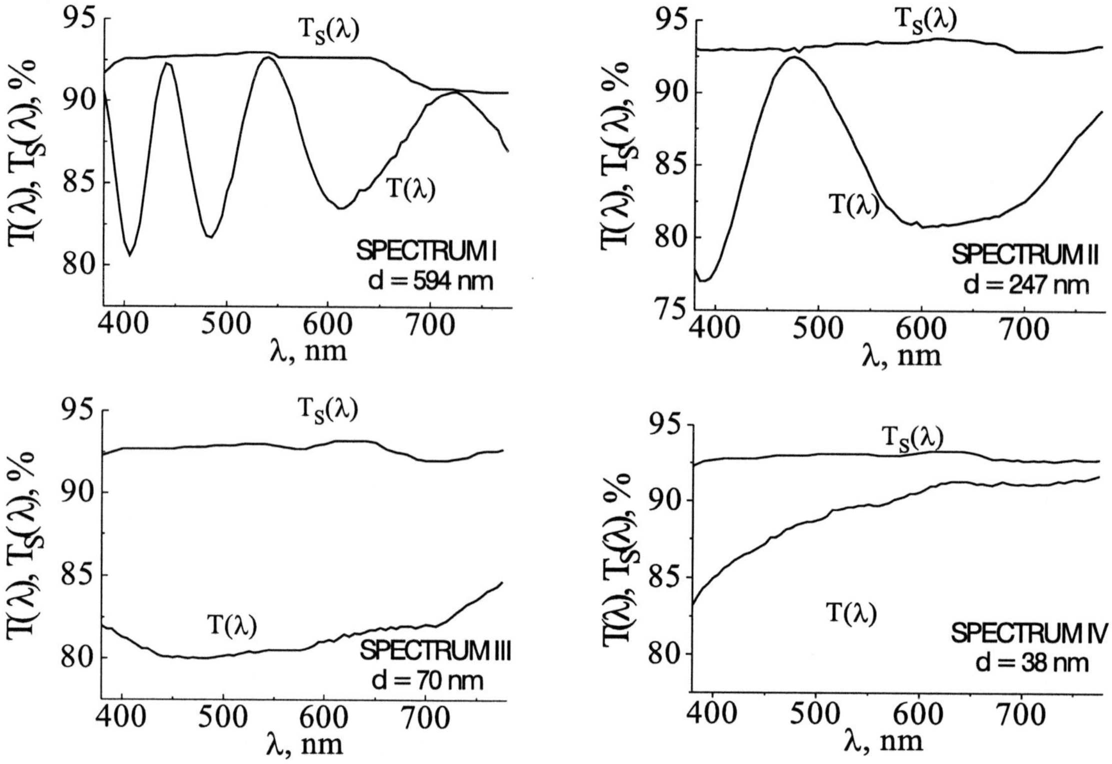

Figure 1. Measured transmittance spectra $T(\lambda)$ of thin barium titanate films of thickness d greater than 200 nm (I), of order of 200 nm (II), and less than 200 nm (III and IV); $T_S(\lambda)$ – transmittance of bare substrates

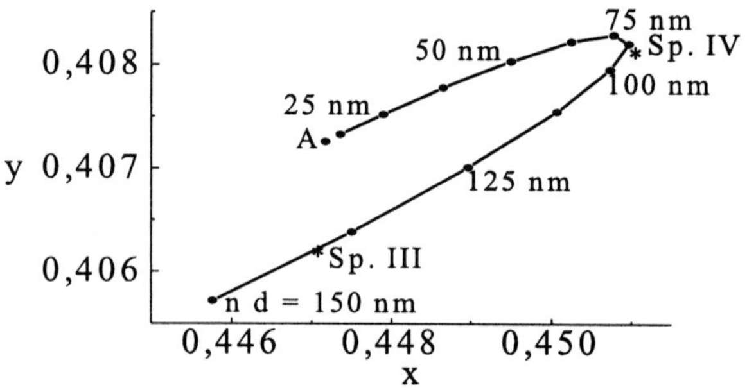

Figure 2: Chromaticity coordinates (A standard illuminant) of calculated transmittance spectra for films with refractive index $n = 2 = \text{const}$ and optical thickness in the range 0...150 nm

The colour coordinates method consists of two steps. In the first step the refractive index n is supposed to be constant value, i.e. $n(\lambda) = a = \text{const}$. It is reasonable for films with weak dispersion in cases where $a \gg b/\lambda^2$ which is valid for barium titanate thin films [3]. In this way, the unknown quantities in this step are constant a and thickness d. They are determined from the following system of two equations:

$$x(a,d) = x_0, \qquad y(a,d) = y_0. \tag{2}$$

In equations (2) $x(a,d)$ and $y(a,d)$ are chromaticity coordinates [3] corresponding to theoretically calculated spectrum of a film with refractive index $n(\lambda) = a$ and thickness d. x_0 and y_0 are numerical values of chromaticity coordinates corresponding to measured transmittance spectrum. The system of equations (2) is a non-linear one and it is solved graphically, Figure 2. The solution of system (2) gives initial values of parameters a and d which are used in the second step.

In the second step the final values of all unknown quantities (a, b and d) are determined from the following system of three equations:

$$X(a,b,d) = X_0, \; Y(a,b,d) = Y_0, \; Z(a,b,d) = Z_0. \tag{3}$$

In equations (3) $X(a,b,d)$, $Y(a,b,d)$, and $Z(a,b,d)$ are colour coordinates [3] corresponding to theoretically calculated spectrum of a film with refractive index $n(\lambda) = a + b/\lambda^2$ and thickness d. X_0, Y_0, and Z_0 are numerical values of colour coordinates corresponding to measured transmittance spectrum. The system of equations (3) is also non-linear one and it is solved by numerical minimisation of the following function:

$$Q = [X(a,b,d) - X_0]^2 + [Y(a,b,d) - Y_0]^2 + [Z(a,b,d) - Z_0]^2. \tag{4}$$

In our minimisation procedure each iteration consists of several steps which number is equal to the number of the parameters to be determined. Each step is one dimensional minimisation carried out by means of the golden section method. This iterative minimisation is terminated when the decrease of Q (between two iterations) is found to be negligible. Evaluation of colour coordinates method accuracy allows to conclude that this method is more precise than envelope methods because it uses not only extrema points but the whole spectrum [3,5].

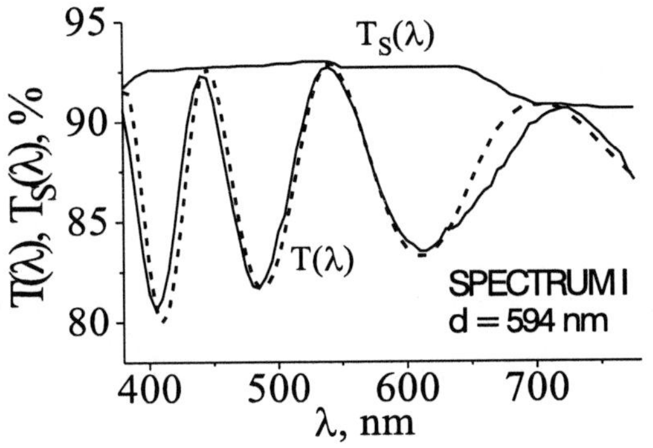

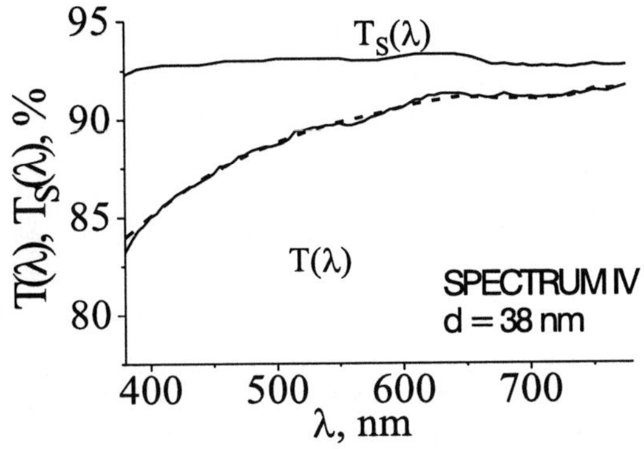

Figure 3. Agreement between calculated and measured transmittance spectra for the thinnest and thickest films presented in Figure 1

Results. Figure 3 shows agreement between calculated and measured transmittance spectra for the thinnest and thickest films presented in Figure 1. One can see that calculations describe measurements very well. The averaged quadratic differences between measured and calculated transmittance in the whole spectral region are not more than 0.5%, which is of the same order as experimental uncertainty of transmittance measurements. All investigated films demonstrate normal dispersion with numerical values of refractive index in the range of 1.8...2.1.

3. Evaluation of minimal film thickness

It can be seen from Figures 1 and 2 that it is possible to decrease the film thickness so that it will be considerable less than 200 nm when in the visible spectral region there is no any interference pattern. A reasonable question is which minimal optical and physical film thickness allows processing by method of colour coordinates using transmittance spectra only? To answer this question we assume that for very thin film (see spectrum IV in Figure 1) difference between film and substrate spectra should be at least 2% in the spectral region under investigation. According to [1,2] it is possible to show that the film transmittance T and substrate transmittance T_S are connected in this way:

$$\frac{1}{T}-\frac{1}{T_S}=\left(\frac{n_S n_0}{n^2}+\frac{n^2}{n_S n_0}-\frac{n_0}{n_S}-\frac{n_S}{n_0}\right)\frac{\pi^2 n^2 d^2}{\lambda^2}. \quad (5)$$

Evaluation results for $d_{\min}$ at $T_S = 92\%$ and $T = 90\%$ and $\lambda = 0.8$ mm are given in Figure 4. Numerical values for $d_{\min}$ given in Figure 4 allow to conclude that the minimal film thickness for application of colour coordinates method depends non-linearly on the film refractive index. The minimal thickness is smaller for films of higher refractive index n. When the difference between film and substrate indices decreases the minimal thickness increases.

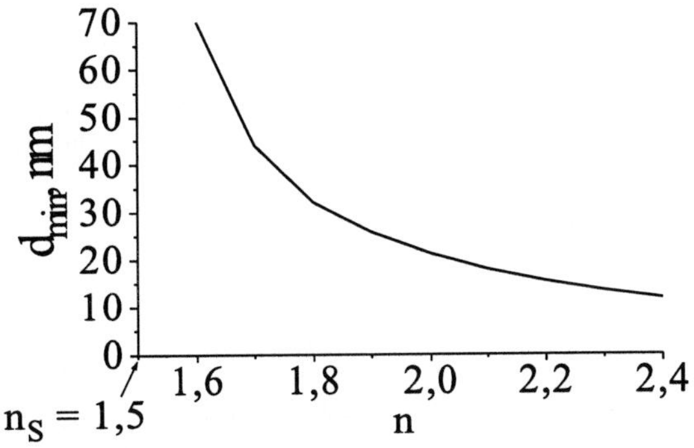

Figure 4. Minimal film thickness $d_{\min}$ for colour coordinates method in the visible for often used values of refractive index n of a film on a glass substrate

4. Conclusions

Brief description and analysis of colour coordinates method for film optical parameters determination developed by the authors is presented.

The measured transmittance spectra of a sequence of barium titanate thin films are analysed by colour coordinates method in order to obtain the thickness and optical dispersion of the films. It is shown that the accuracy of proposed method is higher than for conventional envelop methods. It is a result of fitting procedure using 80 points of the spectra. For often used optical materials evaluation of minimal film thickness, for which the developed methods for film optical parameters determination are possible to be applied, is performed.

References

1. Macleod H. A. "Thin Film Optical Filters", Macmillan, New York (1986) 590.
2. Swanepoel R. Determination of the thickness and optical constants of amorphous silicon. J. of Physics, E: Scientific Instruments 16 (1983) 1214–1222.
3. Pencheva T. and M. Nenkov, "Some Possible Applications of Chromatic Parameters to Thin Films Optical Constants Determination", J. Mod. Opt., 1996, vol. 43, pp. 2449–2462.
4. Dobrowolski J. A., F. C. Ho and A. Waldorf. Determination of optical constants of thin film coating materials based on inverse synthesis. Appl. Opt. 22 (1983) 3191–3200.
5. Nenkov M. and T. Pencheva, "Calculation of Thin-film Optical Constants by Transmittance Spectra Fitting", J. Opt. Soc. Am. A, 1998, vol. 15 pp. 1852–1857.

Nanoscience & Nanotechnology
eds. E. Balabanova, I. Dragieva, Heron Press, Sofia, 2001

SOL-GEL GLASSES AND NANOMATERIALS ON THE BASE OF SiO_2

B. Samuneva, Y. Dimitriev, V. Dimitrov, E. Kashchieva, G. Encheva

University of Chemical Technology & Metallurgy, blvd. Kl. Okhridski 8, Sofia 1756, Bulgaria

Abstract. Silica gels containing Ag_2O and PtO_2 have been synthesized. The gels have been heat treated in the 25–1000°C temperature range. It has been established that the temperature of gel to gel-glass transition decreases when the Ag_2O and PtO_2 content in the gels increases.

Special attention is paid to the formation of silver and platinum metallic particles in amorphous materials. The sizes of the metal particles are determined to be from 3 to 25 nm. The crystallization processes in the gels heat treated at 1000–1200°C are carried out and besides the silver and platinum particles the crystalline phase of α-crystobalite is separated.

It has been shown that nanocomposite materials containing ultrafine silver and platinum particles in SiO_2 amorphous or polycrystalline matrixes can be obtained.

Keywords: silica gel, nanoparticles, silver, platinum

1. Introduction

During the last years the reasurch in the field of sol-gel synthesis of nanocomposites containing ultrafine particles of noble metals (Au, Ag, Pt and Pd) and oxide matrixes (SiO_2, TiO_2, V_2O_5 and WO_3) has been rapidly developed [1–7]. Many authors have studied the preparation and properties of thin coating films of such composite materials containing metal particles [8–18].

Using the low temperature sol-gel technology the nanocomposites with high homogeneity, high purity and high concentration of metal particles can be obtained. It is proved that the silver nanoparticles prepared by sol-gel route in SiO_2 matrix show high antibacterial activity [19].

The purpose of this study is to obtain bulk gels in the SiO_2-Ag_2O and SiO_2-PtO_2 systems in wide composition range and to study the structural and phase evolution with increasing of the temperature.

2. Experimental

The gels of the SiO_2-Ag_2O and SiO_2-PtO_2 systems were prepared using TEOS, C_2H_5OH, H_2O, HNO_3, HCL, $AgNO_3$ and $PtCl_4$ as precursors. Volume ration of the main precursors TEOS: C_2H_5OH: H_2O: $HCl(HNO_3)$ was 1:0.72:0.77:0.33.

The content of Ag_2O in the gel composition is: 0.05, 0.1, 0.2, 0.5, 1.0, 2.0, 3.0, 5.0, 10.0, and 20.0 mol.%. The content of PtO_2 in the gel composition is: 0.1, 0.2, 0.5, and 1.0 mol.%. The content of Ag and Pt in the gels and gel glasses is presented in mol.% of their oxides Ag_2O and PtO_2 because of two reasons: – usually the composition of the glassy materials is presented with the content of the oxides of the chemical elements; – at termal treatment of the gels in most of the cases paticles of Ag and Pt are formed, but for some of the samples the ratio of the quantities of the metal in ionic and atomic form is different and in the present work is not determined.

The starting solution were obtained by sequentially addition and stiring of TEOS and C_2H_5OH, H_2O and HCL and solution of $AgNO_3$ or $PtCl_4$. The gelling time was from 1 to 8 days depending of the metal oxide concentration. It was establiched that the gelling time increases with the increasing in the metal oxide content. The gels were treated at 100°C for 12 hours and at 200°C, 400°C, 600°C, 800°C, 1000°C and only for the gels containing Pt at 1200°C for two hours.

The IR-spectra were recorded by SPECORD-M-80 spectrofotometer in the 1300–300 cm^{-1} range.

X-ray diffraction patterns were obtained using DRON-IM diffractometer (CuKa-radiation, Ni-filter).

The microstructure of the gels was studied using electron microscope EM-400 Philips.

In the present work are presented results, which have not been published in previous article [20], concerning the same problem.

3. Results and Discussion

Colorless silica gels containing up to 20.0 mol.% Ag_2O and light yellow colored gels containing up to 1.0 mol.% PtO_2 have been syntesized. The gelation time was varied from 1 to 8 days. During the heat treatment the color of the gels changed.

Figure 1 shows IR-spectra of four sample containing 0.1, 0.2, 0.5, and 1 mol.% PtO_2, heat treated at 1200°C. In all cases the bands at 1080 cm^{-1}, 800 cm^{-1}, and 460 cm^{-1} are presented in IR-spectra. These bands are assigned to ν^{as}_{SiO}, ν^{s}_{SiO} and ν^{δ}_{SiO}.

It is seen that the band at 960 cm^{-1}, due to the Si-OH vibration is not present. In this way it can be concluded that the processes of network formation are completed at temperatures lower than 1200°C.

This band disappeares at 1000°C or lower temperatures in the cases when SiO_2 gels contain Pt and Ag. This temperature depends on the concentration of the noble metal and the band at 960 cm^{-1} disappears at lower temperatures when the content of Ag_2O encreases [20]. For example, for a gel with 0.05 mol.% Ag_2O this band disappears at

1000°C and for 20 mol.% Ag_2O – at 400°C. The temperature difference of the band disappearence of silica gels containing 0.1 and 1 mol.% PtO_2 is 400 degrees.

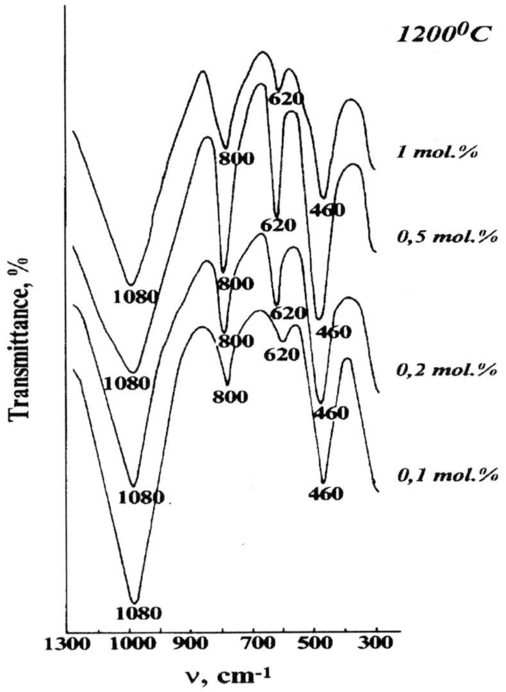

Figure 1. IR spectra of silica gels with 0.1, 0.2, 0.5, and 1.0 mol.% PtO_2 heat treated at 1200°C

The absorption band at 620 cm^{-1} is known to be due to the presence of non metallic crystalline phase – crystobalyte. From the data shown in Figure 1 it can be seen that this band is present in all of the concentration of the PtO_2 in silica gels treated at 1200°C. With increasing the metal concentration up to 0.5 mol.% this band increases, which is an indication for the formation of a bigger amount of cristobalyte. An eception of this case is the sample containing 1 mol.% PtO_2, in which a considerable decrease of this non metallic phase is observed. This is also proved from the X-ray diffraction data of these samples. Having in mind all this, we can conclude that the catalitic role of the Pt for the crystallization processes is limited from the concentration of the metal in the gel glass. At concentrations of 1 mol.% this activity is the lowest in the cases of crystallization at 1200°C.

In Figure 2 X-ray patterns are presented for gel glasses containing 0.2 mol.% PtO_2 thermal treated in the temperature range 200–1200°C.

These results prove the published up to now investigations (X- ray data of gel with 0.1 mol.% PtO_2) that in comparison with Ag non metallic crystalline phase does not appear before 1200°C [20]. From the same figure it has been concluded that in the non metallic phase except the crystoballyte some quantity of alpha quartz also presents. The metallic phase – Pt can be seen clearly at 1200°C, where at lower temperatures the bands are considerably less. This indicates that the above mentioned samples are amorphous in X -ray.

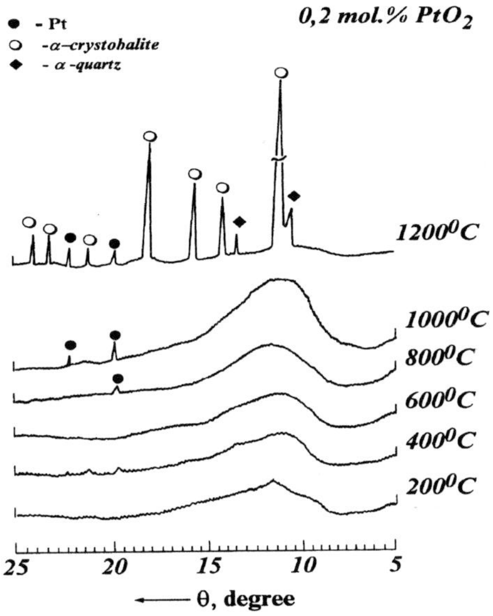

Figure 2. XRD of silica gel with 0.2 mol.% PtO_2 heat treated in 200–1200°C range

The microstructure of some gels, observed by TEM, are given in Figure 3.

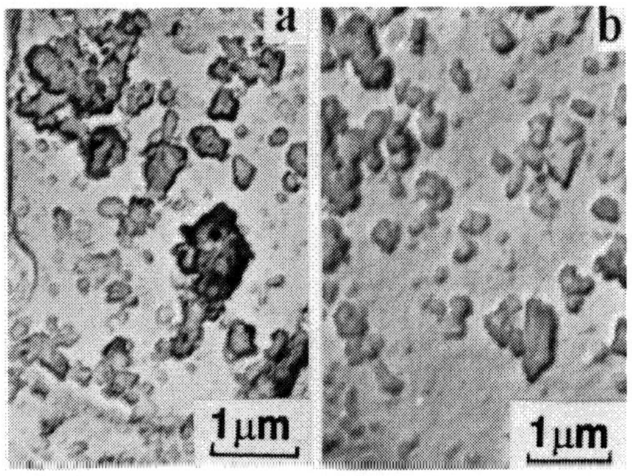

Figure 3. TEM micrographs of silica gels with 0.2 mol.% PtO_2 after heat treatment at 800°C (a) and 1000°C (b)

The observation of the gel powder particles containing 0.2 mol.% PtO_2 allows to determ their sizes beeing between 100 nm and 500 nm. They keep these values at different temperatures – 800°C (Figure 3a) and 1000°C (Figure 3b). In this case separated platinum particals are not observed, that can be related to the small PtO_2 content in the gels and their probable ultrafine sizes that can explain also the presence of small amount of the platinum phase established by X-ray analysis data.

In previous article [20] the silica gels containing 2.0 mol.% Ag_2O at 100°C the presence of silver particles extracted in the C+Pt replicas are found. The increase of the temperature up to 200°C leads to the increase of the concentration and the size of the silver particles. It was found that the particle size in this case in between 3 and 10 nm. At higher concentration of Ag_2O the size of silver particles increase to 25 nm. This results are in agreement whit the data published by G. De et al. [13].

4. Conclusion

Silica gels containing from 0.05 to 20.0 mol.% Ag_2O and from 0.1 to 1.0 mol.% PtO_2 have been obtained at room temperature. It was found that the gel-glass transition temperature decreases whit the increasing of metal oxide concentration. The silver stimulate the formation of the α-crystobalite phase at lower temperature.

According to the XRD results the platinum crystalline phases are found at 800°C and higher temperatures. The nonmetallic crystalline phases (mostly crystobalite) are formed only at 1200°C. By TEM observation it was proved that silver particles are formed and uniformly distributed in matrix. The size of the particles is in the 3–25 nm range and depends from the initial concentration of the silver oxide and the heat treatment temperature.

The observation of the gel powder particles containing 0.2 mol.% PtO_2 allows to determ their sizes beeing between 100 nm and 500 nm. In this case separated platinum particals are not observed, that can be related to the small Pt content in the gels and their probable ultrafine sizes.

References

1. R. Reisfeld, M. Eyal, and D. Brusilovsky, Chem. Phys. Lett., 153, 210 (1988).
2. A. Tour, S. Pendalwar and J. Cooper, Chem. Mater., 2, 647 (1990).
3. Z. Spanhel, M. Menning, and H. Schmidt, Proc. of 17-th Inter. Congress on Glass, Madrid 7, (1992), 9.
4. M. Menning, Z. Spanhel, H. Schmidt and S. Bedholz, J.Non-Cryst. Sol. 147-148, 326 (1992).
5. D. Korvin, and Z. Pye, Glastech. Ber. 67 C, 296 (1994).
6. T. Tsuchiya, T. Sei. Y. Watanabe, M. Mori, and T. Kinari, J. Sol-Gel Sci. Techn. 8, 685 (1997).
7. H. Kozuka, and G. Zhao, Bull. Inst. Chem. Res., Kyoto Univ. 72, 209 (1994).
8. P. Innocenzi, H. Kozuka, and S. Sakka, J. Sol-Gel Sci. Techn. 1, 305 (1994).
9. H. Kozuka, and S, Sakka, Chem. Mater. 5, 222 (1993).
10. H. Kozuka, G. Zhao, and S. Sakka, J. Sol-Gel Sci. Techn. 2, 741 (1994)
11. S. Sakka, H. Kozuka, and G. Zhao, Proc. of Sol-Gel Optics III, San Diego, SPIE v. 2288(1994)p.108
12. B. Breitscheidel, J. Ziedel, and M. Schubert, Chem. Mater. 3, 559 (1991).
13. G. De, A. Licciulli, C. Massaro, L. Tapher, M. Gatalano, G. Bayyaglin, C. Meneghini, and P. Mazzoldi, J. Non-Cryst. Sol. 194, 225 (1996).
14. G. Battaglin, F. Caccavale, F. Gonella, G. Matei, P. Mazzoldi, F. Spizzo, A. Quaranta, F. Garrido, and G. De, Proc. Inter. Symp. On Glass Problems, Vol. 1, Ed.: R. Akcakaya, N. Erinc, G. Albayrak, S. Isevi, Sept.4-6, 1996, Istanbul, Turkey.
15. M. Suzuki, and Y. Taga, J. Non-Cryst. Sol. 150, 148 (1992).
16. B. Ritzer, M. Villegas and J. Fernandez Navarro, J. Sol-Gel Sci. Techn. 8, 917 (1997).
17. M. Epifani, C. Giannini, L .Tapfer and L. Vasanelli, J.Am. Cer. Society 83, 10 (2000)
18. S. Tamil Selvan, M. Nogami, Ar. Nakamura and Y. Hamanaka, J. Non-Cryst. Sol. 254-258, (2-3) 255 (1999)
19. C. Trapalis, S. Spartalis, N. Boukos, A. Kontogeorgakos, P. Bolanis and G. Kordas, Proceedings of the First Balkan Conference on Glass Science and Technology, Volos-Greece, 2000, 516.
20. B. Samuneva, Y. Dimitriev, V. Dimitrov, E. Kashchieva and G. Encheva, J. Sol-Gel Sci. Techn. 13, 969 (1998)

Nanoscience & Nanotechnology
eds. E. Balabanova, I. Dragieva, Heron Press, Sofia, 2001

PREPARATION OF NANOSIZED TiO_2 BY SOL-GEL METHOD AND POSTSYNTHESIS HYDROTHERMAL TREATMENT

P. Stefchev[1], M. Machkova[2], L.Dimitrov[3], V. Blaskov[4], K. Ivanova[1],V. Kozhukharov[2], P. Vitanov[1]

[1]Central Laboratory of Solar Energy and New Energy Sources, Bulg. Acad. Sci., 1784-Sofia, Bulgaria;
[2]University of Chemical Technology and Metallurgy, Sofia-1756, Bulgaria;
[3]Institute of Catalysis, Sofia-1113 Sofia, Bulgaria;
[4]Institute of General and Inorganic Chemistry - BAS, 1113 Sofia, Bulgaria

Abstract. The nanosized TiO_2 is a very promising compound applied for: the preparation of ceramic for high technologies, support in the heterogeneous catalysis, photocatalysis, as a sensor for gas indication, membrance etc. Nanocrystalline TiO_2 has been obtained by sol-gel method. Two kind of stabilized titania sol are prepared. The process of hydrolysis is regulated by addition of acetylacetone to one of the solutions. The final nanocrystalline powder is obtained by two types of treatments: i) heating in the range 80–700°C, ii) postsynthesis hydrothermal treatment of dried powder. The size of the crystallite of obtained powders is in the region 8.7–31.3 nm and specific surface area varies from 11–210 m^2/g. A special attention is drawn to the formation of anatase crystalline modification which is a very perspective as photocatalyst .The characterization of the obtained samples is carried out by the methods of: thermal analysis, XRD (including structural analysis), UV, VIS spectral analysis, and modified BET method.

Keywords: nanocrystaline TiO_2, sol-get method, hydrothermal treatment

1. Introduction

It is well known that the TiO_2 has shown a great promise including useful properties as:

- ability to catalyze the destruction of the wide variety of organic toxic contaminants;
- removing heavy metal ions from solution and converting them into less toxic ones;
- completing the oxidation process from water organic contaminants to CO_2 and diluting mineral acids
- a lack of inherent toxicity;
- high resistance to photodegradation and water hydrolysis processes and
- low cost [1-4].

These useful properties offer a front line of TiO_2 as an advance material in the new concepts of nanotechnology ecology. The problem can be resolved in a future progress in both directions: a better catalyst and a better photoreactor design. The nanocrystaline TiO_2 powder has an advantage in relation to photocatalytical properties. The above cited small particles of TiO_2 photocatalyst can be synthesized through several methods, e.g. the sol- gel organic method [5,6], the hydrothermal method [7] and via hydrolysis of inorganic salts [8].

The aim of the present work is to obtain nanosized TiO_2 powder via the sol- gel route; and postsynthesis hydrothermal treatment, to characterize the powders obtained having in mind their phase evolution in the course of thermal treatment up to 700°C and potential usage.

2. Experimental

Methods of preparation. Two initial samples were an object of preparation and next studies. Sample 1 was prepared by a hydrolysis process of titanium tetraisopropoxide [$Ti(C_3H_7O)_4$] in water solution of HNO_3. The obtained precipitate was peptized by heating at 80°C for 8 hours. Stable sol containing 5 wt.% TiO_2 was formed. Acetyl acetone (AcAc) in addition was added to titanium tetraisopropoxide [$Ti(C_3H_7O)_4$] for sample 2 synthesis. The obtained soles were dried at isothermal conditions (80°C for 24 h). Part of dried powder is hydrothermal treated in autoclave at pH=1 (sample 3) and pH=8,5 (sample 4). After hydrothermal treatment the samples were heated at 230°C. The xerogels achieved were heated from 370°C to 700°C for 4 h. to identify the most effective crystallization temperature.

Methods applied for characterization. Thermal analysis was performed up to 1000°C at heating rate 10°C/min; A_2O_3 heated up to 1200°C was used as inert sample. X-ray diffraction (XRD) patterns are collected at 295 K using Cu K_α ($\lambda = 1.5418$ Å) radiation. Infrared (IR) spectroscopy measurements was carried out in KBr tablets in the region from 400 cm^{-1} to 4000 cm^{-1}. Optical measurements (Transmittance (T), Reflectance (R) and Absorption (A)) were performed by Spectrophotometer UV-VIS-NIR, Perkin Elmer 330, USA; standard of $BaSO_4$ was used. Two kinds of samples were prepared for optical measurements as follows: i) sol solutions in a quartz cuvette, and ii) TiO_2 thin films deposited on quartz plate by dipping sol- gel process; regularly dipping process of quartz plates ($1.0 \times 2.5 \times 5$ cm) with drawing speed of 14 cm/min. After treatment at 120°C for 10 min the thin coating samples were heated at

500°C for 1 h. The specific surface area (SSA) was determined by Klyachko-Gurvich apparatus.

3. Results

Figure1 shows heating curves recorded at polythermal scanning regime of powder Sample 1 up to 1000°C.

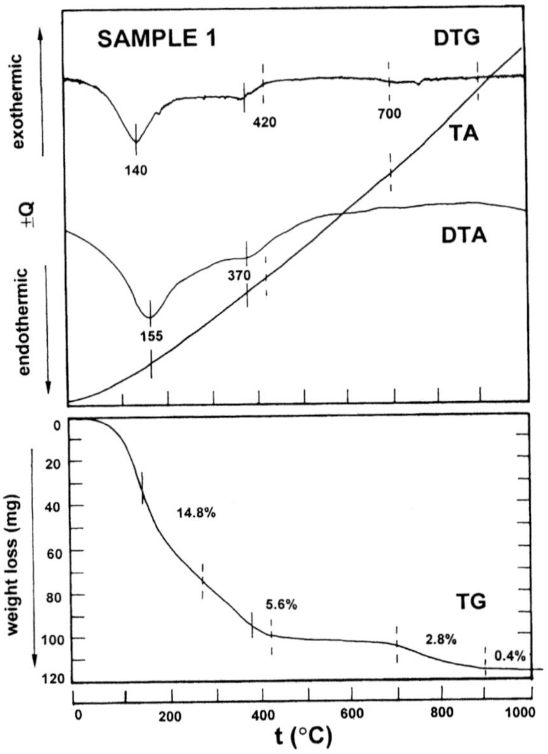

Figure 1. Heating curves (TA, DTA, TG and DTG) for sample 1 in air after primary heating at 80°C for 24 h; sample weight 500 mg

Details about the phase distribution and stability vs. temperature can be extracted from the XRD measurements. Figure 2 shows XRD pattern evolution of powder samples heated in the range from 80°C to 700°C. The aim was to detect if any influence exist after AcAc additive on the phase progress.

The average dimensions of TiO_2 particles varied 8.7–8.8 nm for samples heated at 370°C; for Sample 1 heated at 500°C to 10.4 nm and for Sample 2 heated at the same temperature to 31.3 nm data was checked. The size has been determined according to Sheerer formulae.

After drying and samples heating at 500°C/1 h. the films thickness was measured and values at about $d \approx$ 33 nm were established in a Tally Step profilometer unit. The absorption value was calculated from the spectrum transmittance and reflectance curves. The absorption value, marked by A, can be checked as well. Figure 3 shows the band gap (E_g) value of a TiO_2 coating.

The graphically checked value of the band gap using equation:

$$\alpha^2 = (A')^2(E_{h\nu} - E_g)$$

where A' is the absorption, $E_{h\nu}$ is the photon energy, E_g is the band gap, and α is the absorption coefficient, is 3.9 eV. The value is different in comparison to the well known value of 3.2 eV for the TiO_2- anatase phase in bulk.

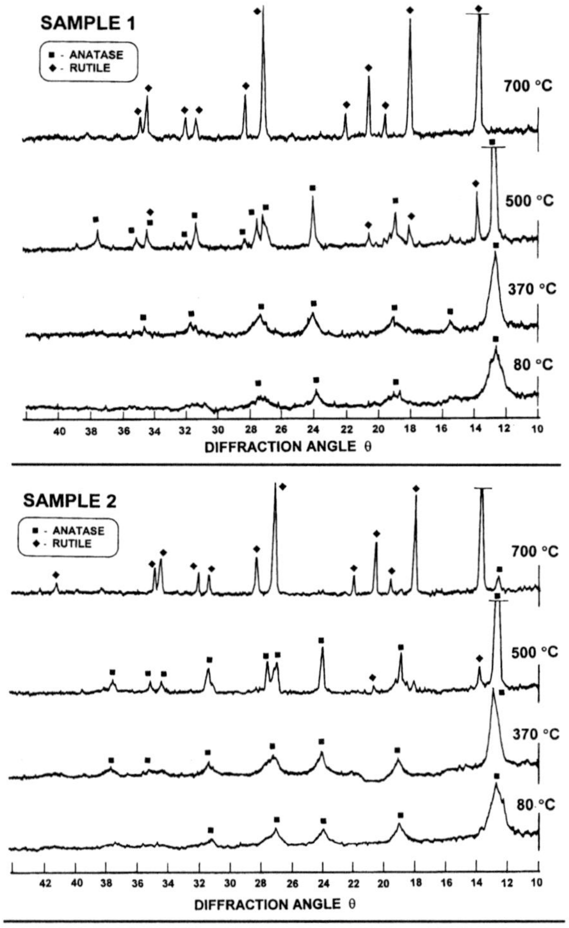

Figure 2. XRD pattern of thermal evolution of Samples 1, 2

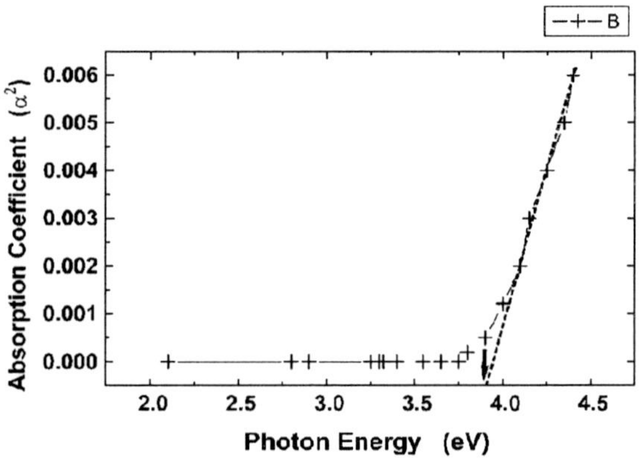

Figure 3. A plot of the absorption coefficient as a function of the photon energy

4. Discussion

The first peak at 155°C is due to a free water content, since the second one at 370°C is a contribution from the organic constituents degradation and next burning process of the C–H units. There are not any thermal effects manifested on the TA, DTA and DTG curves in the heating region from

400°C to 700°C. Above 700°C sample weight loss is 2.8% (700–900°C) and 0.4% (900–1000°C), respectively.

XRD shows that at low temperature region only anatase phase is registered. At 500°C for 4 h a primary crystallization of rutile is manifested. The last phase is dominant at 700°C. After a correlation of XRD data the rutile phase is less in Sample 2. At 700°C, except the rutile phase, a small amount of residual anatase phase is identified. In the present study we establish an anatase to rutile transition at 420°C for Sample 1 and anatase to rutile transition at 440°C for Sample 2., i.e. we have a stable phase that can work up to 450°C.

Table Specific surface areas of the obtained powders

Samples	Temperature (°C)	SSA (m^2/g)
Initial sample 1	80	167
Initial sample 2	80	219
Sample 3	230	155
Sample 4	230	140
Sample 3	370	116
Sample 4	370	120
Sample 1	370	113
Sample 2	370	104
Sample 2	440	79
Sample 3	440	61
Sample 4	440	100
Sample 2	500	11

Sample 1 – precursor without AcAc; Sample 2 – precursor with AcAc; Sample 3, obtained by hydrothermal method at pH =1; Sample 4, obtained by hydrothermal method at pH= 8,5.

Comparison of the experimental transmittance values of the starting sol samples shows that for TiO_2 sol Sample 2 with AcAc the transmittance is higher than in the water base sample preparation. This is well illustrated in the region from 500 nm to 700 nm where ≈ 20% difference exists. The higher E_{gD}-value is based on the fact that some SiO_2/TiO_2 interface influence exist. This is due to silica to titanium oxide surface binding nature and to TiO_2 unit parameters in the film.

5. Conclusions

Nanocrystalline TiO_2 powders are successfully prepared via the sol- gel route and post synthesis hydrothermal treatment.

- The thermal analysis carried out shows that the powder samples possess only endothermal effects. There is a stability state registered on the curves in the region above 400°C.
- XRD powder diffraction measurements at several temperatures indicate that the TiO_2- anatase phase is stable up to 420–440°C. At high temperature treatment of the samples an regular phase transition to rutile phase is registered. The last one is dominant at 700oC. The present studies demonstrate that the sol-gel route as well as the posthydrothermal tretment are perspective methods for synthesis of nanocrystaline pure anatase phase.
- High nanocrystaline TiO_2 powders with specific surface area up to 220 m^2/g have been obtained. The nanoparticles with dimensions from 8.7 to 31.3 nm have been measured.
- The optical parameters of soles and thin films samples have been investigated and was established that the sol sample 2 possesses higher transmittance values than water base sample preparation. The phenomenon is attributed to the smaller particles size. The band gap (E_{gD}) is 3.9 eV.An interface SiO_2/TiO_2 effect exists at silica to titanium oxide layer surface binding.

Acknowledgments

The authors thanks to the Bulgarian National Science Fund for the financial support of the proposal for application to Fifth RTD Framework Programme of EC: Thematic Programme 1.1.4. Energy, Environment and Sustainable Development; special thanks to Dr. M. Mitov for the XRD experiments carried out.

References

1. M. Hoffmann, S. Martin, W. Choi and D. Bahnemann, Chem. Rev. 95 (1995) 69
2. M. Fox and M. Dulay, Chem. Rev. 93 (1993) 341
3. A. Linsebigler, G. Lu, Chem. Rev. 95 (1995) 735
4. T. Thurston, J. Wilcoxon, J. Phys. Chem.B 103 (1999)11
5. S. Music, M. Gotic, M. Ivanda, S. Popovich, A. Turkovich and R. Trojko, Mater. Sci. Eng. B 47 (1997) 33
6. G. Ingo, G. Scavia, A. Bonapasta and G. Padeletti, Procced. vol. 339, Spring Meeting April 4-8, 1994, San-Francisco, Ca., Symposium N Paper N5.56.
7. H. Kominami et al., Catal. Lett. 56 (1998) 125
8. J. Zhao and Z. Wang, Mat. Chem. Phys. 63 (2000) 9

Nanoscience & Nanotechnology
eds. E. Balabanova, I. Dragieva, Heron Press, Sofia, 2001

INVESTIGATION OF OPTICAL INHOMOGENEITY OF THIN FILMS BASED ON TRANSMISSION SPECTRA MEASUREMENTS

Tamara Pencheva and Milen Nenkov

Rousse University, Physics Department; 8 Studentska Str., 7017 Rousse, Bulgaria

Abstract. The investigation deals with correlation between deposition conditions and refractive index depth profile (optical inhomogeneity) of weakly absorbing thin films in the visible spectral region. A new approach based on films interference transmittance spectra has been applied to analyse the optical inhomogeneity of RF sputtered barium titanate. Coatings deposited in different sputtering gas composition at different substrate temperatures are found to demonstrate different nonlinear refractive index depth profile.

Keywords: thin films, refractive index inhomogeneity

1. Introduction

Thin film optical parameters (refractive index n and extinction coefficient k) are often studied after film deposition by various methods which may be divided into two main groups: (1) methods for investigation of homogeneous layers only [1–4] and (2) methods for investigation of variations of the refractive index in the depth of the layer (inhomogeneous layers) [5–7]. The last group may be subdivided into two subgroups: (2a) methods based on films reflectance spectra for linear inhomogeneity only and (2b) methods based on spectroscopic ellipsometry for nonlinear inhomogeneities.

In our previous investigation [7] a spectrophotometric method for determination of the thickness t, refractive index dispersion $n(\lambda)$ and its nonlinear variations in the film depth have been proposed. It is based on a curve fitting procedure for the measured transmittance spectra in the visible spectral region. Two models for the refractive index variations in the film depth (linear and non-linear) together with a homogeneous layer model have been used to describe the films [7]. It has been shown that the measured transmittance spectra are better fitted with an assumption for quadratic variations of the refractive index in the film depth [7]. This method for investigation of thin film optical inhomogeneity is applied in the present study and some important details are given below. The purpose of this investigation is to study correlation between sputtering gas composition and substrate temperature on one hand and the refractive index depth profile of coatings on the other hand. To analyse this correlation, several series of barium titanate thin films were deposited at different substrate temperatures in different gas composition (pure argon and mixtures of argon and oxygen) by means of RF sputtering.

2. Method for Investigation of Optical Inhomogeneity

Analysis of transmittance spectra of barium titanate thin films deposited at substrate temperature 200°C shows one interesting feature – film transmittance is slightly more than uncoated substrate transmittance near interference maxima, see Figure 1. This feature indicates that these films are optically inhomogeneous – their refractive index changes in the film depth [5,6].

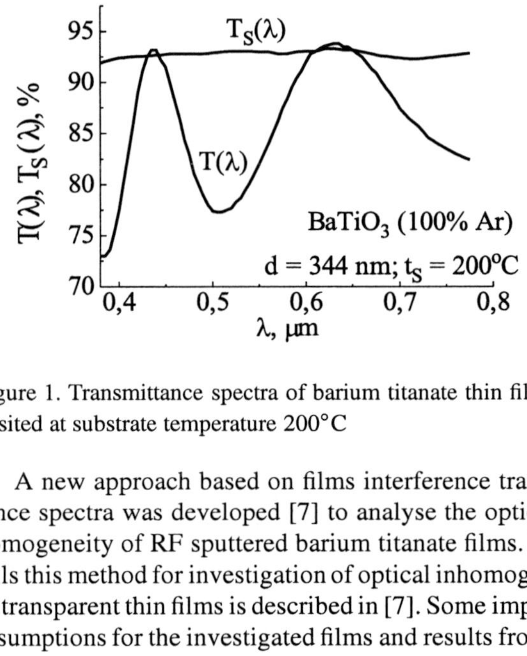

Figure 1. Transmittance spectra of barium titanate thin films deposited at substrate temperature 200°C

A new approach based on films interference transmittance spectra was developed [7] to analyse the optical inhomogeneity of RF sputtered barium titanate films. In details this method for investigation of optical inhomogeneity of transparent thin films is described in [7]. Some important assumptions for the investigated films and results from calculations are given below.

Two main assumptions for a layer are done in the beginning of the calculation procedure. One of them concerns the dispersion equations which is used to describe the wavelength variations of the real and imaginary parts of the complex refractive index [7]. Cauchy-type expressions are chosen to present the dispersion of the refractive index n and the extinction coefficient k:

$$n(\lambda) = a_n + b_n/\lambda^2 + c_n/\lambda^4, \qquad k(\lambda) = a_k + b_k/\lambda^2 + c_k/\lambda^4. \tag{1}$$

We do not give the constants a_n, b_n, c_n, a_k, b_k, and c_k in Eqs. (1) any physical meaning, these constants are used only to describe the wavelength variations of the complex refractive index.

Another assumption is made for the refractive index profile: Every film is divided into 10 sublayers with constant value of refractive index. The physical thickness of each sublayer is – $d/10$. Each sublayer is given an integer number j – from 1 (for the sublayer at the film-substrate boundary) to 10 (for the sublayer at the film-air boundary). The refractive index of each sublayer depends nonlinearly on its number j:

$$n(\lambda)_j = n(\lambda) + \alpha(5.5 - j) + \beta(5.5 - j)^2 \quad (2)$$

where α and β are unknown constants. It may be shown that Eq. (1) presents the refractive index at the middle of the whole film and Eq. (2) – indices at the middle of each sublayer. Nine film parameters are unknown – the constants in dispersion equations (1), in equation (2) and the film thickness d.

The main idea for determination of the unknown film parameters is to find their numerical values, which minimise the merit function δT:

$$\delta T = \left[\frac{1}{80}\sum_{i=1}^{80}\left(T(\lambda_i)_{\mathrm{CALC}} - T(\lambda_i)_{\mathrm{MEAS}}\right)^2\right]^{1/2} \quad (3)$$

where $T(\lambda_i)_{\mathrm{CALC}}$ is calculated transmittance (in %), $T(\lambda_i)_{\mathrm{MEAS}}$ is measured transmittance and $\lambda_i = 380 + (i - 1).5$ in nm are measurement wavelengths. The merit function δT has a simple physical meaning – an average quadratic difference between measured and calculated transmittance values (in %) [2,5,6].

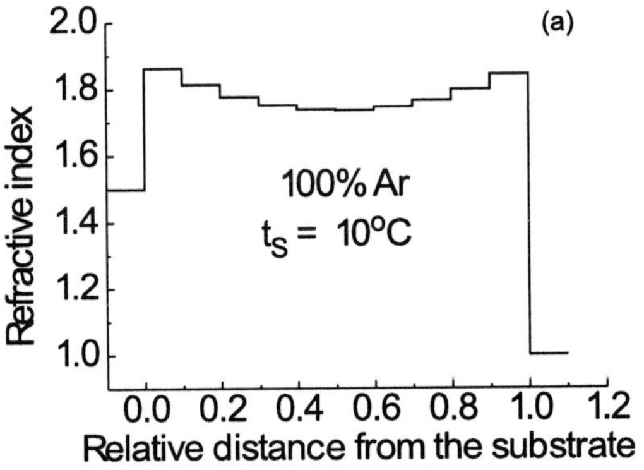

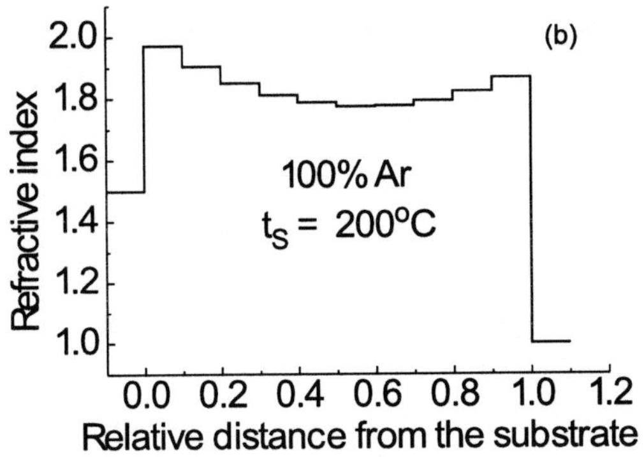

Figure 2. Typical refractive index profiles of films deposited in 100% Ar; (a) – 10°C substrate temperature, (b) – 200°C substrate temperature; relative distance 0.0 corresponds to the substrate-film boundary and relative distance 1.0 – to the film-air boundary

It is known that every optimisation procedure needs some initial values of the parameters, which have to be optimised. In the beginning the films are supposed to be homogeneous, without dispersion: $n(\lambda) = \mathrm{const} = a_n$ and without absorption: $k(\lambda) = \mathrm{const} = 0$. The initial estimates of the constant an and the film thickness d may be extracted from chromaticity coordinates of the corresponding transmittance spectrum [3]. All remaining unknown constants have initial values equal to zero. The transmittance of all sublayers is calculated by means of Abeles matrices [8–10].

3. Results

Figures 2 and 3 shows typical values of the sublayer refractive indices for $\lambda = 555$ nm. Figure 2 presents profiles of films deposited in 100% Ar and Figure 3 – profiles of films deposited in 90% Ar + 10% O_2. All profiles are given in units of relative distance from the substrate-film boundary. Relative distance 0 corresponds to the substrate-film boundary and relative distance 1 – to the film-air boundary. The substrate index in Figures 2 and 3 is 1.5 and index of air is 1. The profiles are calculated using Eqs. (1,2).

The difference between profiles of films deposited at different conditions may be attributed mainly to the difference in the sputtering gas composition and the substrate temperature [11]. Figures 2 and 3 allow to conclude that the

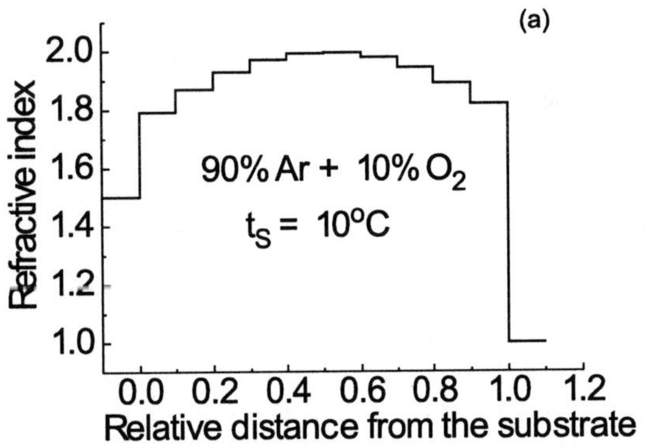

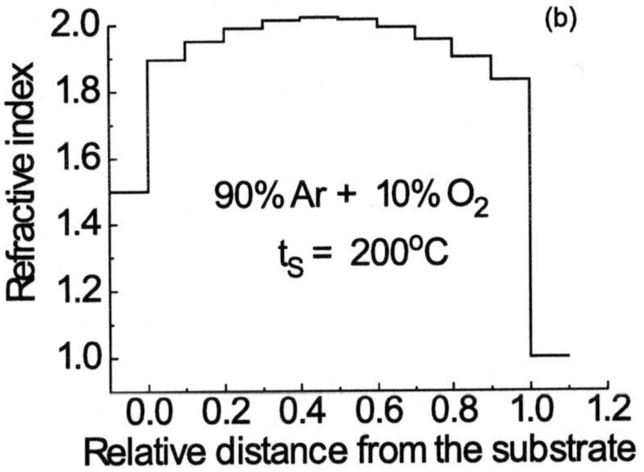

Figure 3. Typical refractive index profiles of films deposited in 90% Ar + 10% O_2; (a) – 10°C substrate temperature, (b) – 200°C substrate temperature

sputtering of barium titanate target in 100% Ar leads to deposition of films which refractive index profile has minimum in the middle of coating and sputtering in 90% Ar + 10% O_2 – to profiles with maximum in the middle of the coating. The comparison of the n-profiles in Figures 2 and 3 shows that the index at the substrate-film boundary is higher than the corresponding value at the film-air boundary for films deposited at substrate temperature 200°C. The profiles of films deposited at 10°C substrate temperature show nearly equal indices at both film boundaries. The smallest variations of the refractive index in the film depth are observed for films deposited in 95% Ar + 5% O_2 sputtering gas composition and at 10°C substrate temperature [12], see Figure 4.

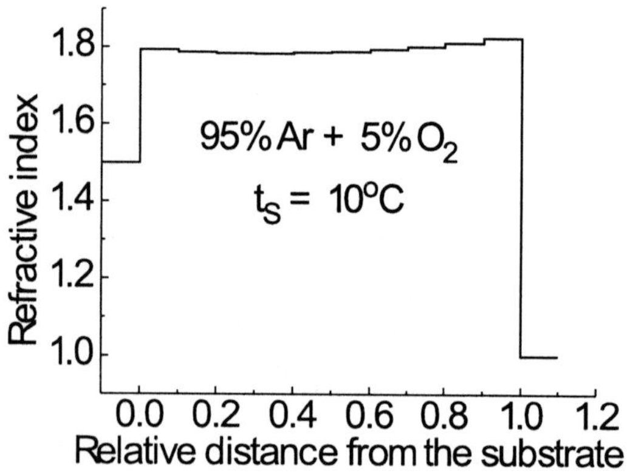

Figure 4. Typical refractive index profile of films deposited in sputtering gas composition 95% Ar + 5% O_2 and at substrate temperature 10°C

4. Conclusions

Barium titanate films deposited at different conditions are studied by means of a new approach for investigation of the optical inhomogeneity of thin films. The coatings deposited in 90% Ar + 10% O_2 and in 100% Ar gas composition demonstrate different nonlinear refractive index profile which weakly depends on the deposition temperature. The refractive indices near the substrate-film boundary for coatings deposited at 200°C are higher than the indices near the film-air boundary. For films deposited at 10°C both boundary film indices are found to be very close. The films obtained in 90% Ar + 10% O_2 have index profiles with a maximum in the middle of the films. The films sputtered in 100% Ar have profiles with a minimum in the central part of the films. The smallest variations of the refractive index in the film depth are observed for films deposited in 95% Ar + 5% O_2 sputtering gas composition and at 10°C substrate temperature.

References

1. Swanepoel R. Determination of the thickness and optical constants of amorphous silicon. J. of Physics, E: Scientific Instruments 16 (1983) 1214–1222.
2. Nenkov M. and T. Pencheva, "Calculation of Thin-film Optical Constants by Transmittance Spectra Fitting", JOSA A, 1998, vol. 15 pp. 1852–1857.
3. Pencheva T. and M. Nenkov, "Some Possible Applications of Chromatic Parameters to Thin Films Optical Constants Determination", J. Mod. Opt., 1996, vol. 43, pp. 2449–2462.
4. Pencheva T. and M. Nenkov, "Estimation of uncertainty in thin film optical constants determination using chromatic parameters", Proc. of 20-th International "Spring" Seminar on Semiconductor and Hybrid Technologies, 1998, Sofia, Bulgaria, Issue 1, 160–166.
5. Borgongo J. P., F. Flori, P. Roche et al. Refractive index and inhomogeneity of thin films. Appl. Opt., 23 (1984) 3567–3571.
6. Borgongo P., B. Lazarides and E. Pelletier, "Automatic determination of the optical constants of inhomogeneous thin films", Appl. Opt., 21 (1982) 4020–4029.
7. Nenkov M. and T. Pencheva, "Investigation of Inhomogeneity of $BaTiO_3$ Thin Films Using Transmittance Spectra Measurements", Thin Solid Films, 1998, vol. 324, pp. 305–310.
8. Nenkov M. and T. Pencheva, "Symmetrical five-layer periods from nonabsorbing high- and low-refractive-index materials", JOSA A, 1997, vol. 14, 686–692.
9. Nenkov M. and T. Pencheva, "Broad-band heat reflector design using five-layer symmetrical periods from nonabsorbing high- and low-refractive-index materials", J. Mod. Opt., 1998, vol. 45, 143–152.
10. Macleod H. A. "Thin Film Optical Filters", Macmillan, New York (1986) 590.
11. Pencheva T. and M. Nenkov, "The Properties of Barium Titanate RF Sputtered in Argon", Vacuum, 1997, vol. 48, 43–49.
12. Pencheva T. and M. Nenkov, "Optical inhomogeneity of $BaTiO_3$ thin films and its correlation with some deposition conditions", Vacuum, 2000, vol. 58, No. 2-3, 358–370.

Nanoscience & Nanotechnology
eds. E. Balabanova, I. Dragieva, Heron Press, Sofia, 2001

OPTICAL AND ELECTRONIC PROPERTIES OF GaAs QUANTUM WELLS EMBEDDED IN AlAs/GaAs SUPERLATTICES

V. Donchev[1], K. Germanova[1], N. Shtinkov[1], H. Brachkov[1], M. Saraydarov[1], I. Ivanov[2], S. J. Vlaev[3]

[1]Faculty of Physics, Sofia University, 5 James Bourchier blvd., 1164 Sofia, Bulgaria
[2]Department of Physics and Measurement Technology, Linkoping University, Sweden
[3]Escuela de Fisica, Universidad Autonoma de Zacatecas, Ap. Postal C-580, Zacatecas 98068, Mexico

Abstract. We present a detailed experimental and theoretical study of nanostructures, oriented for advanced device fabrication and revealing interesting fundamental properties. They consist of GaAs quantum wells embedded in AlAs/GaAs superlattices.

Keywords: GaAs embedded wells, electronic structure, tight binding, envelope function, photoluminescence, excitons, line-shape analysis, graded interfaces

1. Introduction

GaAs quantum wells (QW) embedded in short-period AlAs/GaAs superlattices (SL) have recently received considerable interest because of their advanced device applications and novel physical properties [1–10]. They show important advantages in comparison with single GaAs QW with homogeneous AlGaAs alloy barriers. In this work we present a detailed experimental and theoretical study of such nano-structures.

2. Electronic structure

Theoretical calculations of the electronic structure are performed for a GaAs QW embedded in an $(AlAs)4/(GaAs)_8$ SL (5.5 periods on both sides), which is surrounded by semi-infinite $Al_{0.33}Ga_{0.67}As$ barriers. Two methods are applied: the envelope function approximation and the semi-empirical sp^3s^* spin-dependent tight-binding model combined with the surface Green function matching technique. The behavior of the electron and hole bound states for a wide range of embedded well (EW) widths (from 2 to 50 monolayers) is studied for the case of sharp interfaces [4,5]. Figure 1 illustrates the energies of the conduction band states found by the tight-binding method. The obtained states are described with respect to their energy, spatial distribution and orbital character. An emphasize is given on the interaction between EW and SL states. This interaction occurs when an EW state is close in energy or inside a SL miniband and holds only for states of equal parity. It is manifested by energy shifts and substantial changes in the wavefunction localization of the interacting states: the EW state delocalizes towards the SL and SL states penetrate in the EW. The application of the tight-binding approach allows to obtain interesting features of the electronic structure which are inaccessible for the simple envelope function model. Such are the existence of conduction band states localized in the barriers of the SL with energies 2.327 and 2.424 eV (see Figure 1). The valence band reveals a complex structure, dominated by the coupling between heavy-hole and light-hole states and the interaction between states of different real space localization (EW or SL).

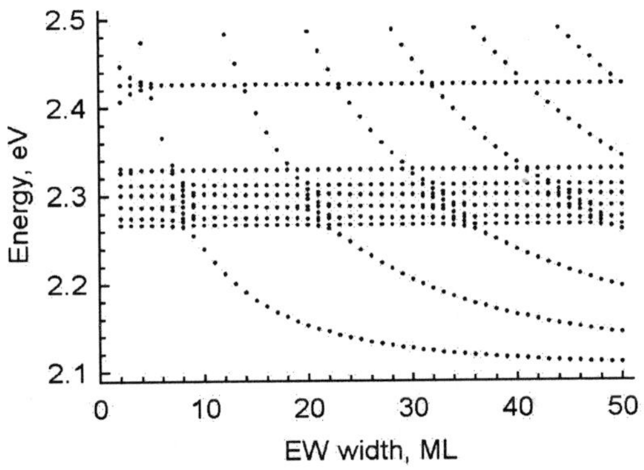

Figure 1. Energies of conduction band states calculated by the tight binding method

The EW could be considered as an effective imperfection in the regular SL. Indeed the coupling-decoupling of an EW state with (from) the SL miniband states (which depends on the EW width, i.e. on the strength of the perturbation) is in principle similar to the formation of localized electronic states in a bulk crystal containing different types of defects. We believe our study gives an insight into this interesting problem.

3. Experimental details

The photoluminescence (PL) method is applied to study two types of samples, using a standard spectroscopic technique [6]. In samples type A, which are grown at 600°C, there is one GaAs QW of 5 nm sandwiched between 26 and 20 periods of an $(AlAs)_4/(GaAs)_8$ SL pepared directly on a GaAs substrate. Samples type B are grown at 550°C. They contain two EWs (12 and 5 nm) and a GaAs buffer layer (400 nm) between the substrate and the SL. There are 5, 24 and 20 periods of SL situated bellow the 12 nm EW, between the EWs and above the 5 nm EW, respectively. PL

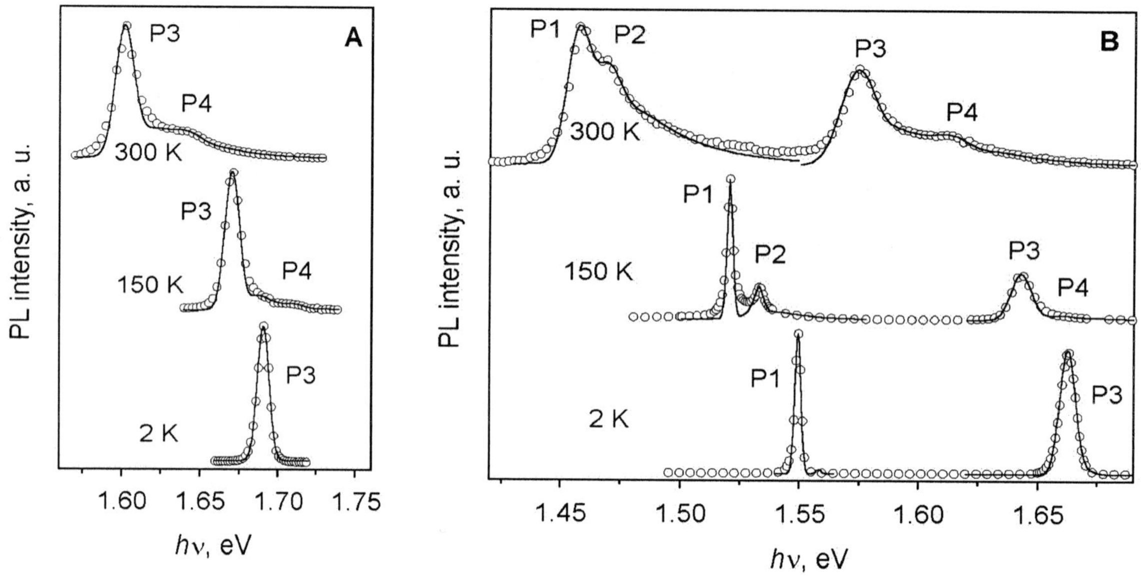

Figure 2. PL spectra (circles) at different temperatures and fits of the PL peaks (lines) for sample type A (a) and B (b)

measurements are performed from 2 to 300 K at different excitation densities between 4 and 400 $\mathrm{W cm^{-2}}$.

Sharp PL peaks related to electron-heavy hole transitions in the EWs are observed at low temperature. They are labeled P3 in samples A and P1 and P3 in samples B (Figure 2). The peaks broaden with increasing the temperature and small peaks or shoulders, P2 and P4, appear on their high-energy sides, corresponding to electron-light hole tran-sitions. Since P1 and P2 are present only in the spectra of samples B, they are attributed to the 12 nm EW. Similarly P3 and P4 are ascribed to the EW of 5 nm.

4. Radiative recombination mechanisms

A detailed analysis of the radiative recombination mechanisms is performed using two approaches [7–9].

(i) The study of the excitation density dependence of the integrated PL lines intensity has shown qualitatively that the recombination at low temperature (2K) is excitonic. Free excitons dominate the recombination up to room temperature, where a contribution of free carriers is also present.

(ii) In the second approach a line-shape analysis of the PL peaks is performed. They are fitted using a simple statistical model, which takes into account both free exciton and free carrier recombination [11]. Typical examples are represented on Figure 2 by solid lines. The results confirm the predomination of free excitons in the radiative recombination. In addition they provide a quantitative estimate of the relative contributions of free excitons and free carriers to the radiative recombination process. It is found that the thermal equilibrium between excitons and free carriers is well described by the mass action law. The temperature behavior of the peak energies and of the broadening parameters determined from the fits follows the expected theoretical trends. The line-shape analysis also gives indirect information about the interface quality.

5. Non-abrupt interfaces and optical transition energies

The experimental energies of the heavy-hole and light-hole exciton transitions are compared with the ones calculated by the envelope function method. When a rectangular interface potential is used the calculated values are lower than the experimental ones. The real interfaces however are rather non-abrupt due to material intermixing. Different sources of this intermixing have been suggested such as surface segregation of Ga atoms and Ga-Al interdiffusion by direct exchange of atoms or via interaction with defects [12]. So, we extend the theoretical analysis towards real interfaces by considering a non-abrupt interface profile. The non-abrupt interfaces are modeled by a concentration profile, obtained by solving the diffusion equation. The diffusion length, L_D, is used as a measure of the interface broadening. The influence of the interface broadening on the electronic structure and the optical transition energies is investigated [10]. The behavior is similar to that observed in a single QW with AlAs barriers, but the transition and states energies are less sensitive to L_D in the EW. The minibands of the superlattice are also affected by the interface broadening. They get wider and move away from the band edges as L_D increases.

The comparison between theory and experiment gives a very good agreement for electron-heavy hole and electron-light hole exciton transition energies in both types of EWs. Besides it provides an estimate of the degree of interface broadening [6,9]. It is of the order of 4.5 and 3.7 monolayers in samples A and B, respectively. The difference is attributed to the higher growth temperature and the absence of a buffer layer in samples type A [6]. These values, although somewhat large (which could be an indication that apart from interdiffusion there are other mechanisms contributing to the interface broadening), provide a rough and quick estimate of the interface quality. The model developed can also be used in the evaluation of complicated systems with a larger interface grading (obtained e.g. by annealing) in order to exploit their new properties attractive for device fabrication.

6. Conclusions

GaAs QW embedded in short-period AlAs/GaAs SL have been studied theoretically and experimentally. The calculations of the electronic structure have shown that an interaction between the localized EW states and the extended SL states with the same parity occurs when an EW state is close in energy or inside a SL miniband. It is manifested by energy shifts and substantial changes in the wavefunction localization of the interacting states. PL measurements are carried out on two types of samples. The analysis of the PL spectra has demonstrated that free excitons dominate the radiative recombination process up to room temperature. A quantitative estimate of the relative contributions of free excitons and free carriers recombination is obtained. An envelope-function based model, which incorporates a non-abrupt interface potential is developed and employed to estimate the degree of interface broadening due to material intermixing. The studies presented in this work contribute to a better understanding of the electronic structure and the optical properties of complicated GaAs/AlAs quantum heterostructures used for fabrication of advanced optoelectronic devices.

Acknowledgement

We thank D. Martin and F. K. Reinhart (EPFL, Switzerland) for supplying the samples, and Prof. V. R. Velasco and B. Hristov for the helpful discussions. This work was supported by the Sofia University Research Fund (contract No. 431/2000) and the Bulgarian National Science Council (contract MU-F-01).

References

1. J. Arriaga and V. R. Velasco, J. Phys.: Condensed Matter 7, 3493 (1995).
2. R. Kucharczyk and M. Steslicka, Solid State Commun. 81, 557 (1992).
3. P. Blood, in: Physics and Technology of Heterojunction Devices, eds. D. V. Morgan and R. H. Williams (Peter Perigrinus, London), 1991, p. 231.
4. N. Shtinkov, V. Donchev, and K. Germanova, in: Thin Films Materials and Devices - Development in Science and Technology, eds. J. M. Marshal, N. Kirov, A. Vavrek, and J. M. Maud (World Scientific, Singapore), 1999, p. 305.
5. N. Shtinkov, S. Vlaev, V. Donchev, and K. Germanova, phys. stat. sol. (b) 220, 153 (2000).
6. V. Donchev, K. Germanova, N. Shtinkov, I. Ivanov, and S. Vlaev, Thin Solid Films 364, 224 (2000).
7. V. Donchev, Tzv. Ivanov, I. Ivanov, M. Angelov, and K. Germanova, Vacuum 58, 462 (2000).
8. V. Donchev, K. Germanova, N. Shtinkov, H. Brachkov, and I. Ivanov, Proc. 4th General conference of the Balkan Physical Union, August 2000, Veliko Turnovo, Bulgaria (to be published in Balkan Phys. Lett.).
9. V. Donchev, N. Shtinkov, K. Germanova, H. Brachkov, and I. Ivanov, Proc. 11th ISCMP, September 2000, Varna, Bulgaria (in press).
10. N. Shtinkov, V. Donchev, K. Germanova, and H. Kolev, Semicond. Sci. Technol. 15, 946 (2000).
11. M. Colocci, M. Gurioli, and A. Vinattieri, J. Appl. Phys. 68, 2809 (1990).
12. A. Ourmazd, M. Scheffler, M. Heinemann, J-L. Rouviere. Mater.Res.Bulletin 27, 24 (1992))

Nanoscience & Nanotechnology
eds. E. Balabanova, I. Dragieva, Heron Press, Sofia, 2001

ROOM-TEMPERATURE PHOTOLUMINESCENCE FROM Tb^{3+} IONS IN SiO_2 AND a-SiC:H THIN FILMS DEPOSITED BY MAGNETRON CO-SPUTTERING

M. Nikolaeva[1], M. Sendova-Vassileva[1], D. Dimova-Malinovska[1], J.C. Pivin[2]

[1]Central Laboratory for Solar Energy, Bulgarian Academy of Sciences, 72 Tzarigradsko Chaussee. 1784 Sofia, Bulgaria
[2]CSNSM, IN2P3-CNRS, Batiment 108, 91405 Orsay Campus, France

Abstract. Thin fims of SiO_2 and a-SiC:H doped with Tb were deposited by magnetron co-sputtering. RBS analysis shows that Tb is homogeneously distributed in the bulk. The laser excited photoluminescence (PL) of the films has been studied. In the case of SiO_2:Tb the PL is due to 4f-4f transitions of Tb^{3+} ions starting from the $^5D_4$ level which is resonantly excited by the 488 nm Ar^+ laser line. The PL in this case has no temperature quenching from 18 K up to room temperature. In the case of a-$Si_{1-x}C_x$:H ($x = 0.5$) the films have an optical gap of 2.5 eV. Optical absorption in the visible is enhanced with increased Tb doping. The PL band is broad and coveres the whole visible spectrum. The position of its maximum coincides with the most intensive PL peak in SiO_2:Tb.

Keywords: photoluminescence, Tb, SiO_2, a-SiC:H

1. Introduction

Rare earth ions in silicon based materials have been the object of much interest recently. This is stimulated by the aim to develop light emitters integrated in the silicon electronics. Er^{3+} has drawn the most attention because one of its most efficient radiative intra-4f transitions is at 1.5 μm which is a very important wavelength for telecommunications. Tb^{3+} is a rare earth ion which has strong radiative intra-4f transitions in the visible. That is why it is a potential candidate for visible LEDs. H. Amekura et al. [1] have demonstrated visible photoluminescence from thin SiO_2 layers implanted with Tb. Magnetron co-sputtering is an alternative method for obtaining rare-earth ion containing films which gives a uniform depth distribution of the impurity atom. We observed for the first time visible photoluminescence in a-$Si_{1-x}C_x$:H:Tb thin films [2]. In this paper results from the study of PL of Tb doped in SiO_2 and a-$Si_{1-x}C_x$:H thin films prepared by magnetron sputtering are presented.

2. Experiment

The thin films of SiO_2:Tb were magnetron co-sputtered in Ar. The target consisted of a suprasil plate with a varying number of Tb chips placed on it. For the a-$Si_{1-x}C_x$:H:Tb films a complex target of c-Si with graphite and Tb chips placed on it was used. The sputtering was performed in Ar and H_2. Crystal silicon and glass substrates were used. The thickness of both structures was 1000 nm. The Tb concentration in the films was measured using RBS. RBS spectra of 2.5 MeV He^{2+} ions were recorded at normal incidence and at a scattering angle of 165°. Films of SiO_2 with Tb concentrations between 0.37 and 13.5 at.% were deposited. The a-$Si_{1-x}C_x$:H:Tb films contained 30 at.% H which was measured by ERDA. The carbon content estimated by NRA was 30 at.%. The Tb concentration deduced from these measurements varies between 0.1 and 7 at.%. For the PL experiments a double SPEX 1403 monochromator and an EMI 9863B photomultiplier were used. The PL was excited with the 488 nm line of an Ar^+ ion laser with power density below 1 W/cm^2 on the sample. Low temperature measurements in the case of SiO_2:Tb were performed using a closed cycle He cryostat. The spectral resolution was 0.25 nm. All the spectra were corrected for the spectral response of the system.

3. Results and Discussion

3.1. SiO_2:Tb

A typical PL spectrum of the SiO_2:Tb thin films is shown in Figure 1. It consists of three characteristic bands which are assigned to the following 4f – 4f transitions of the Tb^{3+} ion: $^5D_4 \rightarrow {}^7F_5$ ($\sim$ 545 nm), $^5D_4 \rightarrow {}^7F_4$ ($\sim$ 590 nm), $^5D_4 \rightarrow {}^7F_3$ ($\sim$ 620 nm). The 488 nm laser line excites resonantly the $^5D_4$ level of the Tb^{3+} ion and the radiative transitions to the levels lying just above the ground level are observed in this experiment. The first of these transitions is the strongest and it has a structure with two maxima at 542 nm and 549 nm. Similar structure has been observed in glasses as well [1]. The predominance of this transition is

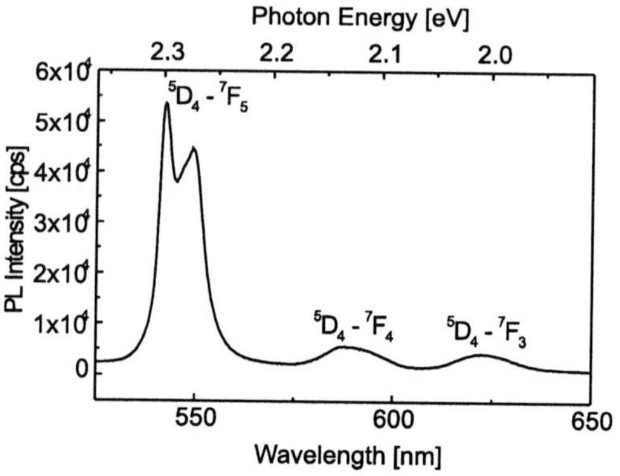

Figure 1. PL spectrum of a SiO_2:Tb sample containing 2.5 at.% Tb at room temperature

explained by the value of the corresponding transition matrix which is highest for the transition according to the Judd–Ofelt theory [1]. In our experiment these emission lines appear on a broad background which we attribute to defect luminescence of the SiO_2 matrix. A decay of the intensity of this background with time under the laser light illumination was detected. We found no such decrease in the Tb^{3+} luminescence. As a measure for the intensity of the Tb^{3+} PL the area under the strongest $^5D_4 \rightarrow {}^7F_5$ band after subtraction of the background is taken.

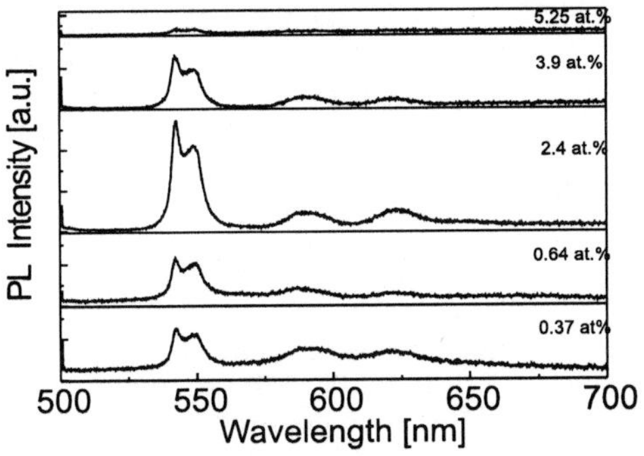

Figure 2. PL spectra of five different SiO_2:Tb samples, containing the indicated concentration of the doping impurity. All spectra are plotted in the same scale

Figure 2 shows the concentration dependence of the SiO_2:Tb PL. Up to 2.4 at.% Tb the PL intensity increases with Tb concentration and then starts to fall. At the highest concentrations we have reached (13.5 at.%) the PL falls by 2 orders of magnitude in comparison with that at 2.4 at.%. The shape of the PL, however, stays essentially the same. The decrease in the PL intensity is probably due to clustering as suggested in [1] and not to cross relaxation. TEM experiments are under way to clarify this point.

The temperature dependence of the SiO_2:Tb PL was studied in the samples showing the strongest luminescence from 18 K up to room temperature. No reduction of the PL intensity was detected as the temperature was raised (Figure 3). This behaviour is similar to that of Er^{3+} in thin films.

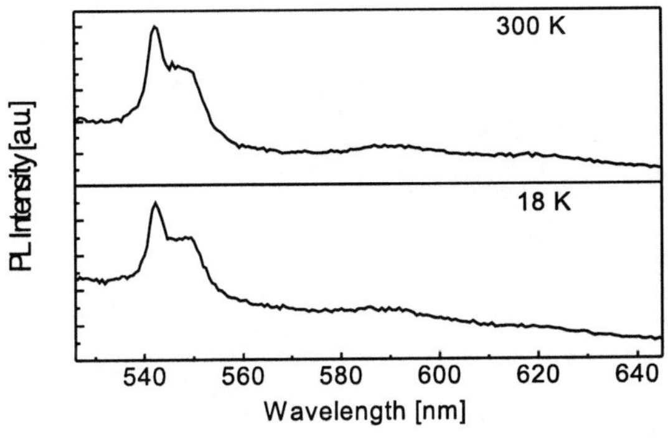

Figure 3. PL spectrum of Tb^{3+} in co-sputtered SiO_2:Tb samples containing 2.5 at.% Tb at 18 K and 300 K. The spectra are plotted in the same scale so that their intensity can be compared

The temperature quenching of its $^4I_{13/2} \rightarrow {}^4I_{15/2}$ transition becomes weaker when the ion has oxygen in its immediate surroundings [3] and is especially weak in SiO_2 [4,5] down to about 20 K when a stronger dependence sets in. Another interesting feature (Figure 3) is that we observe no narrowing of the lines comprising the transitions between the different multiplets of Tb^{3+} when the temperature is lowered. This narrowing is expected because of the smaller homogeneous broadening due to phonons at low temperatures [6]. A narrowing of the emission lines of trivalent rare-earth ions is observed in crystals and to a smaller extent in glasses [7] because there the inhomogeneous broadening due to the wide distribution of site symmetries is greater. Obviously in the case of Tb^{3+} in amorphous SiO_2 the inhomogeneous broadening predominates over the homogeneous one down to 18 K.

3.2. a-$Si_{1-x}C_x$:H:Tb

Figure 4 shows the RBS spectra of a-$Si_{1-x}C_x$:H samples with different Tb concentrations. As it can be seen from the figure the Tb concentration in some samples has a small gradient but for most films it is homogeneously distributed throughout their thickness. The spectra were fitted using the RUMR algorithm [8] in order to extract the concentrations of Tb and Si. In the fitting the data for the C and O concentrations from NRA and the H concentration from ERDA were used. The C concentration in all films is about 30 at.%, the H content is about 38 at.% and Tb varies between 0.1 and 7 at.%. Oxygen is less than 2 at.%. A high C concentration was deliberately aimed at so that a material with a wider optical gap is achieved [9]. In our case the films contain equal amounts of C and Si and the optical gap of the undoped sample is 2.5 eV Figure 5 shows the raw optical transmission spectra of films with increasing Tb content. The transmission of a-$Si_{1-x}C_x$:H without Tb is presented for comparison. The presented experimental data show that the slope of the absorption edge decreases with increasing Tb concentration. Probably this is due to defect states in the band formed by the introduction of the impurity.

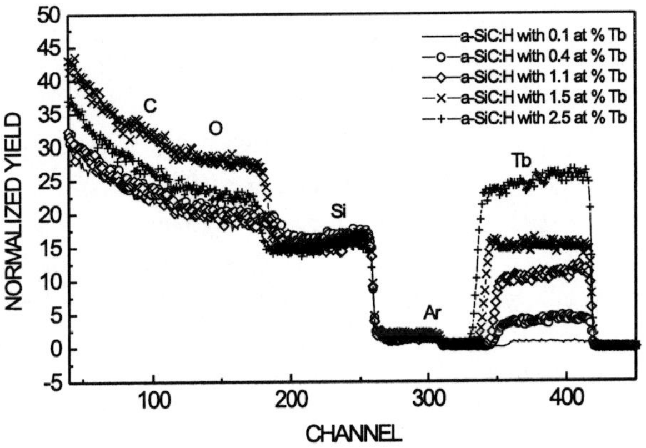

Figure 4. RBS spectra of a-$Si_{1-x}C_x$:H with different Tb concentration

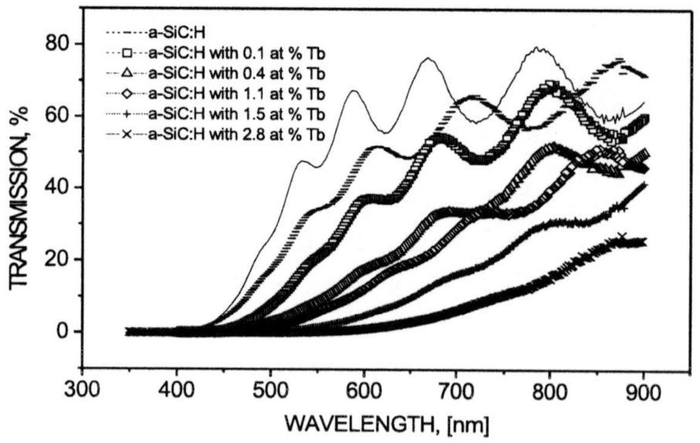

Figure 5. Optical transmission of a-$Si_{1-x}C_x$:H without Tb and containing different Tb concentrations

Fig. 6 shows the PL of a sample of a-$Si_{1-x}C_x$:H:Tb. Its luminescence spectrum is quite different from that of SiO_2:Tb. It consists of a broad structureless band centered at 545 nm. The spectrum of a sample prepared under the same conditions but without terbium is given for comparison. It exhibits the PL band of the a-$Si_{1-x}C_x$:H matrix. Its maximum is in the red region of the visible spectrum and it is weaker. The broad band at 545 nm of the sample containing Tb is obviously connected with the presence of Tb ions. Because of the much narrower band gap of a-$Si_{1-x}C_x$:H in comparison with SiO_2 the mechanism of luminescence in this case is probably different. If we compere the positions of the maxima of PL peaks in SiO_2:Tb and a-$Si_{1-x}C_x$:H:Tb is easy to see that they are at the same wavelength. That is why we consider that this PL peak of a-$Si_{1-x}C_x$:H:Tb is due to the same transitions in Tb^{3+} ions. The greater wight is probably due to the disorder of the material.

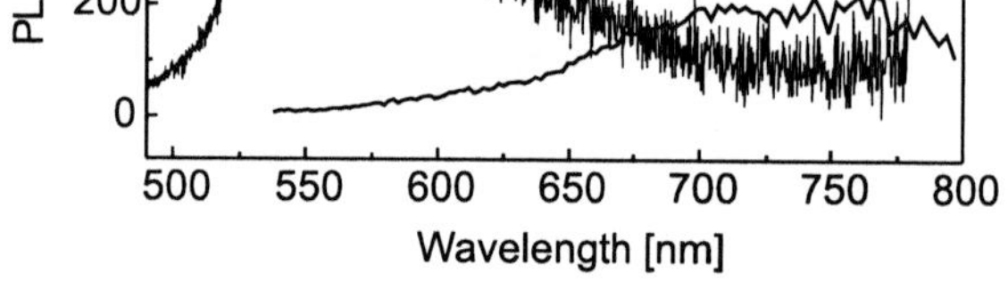

Figure 6. PL of an a-$Si_{1-x}C_x$:H:Tb film (1) and that of an a-$Si_{1-x}C_x$:H film without Tb (2) for comparison

4. Conclusion

We have demonstrated that thin films of SiO_2 and a-$Si_{1-x}C_x$:H containing Tb and exhibiting room temperature photoluminescence can be prepared by magnetron co-sputtering. The room temperature visible PL of the SiO_2:Tb samples increases up to concentrations of 2.4 at.% Tb and then starts to fall. It shows no temperature quenching from 18 K to 300 K. It is due to intra-4f transitions in the Tb^{3+} ions. Optical measurements show that the transmission in the visible decreases with increasing the Tb concentration. The PL peak of Tb^{3+} ions in a-$Si_{1-x}C_x$:H:Tb thin films is centered at 545 nm and its shape is broad and structureless. Probably this shape is due to the disorder of the material.

Acknowledgment

This study was supported by the Bulgarian National Scientific Fund under Contract X-903.

References

1. H. Amekura, A. Eckau, R. Carius, Ch. Buchal, J. Appl. Phys. 84 (1998) 3867.
2. M. Sendova-Vassileva, M. Nikolaeva, D. Dimova-Malinovska, M. Tzolov, J.C. Pivin, Mat.Sci. & Eng. in press
3. S. Coffa, F. .Priolo, G. Franzo, A. Polman, S. Libertino, M. Saggio, A. Carnera, NIM B 106 (1995) 386.
4. J.C. Pivin, N.V. Gaponenko, A.V. Mudryi, V.V. Shushunova, B. Hamilton, Mater. Sci. Eng. B 69-70 (2000) 215.
5. J.H. Shin, W. Lee, H. Han, Appl. Phys. Lett. 74 (1999) 1573.
6. W.M. Yen, W.C. Scott, A.L. Schawlow, Phys. Rev. A (1964) 271.
7. L. Doolittle, Nuc. Inst. and Methods B45, (1990), p1
8. G. Amaranath, S. Buddhudu, F.J. Bryant, J. Non-Cryst. Solids, 122 (1990) 66.
9. N. Tzenov, Ph.D. thesis Bulgarian Academy of Sciences (1996)

SECTION 3:
Synthesis and Processing (Magnetic/Electrical Properties)

Nanoscience & Nanotechnology
eds. E. Balabanova, I. Dragieva, Heron Press, Sofia, 2001

RHEOLOGICAL AND DIELECTRIC MEASUREMENTS OF EPOXY/SMECTITE HYBRIDS – A TOOL FOR EVALUATION OF THE MOLECULAR DYNAMICS AT THE NANOSCALE INTERFACES

R. Kotsilkova[1], P. Pissis[2], A. Kanapitsas[2], S. Rousseva[1]

[1]Central Laboratory of Physico-Chemical Mechanics, Bulgarian Academy of Sciences
[2]National Technical University of Athens, Greece

Abstract. Organic/inorganic hybrids are synthesized as a molecular level dispersion of smectic clay layers in epoxy matrix. Viscosity and viscoelasticity of the epoxy/clay dispersions are studied in order to understand the kinetics of nanocomposite formation and the relaxation dynamics of the polymer chains in the liquid hybrids. Dielectric measurements provide information about local and segmental motions of polymeric chains in the solid hybrids. It thus appears that interactions of the polymer molecules to the inorganic sheets alter the relaxation dynamics of the polymer chains in both liquid and solid state hybrids.

1. Introduction

The recent surge of active interest in polymer-layered silicate (PLS) hybrids have been directed to a tendency of better understanding the structure formation and properties due to the interfaces. In our previous studies we have discussed time-dependent rheological behavior of PLS hybrids in shear and elongation flow [1,2], as well as improved mechanical properties and thermal stability based on the nanolayered structure organization [2,3]. On the other hand, electrical and dielectric measurements are used as a successful way for investigating effects of confinement of polymer and of tethering of polymeric chains to "rigid walls" [4].

In the present study we stress on understanding the effect of interfaces on the molecular dynamics in polymer-layered silicate nanocomposites. Rheological and dielectric methods are used as a tool for studying such effects in liquid and solid state.

2. Materials and Methods

Nanocomposites have been prepared by dispersing of organically modified smectic clay in a low viscosity epoxy resin – Araldite LY556 (CIBA). A smectite (CO-OP Chemical Co., LTD, Japan) consists of an octahedral Al_2O_3 sheet sandwiched between two SiO_2 tetrahedral layers (of ~ 1 nm thick and ~ 50 nm length) with the charges being adjusted by substituting Al^{3+} or Si^{4+} with Mg^{2+} and/or Fe^{2+}. The depressed charges are neutralized with alkaline cations intercalated into the interlayer spaces leading to a laminate structure of several hundred layers. The following type of alkaline cations are used: SAN – $[(C_{16}H_{33})_x (C_{18}H_{37})_y N^+(CH_3)_2]$ ($x = 0.5$; $y = 1.5$); STN – $[(C_8H_{17})_3 (CH_3)N^+]$ and SPN – $[(C_2H_5)_2(CH_3)N^+ (O\text{-}iPr)_{25}]$. The samples investigated are nanocomposites of 2 v%, 5 v%, and 10 v% smectite in epoxy resin. The synthesis method includes mechanical mixing and ultrasonic treatment of the epoxy-clay dispersions. Then, the compositions are cured by two-stage curing process (2 h at 75°C and additional 2 h at 130°C) in the presence of amine curing agent. Rheological properties of the epoxy-smectite dispersions are verified by measurements in oscillatory and shear mode using a cone-plate Brabender Rheotron. Dielectric properties are measured [4,5] by broadband dielectric relaxation spectroscopy (DRS), $10^{-2} - 10^6$ Hz, and by thermally stimulated depolarization currents (TSDC) techniques.

3. Results and Discussion

3.1. Rheology of epoxy-smectite dispersions

Viscoelasticity. The examination of the viscoelastic properties is performed by low amplitude dynamic measurements of epoxy-smectite clay dispersions at 25°C. Figure 1 presents the frequency dependence of storage $G'(\omega)$ and loss $G''(\omega)$ moduli of 2 v%, 5 v%, and 10 v% STN-epoxy

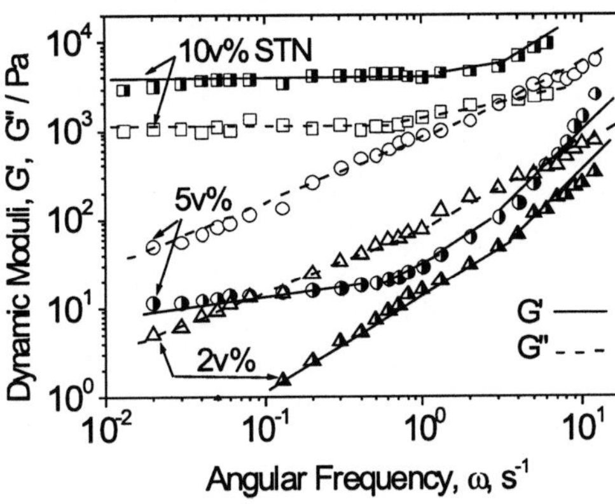

Figure 1. Low amplitude dynamic moduli G' (full symbols) and G''(open symbols) vs. ω of 2 v%, 5 v%, and 10 v% STN-epoxy dispersions at 25°C. Full (G') and dash (G'') lines show the calculated mechanical spectra

dispersions. It is expected that in the frequency range used, the polymer chains should be fully relaxed and exhibit a characteristic terminal behavior, i.e. $G' \propto \omega^2$ and $G'' \propto \omega$.

In contrast to the Newtonian epoxy matrix, nonterminal flow behavior is found for the smectite-epoxy dispersions at frequencies below $\omega = 1\,s^{-1}$. An increase of the smectite concentration from 2 v% to 10 v%, leads to a gradual change of the material's behavior from liquid-like to solid-like. As it is shown in Figure 1, a second plateau appears in both moduli and the values of $G'(\omega)$ become higher than $G''(\omega)$. The nonterminal flow behavior observed in epoxy-smectite clay dispersions could not be attributed to the delamination alone, but mainly could be associated with retardation of molecular relaxation processes produced by the tethering of molecules to the smectite surface.

To confirm the upper suggestion, the discrete relaxation spectrum $H(\tau)(= G_i\,\tau_i)$ has been calculated from the experimental dynamic moduli by using a generalized Maxwell model, which presumes a presence of second plateau in both moduli, where $G'_0, G''_0 \geq 0$.

$$G'(\omega) = G'_0 + \sum G_i \omega^2 \tau_i^2/(1 + \omega^2 \tau_i^2) \quad (1)$$

$$G''(\omega) = G''_0 + \sum G_i \omega \tau_i^2/(1 + \omega^2 \tau_i^2) \quad (2)$$

In Figure 1, the calculated mechanical spectra have been compared to the experimental data of dynamic moduli. The mean relaxation time is calculated from the spectra, $\theta = \sum G_i \tau_i^2 / \sum G_i \tau_i$, for various STN-epoxy dispersions. As shown in Table 1, the slope of G' and G'' vs. ω decreases by increasing the smectite loading and reaches plateau values, while the mean relaxation time strongly increases. It thus appears that the interactions of the polymer molecules to the inorganic sheets shift the relaxation towards the longer relaxation times and the terminal zone – to very low frequencies.

Table 1. Terminal characteristics of G' and G'' vs. ω

Sample	G' slope	G'' slope	$\theta = \sum G_i \tau_i^2 / \sum G_i \tau_i$ [s]
2 v% STN	1,2	0,7	0,6
5 v% STN	0	0,5	3,5
10 v% STN	0	0	46,0

Viscosity. Figure 2 shows the steady-state viscosity, η vs. shear rate, γ of 10 v% smectite-epoxy dispersions of various organically modified smectic clays (SAN, STN and SPN) at 25°C. To evaluate the kinetics of the structure formation, the viscosity curves are compared in Figure 2, before and after 75°C treatment for two hour, without curing agent. The temperature treatment initiates the intercalation process and the tethering of polymer molecules on the smectite surfaces. The reaction between the epoxy resin and the alkaline exchanged forms of clay results in the polymerization of the epoxide and the concomitant delamination of the clay structure. The process was found to be different for STN, SAN and SPN dispersions.

Before the temperature treatment, the viscosity vs. shear rate curves of the smectite-epoxy dispersions show weak plastic behavior. Whereas, after two hours at 75°C, a liquid crystalline-like behavior appears, so the viscosity curves exhibit three characteristic regions: I-shear thinning, II-viscosity plateau, and III-second shear thinning. The viscosity plateau (II) gives an evidence of a shear induced ordering of layers in dispersion at steady shear flow. Thus, it can be expected that there are domains of layers oriented in some preferred direction in the smectite-epoxy dispersions. Probably, the domain structure of ordered mesophases is responsible for the nonterminal flow behavior in the low-frequency zone.

The region III of a superanomalous viscosity, which follows the plateau region, shows that the steady shear flow becomes metastable and a flow spurt occurs. Such property could be related to nanorheology, e.g. to friction in lubricated surfaces at nanoscale separations.

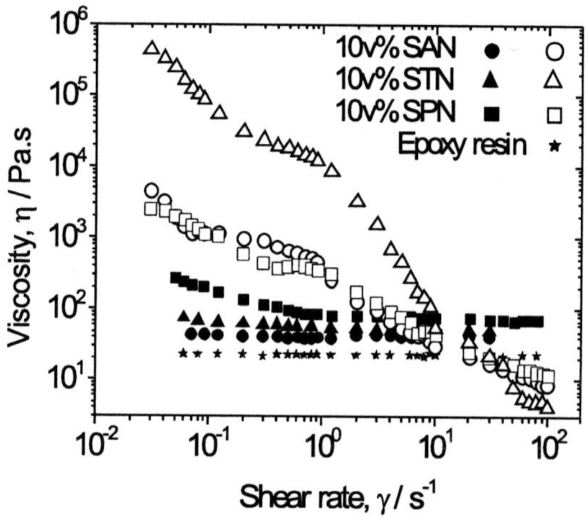

Figure 2. Viscosity vs. shear rate of 10 v% smectite-epoxy dispersions of SAN, STN and SPN at 25°C – before (full symbols) and after two hours at 75°C (open symbols)

For a visualization of the microstructure of the synthesized hybrids, transmission electron microscopy is used. TEM micrograph of the solid 5 v% SPN-epoxy nanocomposite in Figure 3, shows that silicate layers in hybrids are highly anisotropic and cannot be placed completely randomly in the matrix polymer. The TEM results confirm the upper suggestion about the presence of long-range domain

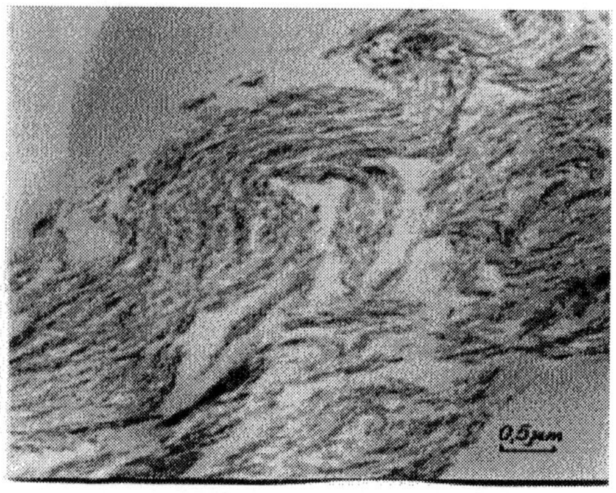

Figure 3. TEM micrographs of 10 v% SPN-epoxy nanocomposites (exfoliated type structure)

structure, which is likely to become better defined at the high silicate content, where the main distance between layers decreases and the majority of polymer chains are tethered to the surface of the silicate layers.

3.2. Dielectric properties of epoxy-smectite hybrids

Figure 4 shows comparative TSDC plots for the solid state hybrids. TSDC measures the depolarization current through the sample and corresponds to measuring dielectric losses against temperature at constant equivalent frequencies in the range $10^{-2} - 10^{-4}$ Hz. Similar to the results of rheological measurements in Figures 1 and 2, distinct differences are observed in the behavior of the samples. The weak peak at about -110°C in epoxy resin is due to the γ relaxation attributed to local motion of unreacted epoxide rings, ether and linkage groups [5]. In the 10 v% SPN and the 5 v% SAN nanocomposites mobility at low temperatures increases with respect to the epoxy resin matrix, whereas the opposite is observed in the 5 v% STN. It is not clear at this stage as to whether the peaks at about -95°C in the 10 v% SPN and -80°C in the 5 v% STN are due to the γ relaxation (which would mean increasing of magnitude and becoming slower) or due to new relaxations. Their temperature position allows bringing them in relation to loss peaks measured by DRS at higher temperatures (not shown here).

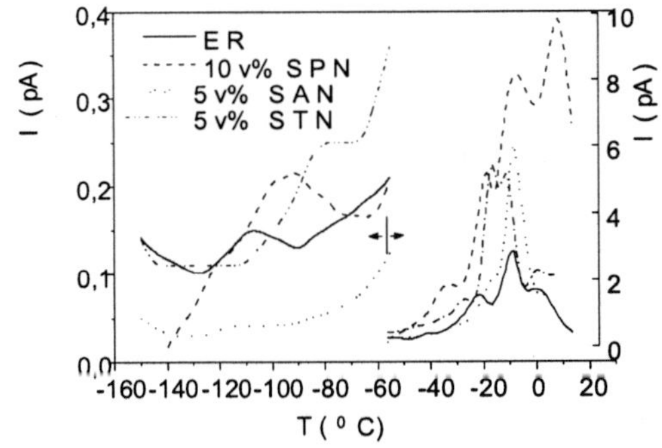

Figure 4. TSDC thermograms obtained with the samples indicated on the plot. Note the change of scale at -50°C

The peaks at higher temperatures in Figure 4 and their DRS counterparts measured at higher temperatures are in the region where in epoxy resin β and ω relaxations have been observed and analyzed [5]. Interfacial effects due to heterogeneity of the samples are also expected to occur in this temperature/frequency region.

The results in Figure 4 suggest that at temperatures between -50 and 20°C mobility increases in the composites with respect to epoxy resin and is highest in the 10 v% SPN hybrid. This might be related with the higher smectite clay content of this sample.

Figure 5 presents results of isothermal DRS measurements replotted as temperature plots of dielectric loss ε'' at 10^5 Hz. The peak at about 90°C in epoxy resin should be due to the α relaxation associated with the glass transition. This peak becomes a shoulder in the 10 v% SPN hybrid, followed by activity at higher temperatures, not present in the pure matrix, probably related with interfacial effects. Further investigation of these effects would reveal details of micromorphology and chain dynamics in the hybrids.

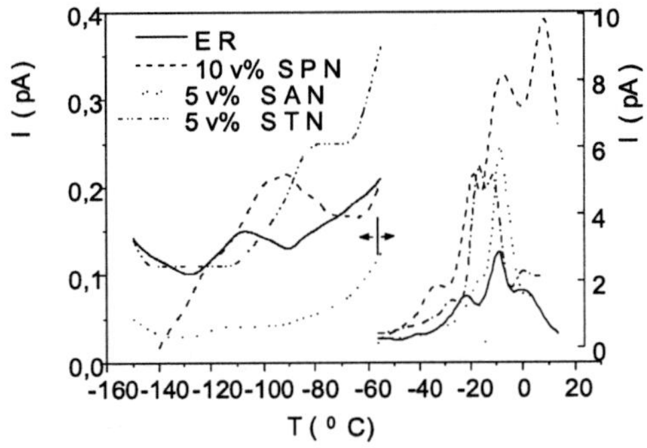

Figure 5. Dielectric losses ε'' against temperature measured at 10^5 Hz on the samples indicated on the plot

Acknowledgement

This work was partly supported by the Greek General Secretariat of Research and Technology, ΠΕΝΕΔ 99ΕΔ52.

References

1. Kotsilkova, R., M. Okamoto, N. Sato, H. Sato, H.Taguchi, T. Kotaka, Shear and elongation flow of hybrids and GFRP. In: Progress and Trends in Rheology V, (ed. I. Emri, R. Cvelbar) (1998) 398-400.
2. Kotsilkova R., Rheology structure relationship of polymer / layered silicate hybrids. Mechanics of Time-Dependent Materials, 2000 (submitted)
3. Kotsilkova R., V. Petkova, J. Pelovski. Thermal analysis of polymer-layered silicate nanocomposites. Proceedings of the 12th International Congress on Thermal Analysis and Calorimetry, August 14-18, 2000, Copenhagen, Denmark, 145-151
4. Bershtein, V.A., L.M. Egorova, P.N. Yakushev, G. Georgoussis, A. Kyritsis, P. Pissis, P. Sysel, L. Brozova, Molecular dynamics in nanostructured polyimide silica hybrid materials and their thermal stability. Macromol. Symp. 146 (1999) 9-15.
5. Maggana C., P. Pissis, TSDC studies of the effects of plasticizer and water on the sub-Tg relaxations of an epoxy resin system. J. Mascrom. Sci. Phys. B36 (1997)

Nanoscience & Nanotechnology
eds. E. Balabanova, I. Dragieva, Heron Press, Sofia, 2001

NANOSIZE DOPING EFFECTS IN POLYCRYSTALLINE HTS

T. Koutzarova[1], S. Miteva[1], M. Ausloos[2], M. Pekala[3]

[1]Institute of Electronics, BAS, 72, Tzarigradsko chaussee, 1784 Sofia, Bulgaria
[2]SUPRAS, Institute of Physics, University of Liege, B-4000 Liege, Belgium
[3]Department of Chemistry, University of Warsaw, Al. Zwirki i Wigury 101, PL-02-089 Warsaw, Poland

Abstract. The present work proposes an analysis and comparison of the results obtained from investigations of the magnetic and resistivity properties of two types of samples: first, with doping in the unit lattice cell (Dy-substituted YBCO), second with nanosize defects on the grain boundaries (K- and Rb- doped YBCO). Substantial differences are seen in the influence of the two kinds of defects in the areas of melt state of HTS.

Keywords: nanosize doping effects, polycrystalline HTS

1. Introduction

The subject of the investigation presented is the nanosize doping effects in polycrystalline HTS of the YBCO type. The defects are artificially created by substitution of the non-magnetic Y^{3+} cation with the magnetic Dy^{3+} cation and additions of alkali metals. Our previous study on Dy-substituted YBCO [1] showed changes in the electromagnetic characteristics of the material which were best expressed for the $Y_{0.4}Dy_{0.6}Ba_2Cu_3O_7$ and $Y_{0.6}Dy_{0.4}Ba_2Cu_3O_7$ samples. In a series of earlier works [2,3], we were also able to show that potassium and rubidium carbonates, introduced as additives to the initial powder used in the ceramic technology for preparation of superconducting compounds, do not participate in the formation of crystallites. At the same time, the existence of the K additive in the synthesis process of the superconducting material allowed a formation of a KCuO-compound as the nanosized thin film remaining selectively on the grain boundaries.

The present work proposes an analysis and comparison of the results obtained from investigations of the magnetic and resistivity properties of two types of samples: first, with doping in the unit lattice cell (Dy-substituted YBCO), second with nanosize defects on the grain boundaries (K- and Rb- doped YBCO).

This was achieved by way of following the behavior of the screening abilities to an a.c. field, the hysteretic losses and the resistivity of the samples near the transition temperature of the doped HTS.

2. Methods

We explored a two-stage classic ceramic-technology procedure for obtaining doped YBCO ceramic oxides and carbonates. The final annealing temperature, which varied for the different levels of doping [1,2,3] was determined by DTA analysis Two Dy-substituted YBCO samples were prepared: $Y_{0.4}Dy_{0.6}Ba_2Cu_3O_7$ and $Y_{0.6}Dy_{0.4}Ba_2Cu_3O_7$.

In stage I, Rb_2CO_3 (samples S1 Rb:YBCO and S2 Rb:YBCO) and K_2CO_3 + CuO (sample KCu:YBCO) were added to the initial powders following the formula $Y_{(1-2x)}Ba_{(2-2x)}Me_xCu_3O_y$ where for rubidium $x = 0.36$ and for potassium $x = 0.95$. In stage II, the S1 Rb:YBCO sample was subjected to a high-temperature isothermic delay for 30 hours; for the samples S2 Rb:YBCO and KCu:YBCO the delay was 120 h. The samples studied were chosen to be with a comparable grain size (about 60 μm). The microstructural investigations were made by SEM and TEM.

3. Results

The electromagnetic characteristics of the HTS pellets were measured mostly at 77 K. To investigate the screening ability of the specimens, a spatially uniform magnetic field was applied perpendicularly to the sample surface and a small coil placed on the surface registered the response signal [4]. The hysteresis loops were obtained using a miniature Hall probe located close to the pellet surface [5].

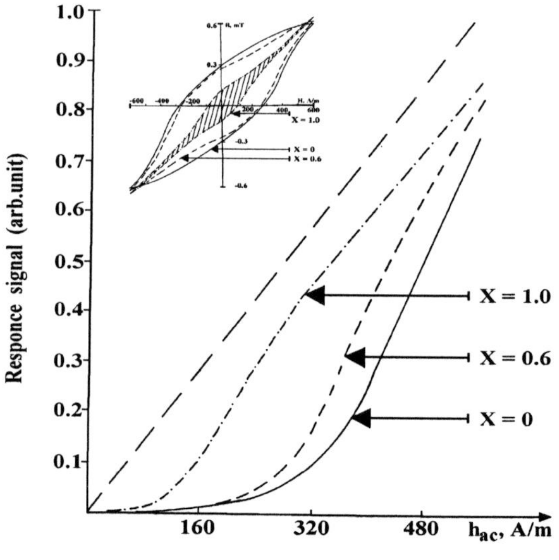

Figure 1. Changes in the screening properties (a) and hysteresis loop behaviour (b) of the samples $Y_{0.4}Dy_{0.6}Ba_2Cu_3O_7$ ($x = 0.6$) in comparison with pure YBCO ($x = 0$) and DyBCO ($x = 1$)

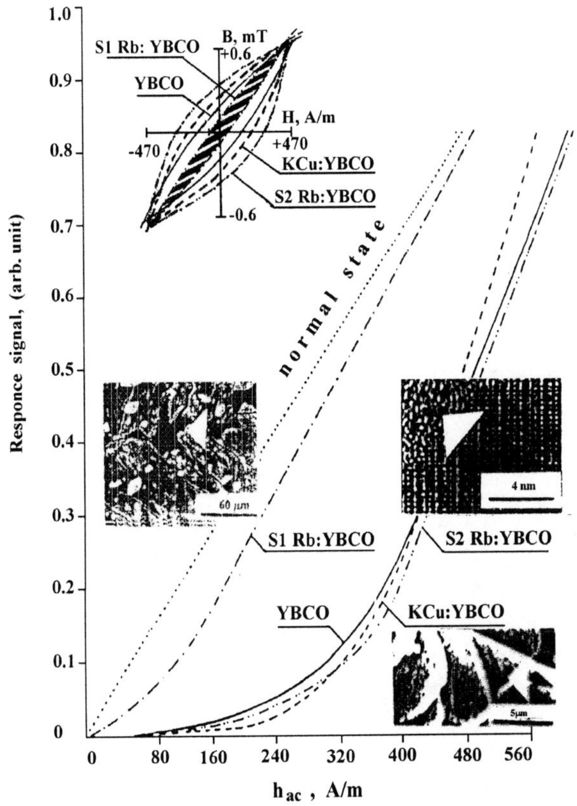

Figure 2. Changes in the screening properties (a) and hysteresis loop behaviour (b) of the samples S1 Rb:YBCO, S2 Rb:YBCO and KCu:YBCO in comparison with pure YBCO

The changes of the screening properties of the $Y_{0.4}Dy_{0.6}Ba_2Cu_3O_7$ in comparison with pure YBCO and DyBCO are plotted in Figure 1a. Figure 1b shows the hysteresis losses for the same samples.

Figure 2a shows the response-signal variation as a function of the a.c. field amplitude h_{ac} ($f = 1$ kHz) for an YBCO sample and for YBCO samples doped with K (sample KCu:YBCO) and Rb (samples S1 Rb:YBCO and S2 Rb:YBCO). At weak fields ($B \leq 0.5$ mT), the samples screened well the magnetic field and were in a state close to Meissner's. It is important to note that pure YBCO, K:YBCO and S2 Rb:YBCO kept their screening properties at greater values of the external field amplitude, while sample S1 Rb:YBCO exhibited a worsening of this characteristic. This tendency is also clearly illustrated by the hysteresis curves in Figure 2b.

The four-probe method was used for precise measurements of the electrical resistivity of the samples. A dc current up to 30 mA was injected as one-second pulses with successively reversed direction in order to avoid Joule and Peltier effects [6]. The electrical resistivity was calculated taking into account the voltage drop and the sample dimensions. The electrical resistivity measurements were performed at several magnetic field strengths between 0 and 0.6 T and the temperature was varied between 20 and 300 K in a closed cycle refrigerator with stabilization by DRC 91 controller with accuracy of 0.02 K [7].

The electrical resistivity versus temperature variation measured at various magnetic field strengths between 0 and 0.6 T is shown in Figure 3a for $Y_{0.6}Dy_{0.4}Ba_2Cu_3O_7$ and in Figure 3b for $Y_{1.81}Ba_{2.81}K_{0.95}Cu_3O_{6.69}$.

4. Discussion

Our investigations demonstrated that the alkali additives did not participate in the growth of the perovskite crystal cell and remained as a second phase. We established that in the process of synthesis the K-additive took part in the formation of KCuO-compound, remaining on the grain boundaries as a thin film. This thin film improves the homogenous growth of the grains of the superconductive phase. For longer isothermic delays, the film has a size in the order of a few tens of nm.

A very different behaviour was observed for the Rb_2CO_3 additive, compared with that of K_2CO_3 additive. The microstructural investigations carried out showed that the Rb additive did not participate in perovskite crystal phase formation, and diffused towards the sample surface in the form of Rb_2O. In that way, the Rb additive changed the phase composition only at the surface layer and no changes were observed in depth. We also investigated the

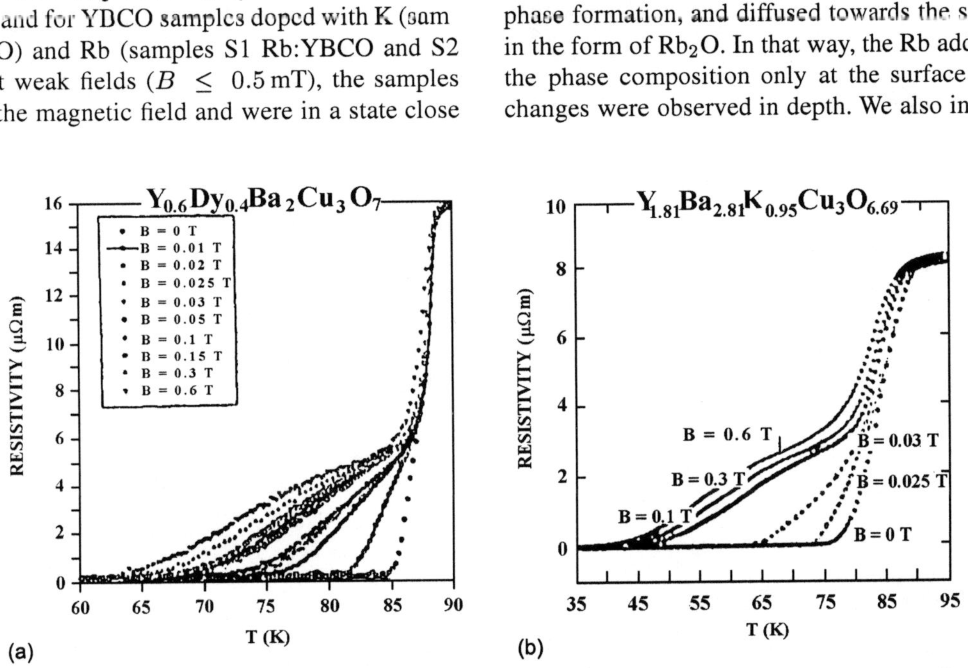

Figure 3. Electrical resistivity vs temperature of a polycrystalline (a) $Y_{0.6}Dy_{0.4}Ba_2Cu_3O_7$ and (b) for $Y_{1.81}Ba_{2.81}K_{0.95}Cu_3O_{6.69}$

S2 Rb:YBCO samples after mechanically removing the surface layer. The sample showed the best screening properties of the whole set of samples studied. The microstructural investigations demonstrated a strong change of the structure on the grain boundaries within a layer of 3-4 nm thickness. The results obtained were compared with those for the sample S1 Rb:YBCO with short isothermal delays. In the sample S1 Rb:YBCO, we established the presence of Rb on the grain boundaries with the size of a few μm. Overall, the screening properties of this sample strongly deteriorated compared with the characteristics of KCu:YBCO with the same level of defects, and of sample S2 Rb:YBCO.

The investigation of the mixed state in HTS changes the simple notions about some superconductivity aspects as derived from low T_c (LTS) superconductors. It appears that superconductor physical parameters present a quite complicated behavior especially in the high magnetic fields, but also at low fields. The thermal fluctuations, vortex dissipation, quasiparticle scattering "interferences" are in fact the main obstacles for well controlling HTS properties and subsequent applications. Thus some fine investigations are still needed.

The study of the electrical resistivity properties of the doping samples exhibited a strong dependence of $R(T)$ in the area of superconducting transfer on the type of additive. The absolute value of the resistivity for sample $Y_{0.6}Dy_{0.4}Ba_2Cu_3O_7$ was 18 $\mu\Omega$m/K at 100 K. In the normal state, the resistivity increased approximately proportionally to the temperature with an average rate of 0.04 $\mu\Omega$m/K. The transition width defined as the difference between the normal state and the resistivity percolation temperature T_p [7] was narrower than 3 K in the absence of an external magnetic field. The transition width became gradually broader up to 37 K at a magnetic field of 0.6 T. Even in a non-zero magnetic field the electrical resistivity dropped down to about 45% of the normal state value in a remarkably narrow interval between 85 and 88 K which is typical for intragrain superconductivity. The intergrain processes are seen to be spread out over a much broader temperature range rising gradually up to 25 K with the magnetic field strength. As can be seen on Figure 3b, the additive on the grain boundaries influences more strongly the polycrystalline sample, where T_p is in the range of 35–40 K.

5. Conclusions

The results obtained demonstrate that the influence of the nanosize doping effect on a unit lattice cell level and on the grain boundaries at the temperatures and magnetic field mentioned have comparable effects on the hysteresis losses and screening properties, when the additive on the grain boundaries is in negligible amounts creating defects in the range of nm. Substantial differences are seen in the influence of the two kinds of defects in the areas of melt state of HTS. The peculiarities in the electrical resistivity behaviour in the mixed state in a highly-nisotropic superconductor, the conditions of the phase coherence transition in the mixed state of a type-II superconductors and information about microscopical parameters and mechanisms of HTS are useful for possible applications, because they rely on a simple assumption, i.e. a distribution of vortex melting temperatures, but they are also of fundamental interest.

References

1. Nedkov I., Koutzarova T., Miteva S., Ausloos M., Bougrine H., Cloots R., Proc. XIV Inter. Conf. Microwave Ferrites 1998, 224-229.
2. Nedkov I., Veneva A., Miteva S., Applied Superconductivity (EUCAS) 1 (1993), 73-76.
3. Nedkov I., Veneva A., Miteva S., Nurgaliev T., Lovchinov V., IEEE Trans. Mag. 30 (1994), 1187-1189.
4. Nurgaliev T., Miteva S., Nedkov I., Veneva A., J. Mag. Mag. Mater. 140-144 (1995), 1305-1306.
5. Miteva S., Nurgaliev T., Nedkov I., Proc. XIII Inter. Conf. Microwave Ferrites 1996, 272-277.
6. Bougrine H., Ausloos M., Rev. Sci. Instrum. 66 (1995), 199
7. Pekala M., Godomski W., Nedkov I., Bougrine H., Ausloos M., p. 301-316, Symmetry and Pairing in Superconductors eds. M. Ausloos and S. Kruchinin, Kluwer Acad. Publ. 1999

Nanoscience & Nanotechnology
eds. E. Balabanova, I. Dragieva, Heron Press, Sofia, 2001

APPLICATION OF NANODIMENSIONAL FUNCTIONAL LAYERS IN LOW POSITIVE TEMPERATURE MEASURING

Bohos Aprahamian[1], Meline Aprahamian[2]

[1]Naval Academy – Varna, Dept. of Electrical Eng.
[2]Technical University – Varna, Dept. of Mathematics

Abstract. Resistive elements, formed upon electroinsulated plates of alumina ceramics with nanodimensional conductive layers are obtained. The prototypes are prepared by magnetron sputtering of the functional layer in the Central Laboratory of Electrochemical Power Sources of Bulgarian Academy of Sciences (BAS). All elements are covered with polyimide insulating nanodimensional film in the Central Laboratory of Photoprocesses of BAS.

Different layer compositions, including amorphous metal alloys are experimented and the change of the electrical resistance due to the temperature rate is studied.

The elements are applied as sensors in electronic bridge circuit for positive temperature measuring up to 773 K.

1. Introduction

The nanotechnologies are interdisciplinary field concerning the synthesis, the research and the development of nanodimensional clusters, structures and wares which dimensions are in the atomic band.

The studies in this field allow the making of nanophase materials and nanocomposites possessing nonordinary mechanical, electrical, optical and magnetic properties.

The object of the proposed research is the study of the possibility for application of nanodimensional functional layers in elaboration of resistive sensors with high change of resistance, caused by the temperature rising. Nanodimensional powder materials elaborated in the Central Laboratory of Electrochemical Power Sources (CLEPS) and in the Central Laboratory of Photoprocesses (CLP) of BAS are used. We expect that the newly developed sensors will have improved characteristics in comparison with the conventionally used, for example posistors.

The authors have many previous studies, related with the application of metal powders, obtained by a borohydride reduction process and possessing an amorphous or ultradispersed structure in resistive sensors with very strong negative dependence of the sensor resistance upon the temperature rising. Many prototypes are elaborated, obtained by screenprinting of resistive pastes with different compositions on electro-insulated plates of aluminum oxide ceramics[1,2].

The proposed research will expend the kinds of the used materials in elaboration of resistive sensors, will contribute the specification of the technological parameters and will study the possibility for application of the newly developed sensors in electrical equipment and automatics, for example in heat protection of electrical machines.

2. Configuration and Preparation Technologies

In developing prototypes of surface resistive sensors on ceramic plates alumina ceramics produced by "Metaloceramics" (Bulgaria), "Rosenthal" and "Rubalit" (Germany) and 22XC type ceramics (Russia) were used.

All specimens were prepared on square ceramic substrates with a side of 2 inch (50.8 mm) and 0.9 mm thick. Having analysed many possible procedures for forming a conducting layer upon a ceramic substrate, we preferred following two of them: magnetron sputtering and screenprinting of a resistive layer, both technologies being applied in integrated circuits production. Though principally different in implementation both of them enable high-grade layers to be obtained with:

— good adhesion with ceramic plate;

— controllable changes of electric resistance before and after layering;

— good corrosion resistance;

— the coefficient of thermal expansion close to that of the substrate.

In forming resistive elements by magnetron sputtering Russian stainless steel of X18H9T – type containing C – 1%, Cr – 18%, Ni – 9% and Ti – 0.5% is used as resistive layer material. Coating is performed in a vacuum-pumping plant of B – 90 type ("Hochvakuum" – Germany) with the following process parameters: cleaning voltage – 2 kV DC, cleaning time – 10 min, operating voltage – 390 V DC, operating gas – Argon, operating pressure – 2×10^{-3} Torr.

In forming resistive elements by screenprinting special resistive pastes are used for resistive layer material, including pastes developed in Bulgaria by the Central Laboratory of Electrochemical Power Sources of BAS, as well as those produced by "Heraeus" – Germany and "DuPont" – USA. DuPont conducting pastes are used for contact areas. Commercial DuPont Model 8010 printer is used, the parameters of the technological process being as follows: speed of the printing panel – 35 mm/s, thickness of the emulsion layer of the screen – 11 μm, mean baking temperature – 850°C, time of maintaining maximal baking temperature – 10 min, paste viscosity – 10^5 Pa.s.

In experimenting the procedures mentioned parameters were changed many times and their effects upon the quality of the layer were studied by statistic analysis.

A protective layer was applied on specimens of resistive elements by screenprinting, this layer enabling operation of the sensor in direct contact with conducting objects without using additional insulation. Dielectric paste of "Heraeus" – Germany production was used for the purpose. Two types of protective layer were tested:

— entire layer covering entire plane of the element on the side of the resistive layer;

— partial layer covering the resistive layer only and following its configuration.

Configuration of surface resistive elements is shown in Figure 1.

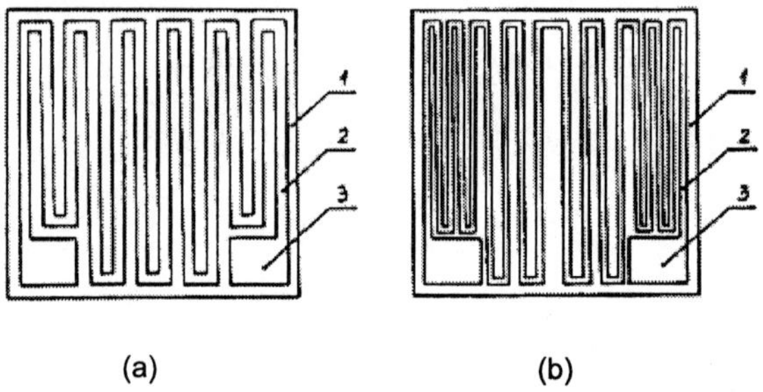

Figure 1. Configuration of surface heater element on ceramic plate. (1) – ceramic plate; (2) – resistive layer; (3) – contact areas; (a) – uniformly distributed resistive layer; (b) – unevenly distributed resistive layer for improvement of the uniformity of the heating element's temperature field

Prototypes are developed with the following parameters: area of the sensor – 2.58×10^{-3} sq.m, maximal power for this area – 60 W, maximal operating temperature – 380°C. Power supply may be of direct or alternating current of up to 250 V. Main technical par-ticulars of the resistive elements are shown in Table 1.

3. Application of Thin Polyimide Layers as Protective Layers of Resistive Sensors

Vacuum deposited polyimide thin layers are a subject of intense investigations. The process of their fabrication comprises the co-evaporation of two precursor monomers. The principle of deposition of these monomers determines their intimate mixing and this predetermines the following process of imidization which takes place at an elevated temperature.

Two types of vacuum deposited polyimide layers differing in their mode of preparation – deposition of the monomer vapours on rotary and linearly moving plates of alumina ceramics have been investigated in the Central Laboratory of Photoprocesses (CLP) of BAS [3]. Main technical particulars of the resistive elements with polyimide protective layers are shown in Table 2.

The experimental results, described in [3], show that the films with the best parameters of the investigated properties are those which monomer vapours precipitate on planetary rotating substrates and the imidization is obtained by curing for 60 min at 170°C and for 60 min at 350°C. The layers obtained in such technological sequence are characterized by higher microhardness and higher stability.

4. Experimental Results and Applications

Through electrical, thermal, mechanical and climatic studies of the prototypes are made in accordance with standards used in Bulgaria. The results obtained prove the sensors to be practicable [1].

Mechanical deformations of the elements due to thermal load are studied in order to choose optimal way of fitting elements in electrothermal devices. Various types of fitting are tested and data for main mechanical stresses in the heater plane are found. Fitting with one fixed point

Table 1. Various types of resistive elements on ceramic plates. CLEPS-BAS – Central Laboratory of Electrochemical Power Sources – Bulgarian Academy of Sciences

No	Ceramic Plate	Resistive Layer	Technology of the Resistive Layer	Protective Layer
1.	96% Al_2O_3 Metaloceramics – Bulgaria	Stainless Steel X18H9T – Russia	Magnetron Sputtering	—
2.	96% Al_2O_3 Metaloceramics – Bulgaria	Resistive Composition "61" of CLEPS-BAS	Screenprinting	—
3.	99% Al_2O_3 Metaloceramics – Bulgaria	Resistive Composition DuPont 1311 – USA	Screenprinting	—
4.	96% Al_2O_3 22XC – Russia	Amorphous Alloy of X18H9T with 13,9% Titanium	Magnetron Sputtering	—
5.	96% Al_2O_3 Rosenthal – Germany	Resistive Composition Heraeus R4005B – Germany	Screenprinting	Entire Layer Heraeus IP9025S – Germany
6.	96% Al_2O_3 Rubalit – Germany	Stainless Steel X18H9T – Russia	Magnetron Sputtering	Partial Layer Heraeus IP9025S – Germany
7.	96% Al_2O_3 Metaloceramics – Bulgaria	Resistive Composition "62" of CLEPS-BAS	Screenprinting	Entire Layer Heraeus IP9025S – Germany
8.	96% Al_2O_3 Rosenthal – Germany	Resistive Composition Heraeus R4010B – Germany	Screenprinting	Partial Layer Heraeus IP9025S – Germany
9.	96% Al_2O_3 Rosenthal – Germany	Amorphous Alloy of X18H9T with 15,5% Titanium	Magnetron Sputtering	Entire Layer Heraeus IP9025S – Germany

Table 2. Various types of resistive elements on ceramic plates with nanodimensional protective layers. CLEPS-BAS – Central Laboratory of Electrochemical Power Sources – Bulgarian Academy of Sciences

No	Ceramic Plate	Resistive Layer	Technology of the Resistive Layer	Nanodimensional Protective Layer	Technology of the Protective Layer
1.	96% Al_2O_3 Metaloceramics – Bulgaria	Resistive Composition "61" of CLEPS-BAS	Screenprinting	Nickel + Polyimide Composition of CLP-BAS	Vacuum Deposition
2.	96% Al_2O_3 Metaloceramics – Bulgaria	Resistive Composition "62" of CLEPS-BAS	Screenprinting	Nickel + Polyimide Composition of CLP-BAS	Vacuum Deposition
3.	96% Al_2O_3 Metaloceramics – Bulgaria	Resistive Composition "62" of CLEPS-BAS	Screenprinting	Nickel + Polyimide Composition of CLP-BAS	Vacuum Deposition
4.	96% Al_2O_3 Metaloceramics – Bulgaria	Stainless Steel X18H9T – Russia	Magnetron Sputtering	Nickel + Polyimide Composition of CLP-BAS	Vacuum Deposition
5.	96% Al_2O_3 Metaloceramics – Bulgaria	Stainless Steel X18H9T – Russia	Magnetron Sputtering	Nickel + Polyimide Composition of CLP-BAS	Vacuum Deposition
6.	96% Al_2O_3 Metaloceramics – Bulgaria	Stainless Steel X18H9T – Russia	Magnetron Sputtering	Nickel Phosphate Composition of CLP-BAS	Vacuum Deposition
7.	96% Al_2O_3 Metaloceramics – Bulgaria	Stainless Steel X18H9T – Russia	Magnetron Sputtering	Nickel Phosphate Composition of CLP-BAS	Vacuum Deposition
8.	96% Al_2O_3 Metaloceramics – Bulgaria	Stainless Steel X18H9T – Russia	Magnetron Sputtering	Polyimide Composition of CLP-BAS	Vacuum Deposition
9.	96% Al_2O_3 Metaloceramics – Bulgaria	Stainless Steel X18H9T – Russia	Magnetron Sputtering	Polyimide Composition of CLP-BAS	Vacuum Deposition

(by pressing) and with 4 fixed points are prove to be most advantageous because dangerous tensile stresses are small and there are no great mechanical loads in the contact areas.

Having analyzed existing types of digital thermometers, we came to a conclusion that it is possible to improve their sensitivity by replacing the thermocouple with a flat sensor consisting of a ceramic plate with a resistive layer of certain configuration. The resistance changes of different sensors caused by the rising of the temperature are shown in Figure 2.

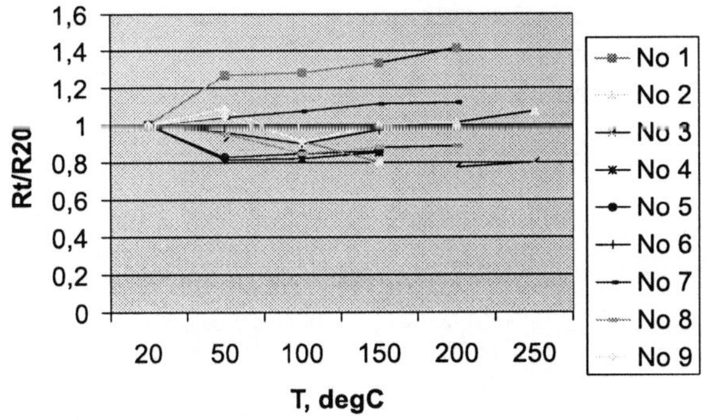

Figure 2. Changes of the resistance of various types of sensors caused by the temperature rising. The numeration is corresponding of the element arrangement in Table 2

The new elements are adopted as sensors in standard bridge circuits for measuring of temperature up to 773 K. Different prototypes were tested as sensors and their accuracy of measuring was studied. Best results were obtained with the prototypes No 5 and 7 shown in Table 1. The changes of the concretely measured temperature with different devices – standard digital thermometer and electronic thermometer with resistive sensor on ceramic plate (No 7 of Table 1) were studied – Figure 3. The accuracy of measuring of the new sensor is worse but admissible. We continue our work for improvement of the new resistive elements characteristics.

Heater elements on ceramic plates with the same configuration are also adopted in practice as thermostatic elements of computer systems. They are considered promising and we have been working on their further modifications.

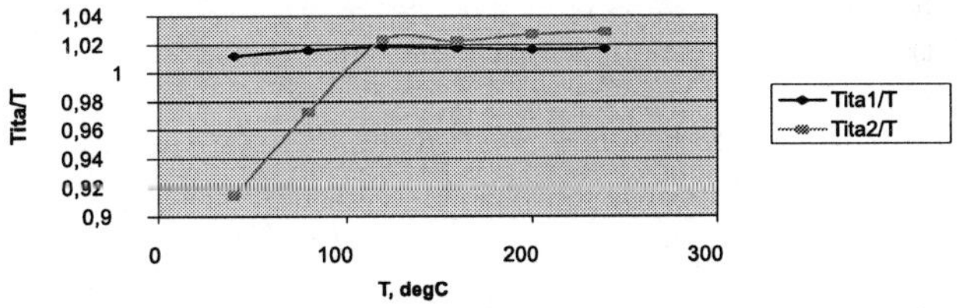

Figure 3. Changes of the concretely measured temperature with different devices caused by the temperature rising. (1) (1) – standard digital thermometer; (2) – electronic thermometer with resistive sensor on ceramic plate

References

1. Aprahamian B., Kontrov S. Surface heater elements on ceramic plates, Papers of the IV-th International Conference on Electron Beam Technologies, Varna, 1994, p. 284–291
2. Aprahamian B. Application of amorphous resistive layers in electrical measuring of speed of fluids, Sbornik atatii "Morski nauchen forum", vol. I, Varna, 1998, p. 301–305
3. Assa J. et al. Thin polyimide layers - preparation and properties, Vacuum, vol. 51, no 2, 1998, p.185-188

Nanoscience & Nanotechnology
eds. E. Balabanova, I. Dragieva, Heron Press, Sofia, 2001

FLEXOELECTRIC MESUREMENTS ON PHOSPHOLIPID MODIFIED POLYMER DISPERSED LIQUID CRYSTALS (PDLC)

T. Angelov[1], Y. Marinov[1], S. Torgova[2], A.G. Petrov[1]

[1]Institute of Solid State Physics, BAS, 72, Tzarigradsko Chaussee, 1784 Sofia
[2]SSC RF "NIOPIK", Organic Intermediates & Dye Institute, B. Sadovaya 114, Moscow, Russia

Abstract. We report on the observation of flexoelectricity of cyanobiphenyl nematic liquid crystal dispersed by ultrasonication in polymethylmethacrilate polymer matrix and studying surface dissipation of orientational energy by the flexoelectric spectroscopy method. Lipid molecules were involved along the polymer dispersed liquid crystals (PDLC) formation to obtain weak homeotropic anchoring at LC droplet surfaces. By the solvent induced phase separation method LC droplet size of 2–5 μm has been reached. The thickness of PDLC films was 10 μm. An linear electro-optic response up to 100 μs has been measured.

Keywords: optoelectronic, flexoelectricity, liquid crystals, polymers

1. Introduction

Polymer dispersed liquid crystals (PDLC) [1] are composed materials consisting of micron-sized liquid crystal droplets dispersed in an optically transparent polymer matrix. PDLC samples are a field-controlled light scattering devices Figure 1 [6]:

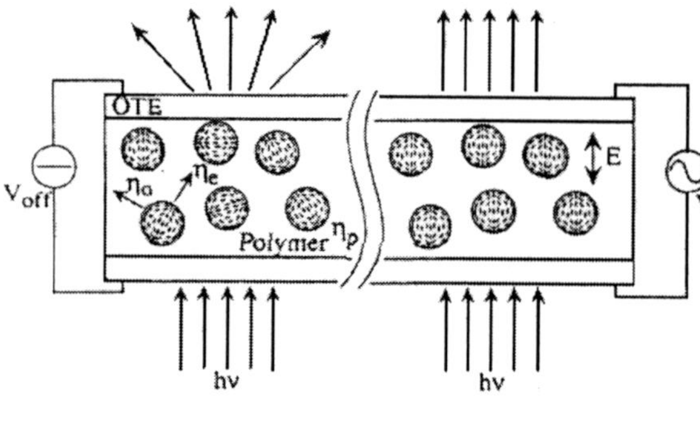

Figure 1. Cross-sectional view of a generalized PDLC thin film device

Approximately spherical droplets of a liquid crystalline material are dispersed within a polymer matrix. The PDLC film is sandwiched between optically transparent electrodes OTE. On the left, the film properties are depicted under zero-field condition. The optical axes of the birefringent droplets (shown here in the "bipolar configuration" [1]) are randomly aligned and the film strongly scatters light because on the mismatch in polymer, n_p and droplet extraordinary, n_c, refractive indices. On the right, under influence of an applied electric field, the liquid-crystalline droplets align with their axes parallel of the field. In cases where the droplet ordinary refractive index, n_0, matches that of the polymer, the film becomes transparent to normally incident light. As in presence of external field initially randomly oriented director of nematic droplets becomes uniformly oriented, thus switching from opaque to transparent state of medium. These liquid crystal systems can be used as light shutters, flexible displays and switchable windows. The electro-optic effect in PDLC scattering devices is dielectric by origin and thus it is quadratic with the field. For such application time response, low field switching and contrast are of substantial importance. Those requirements are compromise satisfied for film thickness higher then 10 μm and droplet size less 1 μm.

The electric field effects on the droplet structure in PDLCs have been studied by several authors [1–8]. Recently, new types of PDLCs utilizing linear electrooptic effects in chiral liquid crystal were presented [9]. In these devices the linear coupling may be dielectric, flexoelectric or ferroelectric. The response time of 100 μs much shorter than those of field controlled scattering devices (millisecond range) was found. The advantages of linear and shorter electrooptic response related to the flexoelectricity presumably will play one of the major roles in future developments of PDLC devices.

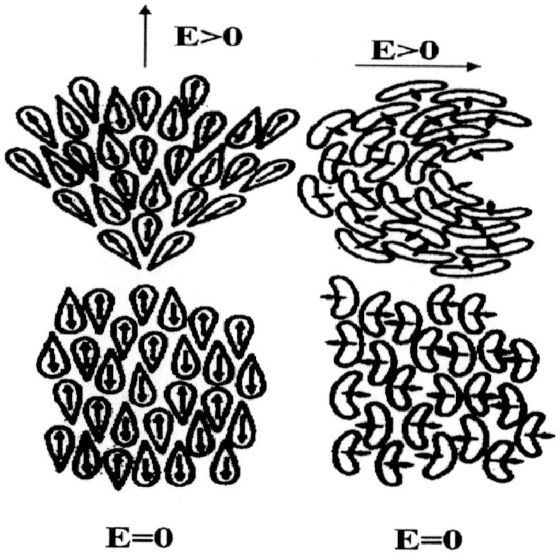

Figure 2

The flexoelectricity Figure 2, is a fundamental property of liquid crystals and gives the linear coupling between electric field and nematic director deformation, expressed by (1):

$$\boldsymbol{P} = e_{1z}\boldsymbol{n}(\mathbf{div}\,\boldsymbol{n}) + e_{3x}(\mathbf{rot}\,\boldsymbol{n} \times \boldsymbol{n}) \qquad (1)$$

where: $\boldsymbol{P}$ – polarization; e_{1z} and e_{3x} – splay and bend flexoelectric coefficients; $\boldsymbol{n}$ – molecular director.

Mainly flexoelectricity is a surface effect strongly depending on surface anchoring. That fact is directly connected to the PDLC, where the nematic interface is fully extended. In this respect the aim of this paper is to investigate the flexoelectric effect in confined nematic liquid crystal systems, under modified droplet boundary conditions.

2. Materials and Methods

Materials: Polymethylmethacrylate (PMMA), L-α-phosphatidylcholine Soy bean lecithin (Sigma) and liquid crystals eutectic mixture cyanobiphenyls LC-807, average molecular size of 20–30 nm – "NIOPIK" – Russia.

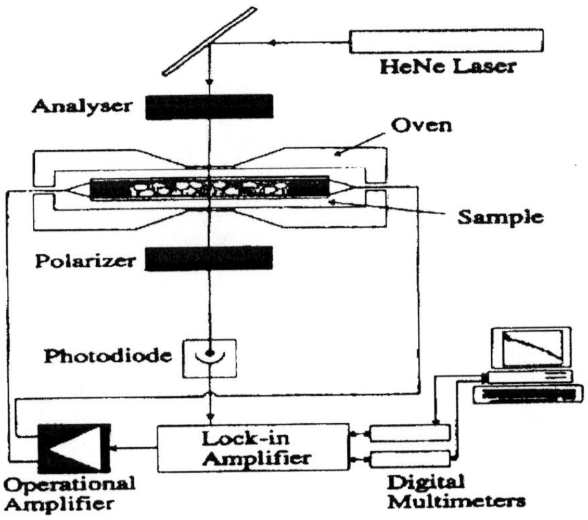

Figure 3. Experimental setup

Preparation of the PDLC: PMMA, Lecithin and LC-807 in weight fractions 0.3416:0.0366:0.6218 in chloroform were dissolved. Homogeneous solution was obtained after mecnanical stirring and sonication by UM-4 "Unitra". Solution was then spread onto a ITO conductive glass substrate. Upon solvent-induced phase separation (SIPS technique) at room temperature, an opaque PDLC film of 10 μm thickness was obtained, subsequently covered by a second ITO glass substrate. Flexoelectric phase sensitive spectroscopy method described elsewhere [10] was used throughout the all experiments. A schematic diagram of our experimental system is shown in Figure 3. Using microscopic observations we determine an average size of the liquid crystal droplets around 2.5 μm.

3. Results and Discussion

Standard qualitative measurements of PDLC light transmittance was carried out. The obtained data for red and white irradiation are shown in Figure 4. The electrooptic responses are in good consistence with the similar PDLC devices [6,7] showing high transparent state without substantial light irradiation selectivity.

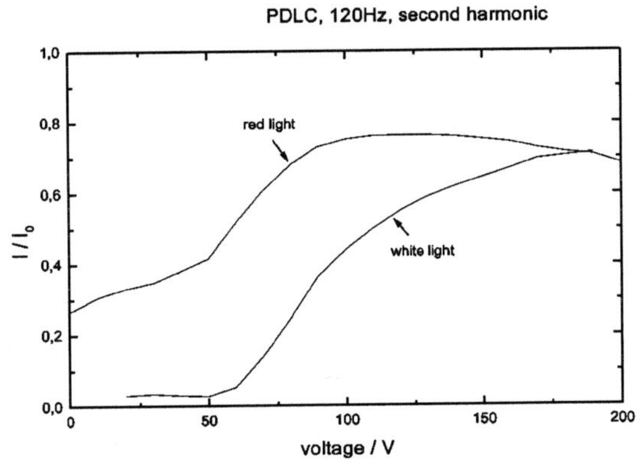

Figure 4. Transmittance as a function of applied voltage upon different light irradiation

The dielectric measurements were followed by flexoelectric one at lower electric field in order a linear coupling to be find. Flexoelectric spectra provide information about surface dissipation of orientational energy in nematics [10]. For homeotropic nematic layers oriented by self-assembled monolayers higher frequency part of viscoelastic spectra reveals important aspects of orientational interaction nematic-substrate. It is worth to mention that surface driving forces are presented within interface of few nanometers. Furthermore strong anchoring strength dependence on surfactant density was found in planar cells [11]. Understanding the above mentioned interaction in confined systems like PDLC is important from fundamental and applied points of view. The non- uniform nematic director field and the need of finding appropriate boundary condition on liquid crystal droplets surfaces makes the studying of such systems more complicated.

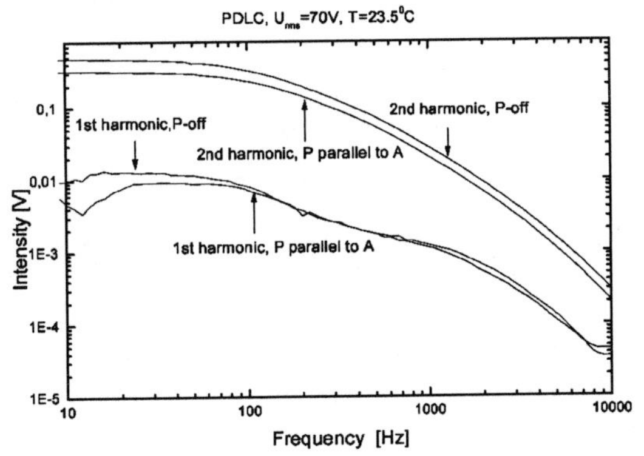

Figure 5. Frequency dependence of the 1st and 2nd harmonic of modulated transmitted light trough PDLC. $U = 70\,V$, $T = 23.5°C$

In Figure 5 the 1st harmonic flexoelectric spectra of LC 807 droplets dispersed in polymethylmethacrylate and weakly anchored on the self-organized phospholipid monolayer at the polymer/LC interface, are displayed. Besides expected 2nd harmonic due to dielectric coupling of director field inside the droplets to the ac electric field, a 1st harmonic response was also obtained. This is a strong evidence of a linear, viz. flexoelectric coupling. The shape of the 1st harmonic spectrum displays a low-frequency plateau, followed by a mid-frequency plateau and, eventually, a crossover frequency of roughly 1000 Hz, which separates bulk from surface dissipation of energy. Frequency range up to 10000 Hz was investigated, which corresponds to steady electro-optical response down to 100 μs.

4. Conclusuon

For the first time 1st harmonic flexoelectric spectra of cyanobiphenyl PDLC modified by phospholipid was obtained. Existence of steady linear electrooptic part in transmitted light due to flexoelectric coupling was proved.

Acknowledgements

Thanks are due to Dr. Hr. Hinov for helpful discussions.

References

1. P.S. Drzaic, Liq. Crystals, vol.3.11, p. 1543, 1988.
2. A.L. Lovinger, K.R. Amundson, D.D. Davis, Chem. Mater. 6, pp. 1726–1736, 1994.
3. J.R. Kelly, P. Palfy-Mahoray, Mol. Cryst. Liq. Cryst. vol. 223, pp. 251-261, 1992.
4. K. Lee, S.-W. Suh, S.-D. Lee, Appl. Phys. Lett vol.64, 6,1994.
5. D.A. Higgins, Adv. Materials, 14, 4, p.251, 2000.
6. S. Zumer, J.W. Doane, Phys. Rev. A, vol. 34, 4, p3373, 1986.
7. J.R. Kelly, P. Palfy-Mahoray, Mol. Cryst. Liq. Cryst., vol.243, pp. 11–29, 1994.
8. R.B. Meyer, Phys. Rev. Lett. vol.22, 918,1969.
9. P. Rudquist, L. Komitov, P. Rudquist Thesis, Note 3,1994.
10. Y. Marinov, N. Shonova, C. Versace and A.G. Petrov, Mol. Cryst. Liq. Cryst., vol. 329, pp. 1145–1150, 1999.
11. Ü. Kühnau, A.G. Petrov, G. Close, H. Schmiedel, Phys. Rev. E, vol. 59, l, pp. 578–585.

SECTION 4:
Synthesis and Processing (Electrochemical/Chemical Properties)

Nanoscience & Nanotechnology
eds. E. Balabanova, I. Dragieva, Heron Press, Sofia, 2001

THE ROLE OF ELECTROCMHEMISTRY IN MODERN NANOSTRUCTURING OF SOLID STATE SURFACES

W. J. Lorenz[1], W. Wiesbeck[1,2], F.X. Kärtner[1,3], and W. Schindler[3]

[1]Institut für Nanotechnologie, Universität Karlsruhe, D-76131 Karlsruhe, Germany
[2]Institut für Hochfrequenztechnik und Elektronik, Universität Karlsruhe, D-76131 Karlsruhe, Germany
[3]Institut fur Hochfrequenztechnik und Quantenelektronik, Universität Karlsruhe, D-76131 Karlsruhe, Germany

Abstract. Nanoelectronic and nanomagnetic devices are expected to play a dominant role in future nanotechnology. At present, nanostructures of metals or oxides on different substrate materials such as foreign metals or semiconductors cannot be fabricated by conventional vapor phase techniques below a lateral width of less than 10–20 nm. Nanostructures of this size, however, show quantum effects even at room temperature. At present, small 0D metal or oxide clusters, 1D metal and oxide wires as well as 2D metal or oxide nanostructures can only be formed by electrochemical means using *in situ* Scanning Probe Microscopy (SPM) and appropriate polarization routines of the substrate and the probe to adjust defined supersaturation conditions for local 0D metal cluster formation and directed growth. The thermodynamic, structural and kinetic properties of such "Low-Dimensional Systems" are still poorly understood. They are expected to behave differently from 3D metal bulk phases since the majority of the metal atoms is located at the interface and surface of the structures. For practical applications in nanoelectronic and nanomagnetic devices, low dimensional systems must exhibit sufficient stability not only at the solid/electrolyte interface, but also at the solid/gas interphase.

1. Introduction

The realization of nanometer scale magnetic and electronic devices which show a novel functionality compared to classical devices, as well as the exploitation of novel catalytic or sensoric properties of nanoscale particles requires the production of metallic (M) or oxidic (O) nanostructures with lateral widths of less than 20 nm on top of different electron-conducting and semi-conducting surfaces, denoted as substrates (S). Such small nanostructures open up the possibility to utilize, e.g., quantum effects even at room temperature, which is substantially different from a simple increase of device density due to down-scaling of the device size. This has been demonstrated at room temperature in electrodeposited point contacts, or in electrochemically prepared devices, if the lateral device widths drop down below a critical size [1–9].

The main difficulty in the preparation of suitable nanostructures is the fact, that nanostructures of only a few nanometer lateral width and height cannot be fabricated at present by the usual vapor phase sputtering and evaporation techniques using lithographic intermediate steps. Besides the limiting parameter of the wavelength of the radiation used for lithography, damage in the structures due to irradiation or liftoff-techniques or other irreversible modifications of the nanostructures due to the large number of processing steps [10] become more pronounced with ever smaller structure sizes. Thus, at present the limit of lithographic techniques is in the range of a few tens of nanometers using electron or ion beam irradiation and advanced photoresists [11,12]. Furthermore, such small lateral widths have been only achieved with Si nanostructures by using oxidation processes for the ultimate thinning of the structures. Metallic nanostructures, however, would be more desirable for applications due to their large conductivity, and due to their larger skin depth at very high frequencies up to the THz-domain.

Periodic structures can be produced by selforganization processes, or by exploiting specific substrate properties like superstructures or surface reconstructions, or even moreless regularly spaced step edges [13], which allow the periodic arrangement of deposits onto these surface sites. Arrays of small magnetic clusters deposited by organic molecules which arrange on surfaces in a regular way, or magnetic clusters deposited onto the herringbone structure of the reconstructed Au(111) surface [14–16] are examples of such techniques. Also the method of cluster deposition by atom beams has been shown to result in regular arrays of nanometer sized clusters [17,18]. Unfortunately, such self-arranged deposits cannot be formed at predefined substrate locations, but only over large substrate areas in a regular pattern.

However, in the last few years it has been demonstrated that electrochemistry is capable to deposit zero-dimensional (0D) M clusters with a diameter and a height of only a few nanometers locally on different substrates using *in situ* Scanning Probe Microscopy (SPM) techniques at a solid/liquid interface, and appropriate polarization routines of both the substrate and the SPM probe [19-35]. It has been shown that critical 0D M clusters consisting of only a few M atoms can be fabricated at precisely predefined positions on foreign metal substrates, graphite, semiconductors, and superconducting oxide ceramic films. The different types of nanostructuring methods for local formation of such "Low-Dimensional Metal Systems" (LDMSs) with different dimensionality $i = 0, 1, 2$ [19–35] will be discussed below. Oxidic nano-structures, denoted as "Low-Dimensional Oxide Systems" (LDOSs), have been obtained by a direct writing process using *in situ* Atomic Force Microscopes (AFM) under electrochemical conditions [8,9]. Atomic contacts, prepared under electrochemical conditions at the solid/liquid interface [4,6] or using *in situ* STM [1–3], have been reported to show quantum effects of the conductivity even at room temperature.

In contrast, metallic nanostructures prepared at solid/vacuum interfaces usually show lateral widths of between 50 and 1000 nm, and exhibit quantum effects only at very low temperatures [36–39]. The observation of quantum effects at higher temperatures, however yet much below room temperature, has been only achieved using unconventional techniques to fabricate point contacts under mechanical strain conditions [40,41].

A significant advantage of the electrochemical fabrication of nanostructures is the fact that supersaturation and undersaturation conditions at both the substrate/electrolyte interface and at the SPM probe/electrolyte interface can be precisely adjusted and rapidly changed via the corresponding electrode potentials. Supersaturation and undersaturation are directly related to the Fermi levels of them. The exact control of the Fermi energies of the substrate and the SPM probe allows to produce M or O nuclei of the desired size and to adjust the growth conditions to form unique 0D M or O clusters, well-defined 1D M or O nanowires and 2D M or O nanostructures from separated 0D M or O clusters. Thus, the SPM may be expected to become a preparation tool, compared to the simple imaging function which SPM's have been used for so far.

It is worth to note, that practical applications require a sufficient time stability of such "Low-Dimensional Systems" (LDSs), i.e. LDSs should not dissolve or corrode over sufficient time periods. However, at present there exists practically only a poor knowledge concerning the energetics and the dissolution and corrosion behavior of LDSs determining the long-time stability of nanostructures. Consequently, fundamental studies of the thermodynamic, structural and kinetic properties of LDSs in selected electrochemical model systems are necessary.

Summarizing this very brief overview, electrochemical techniques, from the present point of view, seem to be capable to grow nanostructures without irreversible modifications due to the preparation process, which are almost unavoidable in nearly all lithographic preparation techniques. Thus, one may expect, that electrochemical growth of nanostructures may contribute substantially to some important fields of nanotechnology, since it allows the investigation of "as grown" nanostructures already in the near future.

2. Delocalized Electrochemical Formation of LDMSs

Electrochemical phase formation and dissolution are related processes that are strongly influenced by the atomic structure of the substrate and by surface inhomogeneities of real substrates. In the past decade, so-called "delocalized", i.e. undirected, metal phase formation processes on native and foreign metal substrates have been intensively studied on an atomic level using *in situ* SPM techniques

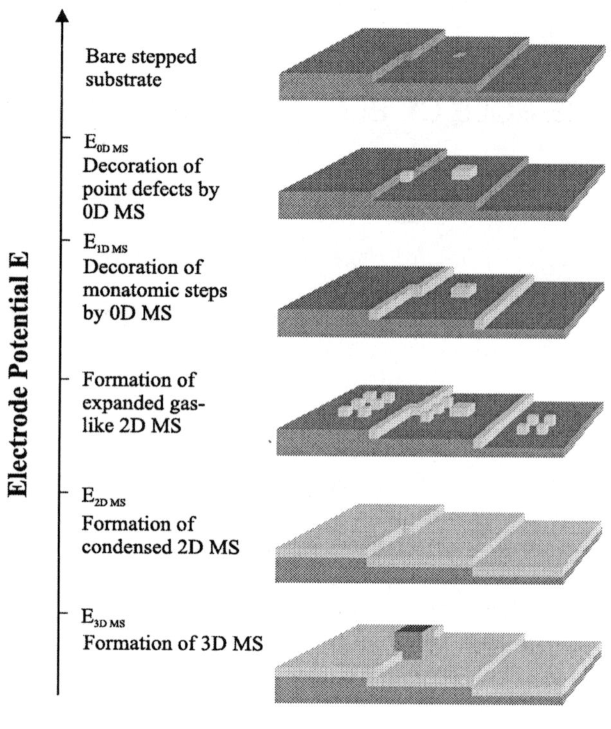

Figure 1. Formation of LDMSs of different dimensionality on foreign substrates depending on the electrode potential E

[19,20 (and references herein)]. The main parameters determining the nucleation and growth steps in phase formation and dissolution processes are the interaction energy between a M adatom and S, the crystallographic M-S misfit, internal strain within LDMSs and/or topmost layers of S, and the influence of substrate surface inhomogeneities [19,20].

In electrochemical systems with a strong M adatom-S interaction it has been found that LDMSs such as 0D M clusters and 1D M phases are preferentially formed at 0D and 1D surface inhomogeneities of S at relatively high undersaturation (underpotential deposition (UPD)) conditions. At lower undersaturation, 2D M overlayers with expanded and condensed structures are formed on atomically smooth terraces of S in such systems. The formation of 2D M overlayers always starts from 0D M clusters and 1D M phases via 2D nucleation and 2D growth processes. The 3D M phase is formed under supersaturation (overpotential deposition (OPD)) conditions on top of LDMSs. In appropriate model systems, stepwise formation of LDMSs and the 3D M phase depending on the electrode potential has been observed. This is schematically illustrated in Figure 1. In such electrochemical systems with a strong M adatom-S interaction, the 3D M growth mechanism in the initial growth stages takes place either layer-by-layer (Franck-van der Merwe growth mode) or by growing metal islands (Stranski-Krastanov growth mode) depending on the crystallographic M-S misfit (Figure 2). Upon increase of the crystallographic strain energy, lattice relaxation of the growing film towards its bulk properties is observed [42]. A variety of relaxation mechanisms, depending on the specific M-S system and its interfacial alloying, have been reported in the recent literature [42-46].

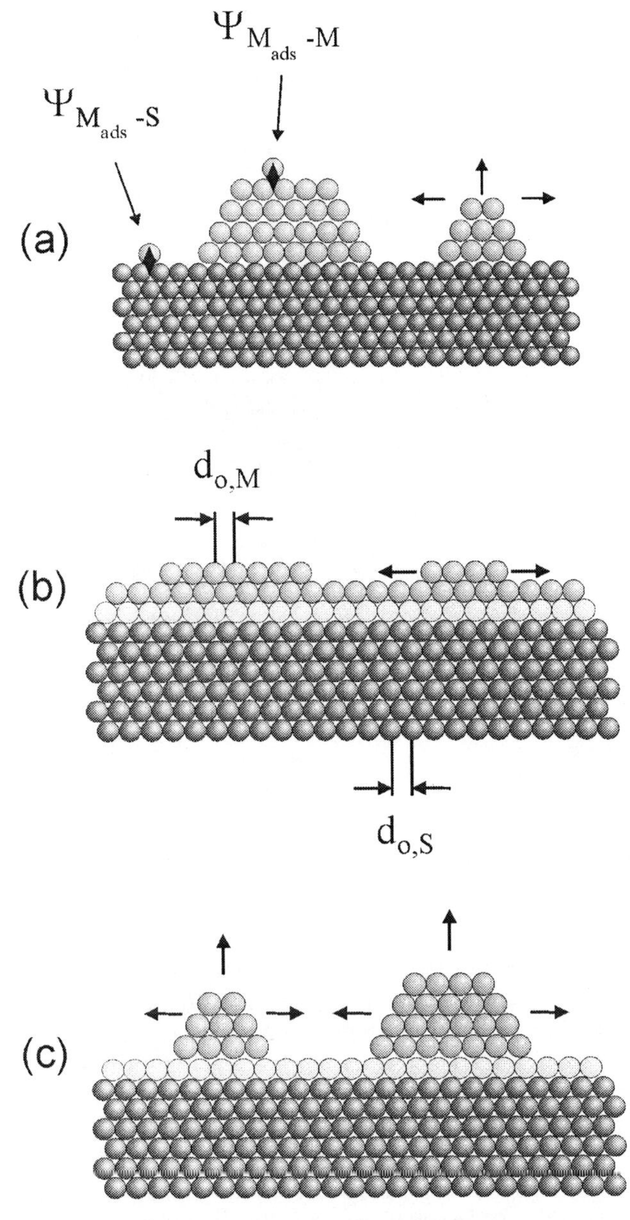

Figure 2. Schematic of the different growth modes of M on foreign substrates S and on the same substrate M, depending on the binding energy between M_{ads} and S, $\Psi_{M_{ads}\text{-}S}$, and on the binding energy between M_{ads} and M, $\Psi_{M_{ads}\text{-}M}$, and depending on the crystallographic misfit characterized by the interatomic distances $d_{0,M}$ and $d_{0,S}$ of the 3D M and S bulk phase. (a) "Volmer-Weber" growth (3D M island growth) for $\Psi_{M_{ads}\text{-}S} \ll \Psi_{M_{ads}\text{-}M}$, independent of $(d_{0,M} - d_{0,S})/d_{0,S}$. (b) "Frank van der Merwe" growth (M layer-by-layer growth) for $\Psi_{M_{ads}\text{-}S} \gg \Psi_{M_{ads}\text{-}M}$ and negligible crystallographic misfit $(d_{0,M} - d_{0,S})/d_{0,S} \approx 0$. (c) "Stranski-Krastanov" growth (3D M island formation on top of a predeposited 2D M_{ads} monolayer on S) for $\Psi_{M_{ads}\text{-}S} \gg \Psi_{M_{ads}\text{-}S}$ and a significant crystallographic misfit

In systems with a weak M adatom-S interaction, a "delocalized" formation of 2D LDMSs has not yet been observed. Obviously, the 3D M phase is nucleated and islands or layers grow on top of an unmodified atomically flat S surface in the OPD range (Volmer-Weber growth mode) independent of the M-S misfit, which can be leveled out by misfit dislocations (Figure 2). The different growth modes schematically illustrated in Figure 2 neglect the important role of substrate surface inhomogeneities and the internal strain of LDMSs and/or of the topmost layer of the S surface in the electrochemical phase formation process discussed above.

Nanotechnological applications, however, require an active structuring of foreign substrates, in particular semiconducting substrates, in a well defined way and at precisely predefined substrate sites. The huge knowledge about "delocalized" formation of LDMSs at substrate surface inhomogeneities and on atomically smooth terraces of real surfaces is certainly very helpful, and will contribute to the solution of this task. However, new growth procedures are necessary, which meet better the requirements of an active nano-structuring at preselected substrate sites.

3. Localized Electrochemical Formation of LDMSs

Therefore, "localized", i.e. directed, metal deposition processes have been developed using *in situ* SPM techniques as an atomic tool and appropriate polarization routines of the substrate and the SPM probe to produce locally well-defined LDMSs on atomically smooth terraces of different substrates independent of surface inhomogeneities, M adatom-S interaction, and crystallographic M-S misfit. Here, "localized" metal deposition means that the site(s) at which deposition occurs are carefully selected instead of the "normal" case where deposition occurs at random initiation sites. In particular, only "localized" nanostructuring of surfaces in systems with weak M adatom-S interaction (negligible metal UPD effects, Volmer-Weber growth mode, Figure 2) is of interest specifically for nanoelectronic devices.

Generally, different techniques for directed nanostructuring can be distinguished: Methods with the probe in tunneling contact to the substrate surface, or methods with a large separation between probe and substrate surface during formation of the nanostructures, i.e. with the probe out of tunneling contact using the probe as an electrochemical electrode. Among the first class of methods, i) probe-induced, ii) electric field-induced, and iii) defect-induced methods and their combinations have been developed [19,20]. An example of a combined tip-induced and field-induced technique is schematically shown in Figure 3. Among the second class of methods the Scanning Electrochemical Microscope (SECM, SICM) [47-50] or scanning probe based electrochemical techniques [29] can be found. The latter methods change the Nernstian equilibrium potentials at S in order to achieve deposition or dissolution (Figure 4).

The pioneering work of Kolb et al. [24–26] showed, that isolated small 0D Cu clusters and well-ordered arrays consisting of several thousands of 0D M clusters on atomically flat terraces of Au(111) substrates can be fabricated using an automated STM-tip-induced nanostructuring technique: First, a certain amount of M is predeposited on a STM tip followed by a transfer of M from the tip to S by approaching the tip towards S via the z-piezo control. The M transfer is characterized by a "jump to contact" and

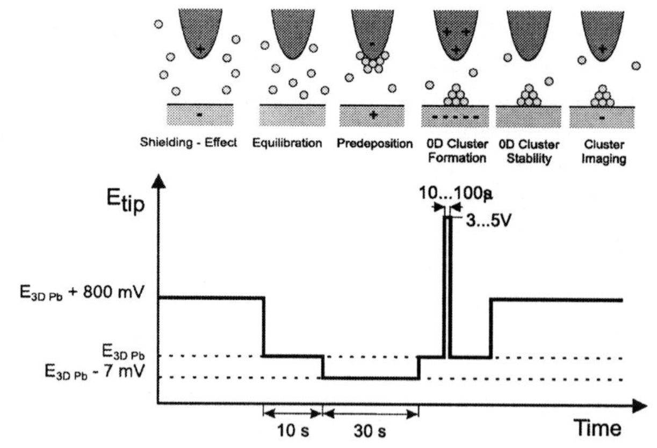

Figure 3. Combined tip-induced and electric field-induced deposition procedure of clusters, as used for the deposition of Pb clusters on n-Si(111):H substrates (see Figure 5)

the formation of a "connective neck" between the M covered tip and S. In this case, the appearance of an electrical short-cut [24], and hence a local increase of the substrate temperature can not be excluded. However, this method can only be applied successfully in model systems with a strong interaction energy between M adatoms and S, i.e. showing significant UPD phenomena in the undersaturation range (see below).

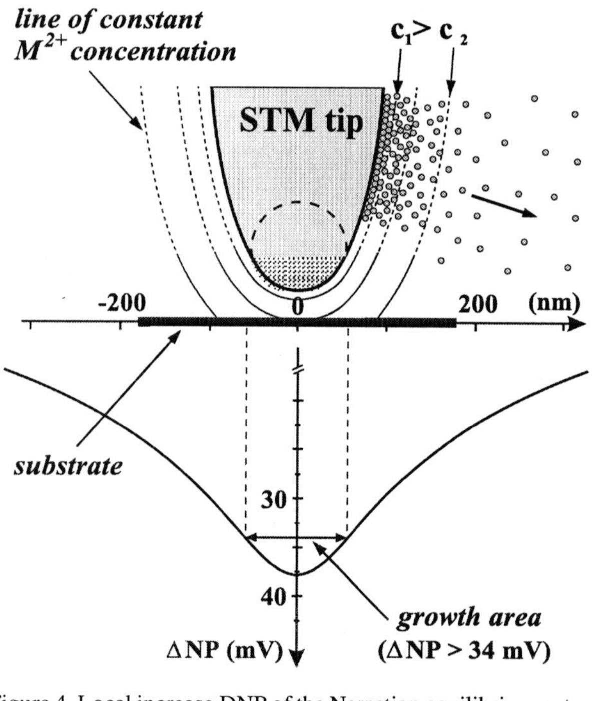

Figure 4. Local increase DNP of the Nernstian equilibrium potential of LDMSs due to an increase of the M^{2+} ion concentration around the STM tip, and in particular in the gap between STM tip and S (with respect to a 1mM M^{2+} ion concentration in the electrolyte). The distance between STM tip and S is more than ten nanometer, resulting in separated double layers around STM tip and S. This leads to a solely electrochemical formation of growth nuclei within a certain area underneath the STM tip, denoted as "growth area" in the figure

On the other hand, Penner et al. [31-33] and Lorenz et al. [19,20] have developed a field-induced nanostructuring method that can also be applied in technically important systems with weak M adatom-S interaction. However, the application of relatively high voltage pulses on the STM tip staying in the tunneling mode produces substrate surface defects which act as nucleation and growth sites for the subsequent local M deposition.

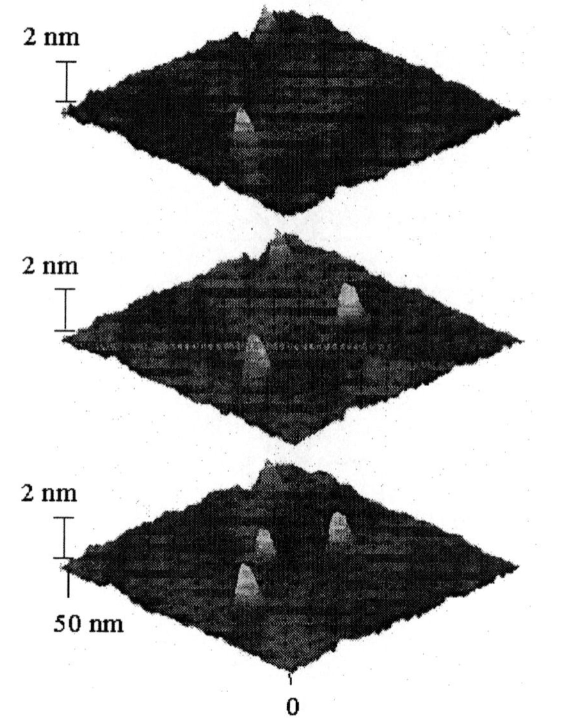

Figure 5: STM images of Pb clusters deposited onto n-Si(111):H, upon a subsequent application of the polarization routine as shown in Figure 3

Electrochemical nanostructuring methods, which operate more universal, were recently developed by Lorenz et al. [21–23] (Figure 3 and 5) and Schindler et al. [27–30] (Figure 4 and 6). Whereas the first method is based on a combination of tip-induced and electrochemical pulse (field-induced) techniques in the tunneling mode of the STM tip, the second method operates solely electrochemically at large distances between STM tip and S avoiding any unwanted mechanical or electrical tip-surface interactions. The principle of both methods is a predeposition of M onto the STM tip and an electrochemical transfer of M from the tip to the underlaying substrate. In this case, the transfer of M is initiated by an anodic pulse polarization applied to the tip dissolving M from the tip. Then, the locally increased M ion concentration between the SPM tip and S leads to a local shift of the Nernstian equilibrium potentials of LDMSs in positive direction. Since the potential of S is held constant, a supersaturation is locally generated at S which leads to a local M deposition and the formation of 0D M clusters.

It was shown that both methods operate in systems with weak M adatom-S interaction, independent of the crystallographic misfit, and without significantly damaging the substrate surface. This has been proven by the localized Pb deposition on n-Si(111):H [23,51] (Figure 5), and the localized Co and Ni deposition on Au(111) [27–30] (Figure 6).

Besides the active nanostructuring techniques based on STM, a variety of experiments have been reported using AFM-cantilevers to modify substrate surfaces more or less mechanically [52,53]. These techniques utilize the well known preferred electrochemical nucleation at defect sites in the substrate surface, which have been pregenerated by the cantilever.

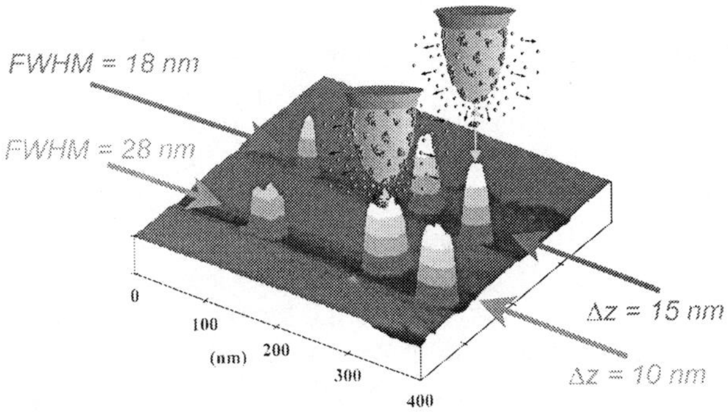

Figure 6. STM image of Co clusters deposited onto Au(111) by a local increase of the Nernstian equilibrium potential at S due to the M dissolution from the STM tip, as shown in Figure 4. The distance between tip and S during deposition is Δz, FWHM denotes the full width at half maximum of the clusters. Each grey scale corresponds to a height of 1 nm. (The lower scanlines across the clusters are an artefact of the STM measurement.)

Another field of promising nanostructuring techniques is the delocalized deposition into patterned masks of monomolecular thick Alkanthiol- or Biphenyl-layers. These techniques also avoid irreversible modifications due to the preparation process of the nanostructures. Whereas large structures with diameters in the mm range have been successfully grown [54], the homogeneous filling of structures with lateral sizes below 20 or even 10 nm still has to be proven. Deposition conditions may be necessary to be chosen far from equilibrium, similar to the conditions during filling of nanowires with diameters of some tens of nanometers [55-57].

4. Properties of LDMSs

The properties of electrochemically fabricated LDMSs are still not well understood due to a lack of reliable investigations. From the few preliminary investigations in the literature, the following aspects may be deduced, and seem to be important in future studies:

i) In contrast to the extensive knowledge concerning the structural properties of 2D M UPD systems, there are practically no structural results for 0D M clusters and 1D M phases. It is unclear if M grows epitaxially, commensurate, or incommensurate on S, which probably depends on the M-S system. The influence of the crystallographic M-S misfit on the M structure is unknown as well as the influence of internal strain in the M structure on the physical and chemical properties. Results obtained in M UPD systems suggest that the structure of LDMSs will significantly differ from that of a corresponding and relaxed 3D M phase as found by Kolb et al. for the Au(100)/Cu_2^+ system [58].

ii) The deposition of LDMSs will influence both the underlying S structure leading to surface relaxation processes, surface alloying processes, and/or the formation of 0D S inhomogeneities at which 0D M clusters may be preferentially deposited. New experimental results concern the formation of a M-S surface alloy as a stabilizing mechanism of LDMSs [59,60].

iii) So far, only limited information exists concerning the stability of LDMSs and nanostructures at room temperature. In contact with an electrolyte it has been found by several investigators [23,29,30,61] that 0D M clusters with a height and a diameter of only a few nanometers are stable for an unusual long time (of about one hour) even in the undersaturation range with respect to the 3D M bulk phase.

Nanometer sized 0D M clusters contain less than 10000 atoms. Most of them are interfacial atoms. Therefore, it must be concluded that these small M clusters behave more as 0D quantum dots than as 3D copper crystallites.

Recently, first attempts were made to explain the observed cluster stability in terms of either a tip-shielding effect [62], kinetic hindrances [61], or quantum-electronic assumptions [63]. On the other hand, a thermodynamic LDMS concept was developed which is based on the assumption that the activity of LDMSs decreases with decreasing dimensionality [20,64]. Consequently, the equilibrium conditions of LDMSs in electrochemical systems can be described by Nernst-like potentials taking into account the dimensionality i:

$$E_{i\mathrm{DMS}} = E^0_{3\mathrm{DMS}} = \frac{RT}{nF} \ln \left(\frac{a_{\mathrm{M}^{z+}}}{a_{i\mathrm{DMS}}} \right) \tag{1}$$

with $i = 0, 1, 2, \text{or } 3$.

The equilibrium potential, $E_{i\mathrm{DMS}}$, is determined by the activity of the M ions in the electrolyte, $a_{\mathrm{M}^{z+}}$, and the activity of the iD metal system, $a_{i\mathrm{DMS}}$, which is constant for a condensed and unstrained iD M system, but depends on the internal strain and the surface coverage for expanded iD M systems according to the corresponding adsorption isotherm. Consequently, the Nernstian equilibrium potentials of iD M systems with $i = 0, 1, 2, 3$ are shifted into positive direction with decreasing i and LDMSs become stable even in the UPD range:

$$E_{3\mathrm{DMS}} < E_{2\mathrm{DMS}} < E_{1\mathrm{DMS}} < E_{0\mathrm{DMS}}. \tag{2}$$

Consequently, different undersaturation (UPD) and supersaturation (OPD) ranges with respect to the corresponding equilibrium potentials of the iD M systems can be defined which determine the thermodynamic stability as demonstrated in Figure 7 using the example of the Ag(111)/Pb^{2+} system [19,65].

The relatively high stability of 0D M clusters can be understood in terms of this LDMS concept. The clusters are stable at potentials more negative than $E_{0\mathrm{DMS}}$, but more positive than , i.e. within the undersaturation range related to the 3D metal phase, but in the supersaturation range related to the 0D metal system.

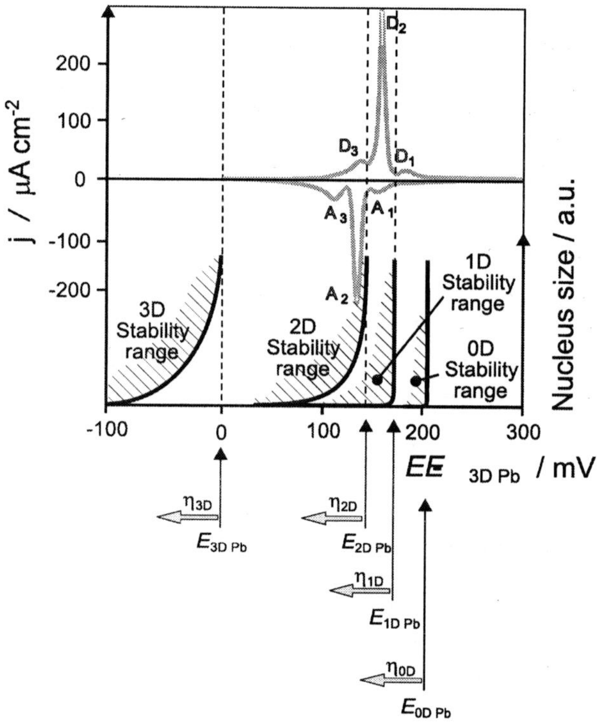

Figure 7. Stability ranges of LDMSs in the system Ag(111)/Pb2$^+$, as deduced from cyclic voltammetry measurements. The 0D stability range is extrapolated – it can not be observed in the measurements

The thermodynamic stability ranges of LDSs are schematically shown in Figure 7 where the equilibrium potentials, $E_{i\mathrm{DMS}}$, of iD metal systems are plotted vs. their nucleus size.

However, the thermodynamic concept, and a possible surface alloy formation give only qualitative interpretations of the unexpected stability of LDMSs on foreign substrates. More advanced theoretical investigations are necessary to explain the electronic behavior of 0D copper clusters on a quantummechanical basis. The appearance of discrete energy levels for electrons in nanometer sized M clusters must be examined using appropriate quantummechanical approaches [59]. An open problem in this case is, how the electronic wave functions of the substrate or a 2D M UPD overlayer or a surface alloy will interfere with those of a 0D M cluster formed on top. For this purpose, also structural information on LDSs becomes important.

iv) The formation of nanometer sized 1D and 2D M systems starting from locally formed 0D M clusters is presently still unsolved. It should be possible under defined supersaturation conditions for M electrodeposition. Another way is the development of a "direct writing method" using the STM tip as a tool and defined electrocrystallization conditions.

v) Initial kinetic results on the deposition/dissolution processes of 1D and 2D metal UPD systems have been obtained combining *in situ* STM and electrochemical impedance spectroscopy (EIS) [20]. In the model system Ag(111)/Pb^{2+}, which is characterized by a strong M–S interaction, it was shown that the exchange current density of kinks of a 1D MS is significantly lower than that of kinks in a condensed 2D MS. This finding is expected theoretically taking into account the binding energies of first next neighbours. However, different results are expected in systems with weak M adatom–S interactions, which are candidates for possible nanotechnological applications.

vi) The anodic dissolution and corrosion behavior of LDMSs has not been studied systematically so far. Previous in situ SPM investigations of corrosion and passive layer formation processes on an atomic level have not treated low-dimensional systems [66–91]. Recently, the LDMS concept was successfully applied to the stepwise formation of low-dimensional oxidic systems (LDOSs) playing an important role in the dissolution and passivation kinetics of transition metals [64]. The equilibrium conditions of LDOSs can also be described by Nernst-like equations corresponding to eq. (1), but the equilibrium potentials (cf. eq. 2) are shifted into negative direction with decreasing dimensionality of LDOSs. Therefore, the stepwise formation of LDOSs is similar to that of LDMSs schematically shown in Figure 1 replacing the electrode potential E by $-E$. The LDOS concept leads to a much better understanding of metal dissolution and passivation processes on an atomic level, but it remains rather hypothetical since LDOSs cannot be imaged with atomic resolution by *in situ* SPM methods so far. Possibly, tunneling spectroscopy may help to solve this problem.

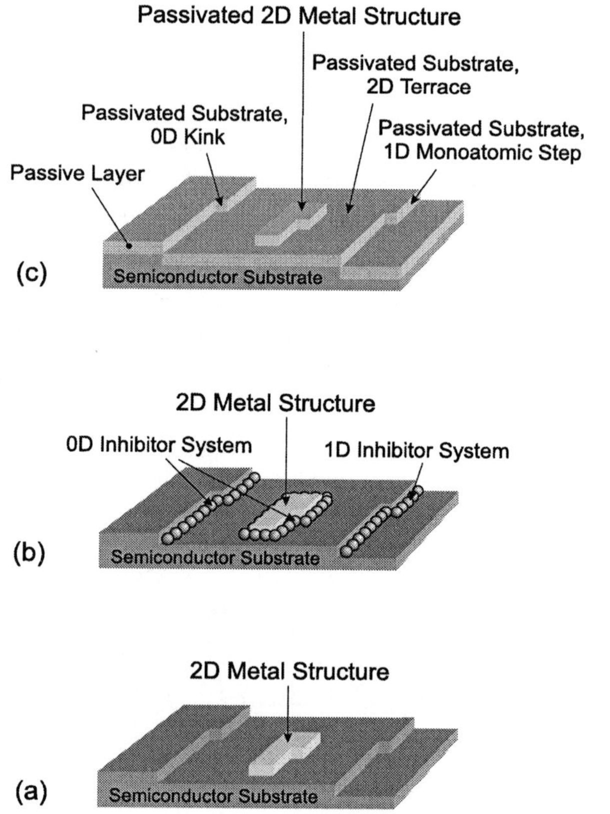

Figure 8. Possible stabilization mechanisms of nanostructures (a) using 1D inhibitor systems (b), or passive layers (c)

Nevertheless, it is clear that a sufficient time stability of nanosystems is required for practical applications.

In other words, LDMSs should not dissolve or corrode over sufficient time periods independent of the ambient phase. Therefore, stabilizing effects of passive layer formation or adsorption of corrosion inhibitors on LDMSs has to be studied. Possible mechanisms for stabilizing nanostructures are shown in Figure 8 utilizing inhibitors or monomolecular layers, which reduce the surface mobility of the metal structures, and which avoid degradation of the structures upon removal from the preparation chamber.

Summarizing, the widely unknown thermodynamic, structural and kinetic properties of LDMSs have to be studied intensively in order to allow for useful applications of LDMSs in future nanomagnetic and nanoelectronic devices.

5. Possible Applications of LDMSs

At present, quantum-electronic devices such as quantum dots, quantum wires, tunnel junctions, and single electron devices have been fabricated with relatively large quantum confinements characterized by 10 nm < lateral width < 100 nm. They were made by classical nanostructuring techniques from the gas phase such as e-beam lithography or ion-beam writing, involving lithographic intermediate steps. Consequently, quantum effects can only be observed in these devices at very low temperatures in the range of milli-Kelvin to a few Kelvin [36–39]. On the contrary, if quantum confinements are used with a lateral width of only a few atoms, quantum effects become observable even at room temperature. As mentioned above, this has already been demonstrated for electrochemically fabricated quantum wires [1–7] and single electron transistors [8,9].

The electrochemical fabrication of LDMSs and their application in nanomagnetic and nanoelectronic devices seems to be a suitable way to overcome the limits of the conventional gas phase techniques. However, the above discussed mainly unknown properties of electrochemically fabricated LDMSs must be intensively studied to guarantee a necessary long-time LDMS stability for practical applications.

The fabrication and the dc- as well as ac- characterization of novel nanoelectronic devices, which are suitable for the development of high speed electronics in the THz-domain, is important for future nanotechnology. These devices will be fabricated on conventional semiconductor substrates by localized electrochemical metal deposition using *in situ* SPM techniques. On the other hand, contact pads, photo-detectors, and photoconductive switches will be conventionally prefabricated on the substrate by optical and e-beam lithography. This hybride fabrication technology allows both the *in situ* and *ex situ* characterization of well-defined quantum wires, single-electron-transistors and Schottky-diodes which will show quantum-blockade and other quantum size effects even at room temperature. These devices should show high-frequency characteristics improved by one or two orders in magnitude when compared to the current state of the art technology.

In the field of catalysis and sensors, a detailed knowledge of the properties of nanostructures will result in a better understanding of the elementary steps in catalytic processes, or in novel sensoric applications. Detection limits may be expected to be decreased to the principal limit of single molecules, as has been shown recently [2,4].

6. Conclusions

The electrochemical fabrication of LDMSs and LDOSs utilizing *in situ* SPM techniques seems to be at present the only way to study the properties of very small LDSs and to consider their application in nanoelectronic and nanomagnetic devices. The appearance of quantum effects even at room temperature in nanoelectronic devices with very small constraints opens up novel applications, in particular for high-frequency devices. Principle concepts derived even from the delocalized metal deposition may be useful for the interpretation of the unusual stability behaviour of SPM generated nanostructures compared to bulk like structures.

Acknowledgements

We would like to thank Martin and Philipp Hugelmann for the image processing of the schematic images. Some of the authors (W.J.L., W.W. and W.S.) would like to acknowledge gratefully financial support by the Deutsche Forschungsgemeinschaft under the contracts WI1044/10 and SCHI492/1.

References

1. C. Z. Li und N. J. Tao, Appl. Phys. Lett. 72, 894 (1998).
2. C. Z. Li, H. Sha und N. J. Tao, Phys. Rev. B 58, 6775 (1998).
3. C. Z. Li, A. Bogozi, W. Huang und N. J. Tao, Nanotechnology 10, 221 (1999).
4. C. Z. Li, H. X. He, A. Bogozi, J. S. Bunch and N. J. Tao, Appl. Phys Lett. 76, 1333 (2000).
5. C. Shu, C. Z. Li, H. X. He, A. Bogozi, J. S. Bunch und N. J. Tao, Phys. Rev. Lett. 84, 5196 (2000).
6. C. Z. Li, H. X. He und N. J. Tao, Appl. Phys. Lett. 77, 3995 (2000).
7. H. X. He, C. Z. Li und N. J. Tao, Appl. Phys. Lett., submitted.
8. K. Matsumoto, M. Ishii, K. Segawa, Y. Oka, B. J. Vartanian und J. S. Harris, Appl. Phys. Lett. 68, 34 (1996).
9. J. Shirakashi, K. Matsumoto, N. Miura und M. Konagai, Appl. Phys. Lett. 72, 1893 (1998).
10. Solid State Technology 42, No.4, 14 (1999).
11. T. Koester, F. Goldschmidtboening, B. Hadam, J. Stein, S. Altmeyer, B. Spangenberg, H. Kurz, R. Neumann, K. Brunner, and G. Abstreiter, J. Vac. Sci. Technol. B 16, 3804 (1998).
12. A. Tilke, R.H. Blick, H. Lorenz, J.P. Kotthaus, and D.A. Wharam, Appl. Phys. Lett. 75, 3704 (1999).
13. J. de la Figuera, M.A. Huerta-Garnica, J.E. Prieto, C. Ocal, and R. Miranda, Appl. Phys. Lett. 66, 1006 (1995).
14. B. Voigtländer, G. Meyer, and N.M. Amer, Phys. Rev. B 44, 10354 (1991).
15. F.A. Möller, O.M. Magnussen, and R.J. Behm, Phys. Rev. Lett. 77, 3165 (1996).
16. O. Fruchart, M. Klaua, J. Barthel, and J. Kirschner, Phys. Rev. Lett. 83, 2769 (1999).

17. Th. Schulze, B. Brezger, P.O. Schmidt, R. Mertens, A.S. Bell, T. Pfau, J. Mlynek, in: Microelectron. Eng. 46, 105 (1999).
18. A. Bard, K.K. Berggren, J.L. Wilbur, J.D. Gillaspy, S.L. Rolston, J.J. McClelland, W.D. Phillips, M. Prentiss, G.M. Whitesides, J. Vac. Sci. Technol. B 15, 1805 (1997).
19. E. Budevski, G. Staikov and W. J. Lorenz, Electrochemical Phase Formation and Growth – An Introduction to the Initial Stages of Metal Deposition, Wiley-VCH, Weinheim, 1996.
20. G. Staikov, W. J. Lorenz und E. Budevski, Imaging of Surface and Interfaces – Frontiers in Electrochemistry, P. Ross and J. Lipkowski, Editors, Vol. 5, p. 1, Wiley-VCH, Weinheim 1999.
21. R.T. Potzschke, A. Froese, W. Wiesbeck, G. Staikov, and W.J. Lorenz, in Proceedings of the Third Symposium on Electrochemically Deposited Films, M. Pounovic and D.A. Scherson, eds., PV 96-19, p.21, The Electrochemical Society Proceedings Series, Pennington, NJ (1997).
22. G. Staikov and W.J. Lorenz, in Electrochemical Nanotechnology - In situ Local Probe Techniques at Electrochemical Interfaces, W.J. Lorenz and W. Plieth, eds., p.13, Wiley-VCH, Weinheim (1998).
23. R. T. Pötschke, G. Staikov, W. J. Lorenz und W. Wiesbeck, J. Electrochem. Soc. 146, 141 (1999).
24. D.M. Kolb, R. Ullmann, T. Will, Science 275, 1097 (1997).
25. D. M. Kolb, R. Ullmann und J. R. Ziegler, Electrochim. Acta 43, 2751 (1998).
26. J.C. Ziegler, G.E. Engelmann, and D.M. Kolb, in: Electrolytic Metal Deposition: Fundamental Aspects and Applications, A. Dakkouri and D.M. Kolb, eds., p.151, R. Oldenburg Verlag, Munich (1999).
27. D. Hofmann, W. Schindler, and J. Kirschner, Appl. Phys. Lett. 73, 3279 (1998).
28. W. Schindler, D. Hofmann, and J. Kirschner, J. Appl. Phys. 87, 7007 (1999).
29. W. Schindler, D. Hofmann, and J. Kirschner, J. Electrochem. Soc. 148, C124 (2001).
30. W. Schindler, Elektrochemie und Magnetismus von Co Monolagen und Nanostrukturen, Habilitationsschrift, Universität Karlsruhe, Germany (2000).
31. W. Li, J.A. Virtanen, and R.M. Penner, Appl. Phys.. Lett. 60, 1181 (1992).
32. W. Li, J.A. Virtanen, and R.M. Penner, J. Phys. Chem. 96, 6529 (1992).
33. W. Li, T. Duong, J.A. Virtanen, and R.M. Penner, in: Nanoscale Probes of the Solid / Liquid Interface, NATO ASI series E: Applied Sciences, A,A, Gewirth and H. Siegenthaler, eds., vol. 228, p.183, Kluwer Acad. Publ., Dordrecht (1995).
34. R. Schuster, V. Kirchner, X.H. Xia, A.M. Bittner, and G. Ertl, Phys. Rev. Lett. 80, 5599 (1998).
35. J.R. LaGraff, A.A. Gewirth, J. Phys. Chem. 99, 10009 (1995).
36. R.J. Haug, Electrochim. Acta 40, 1283 (1995).
37. D. Berman, N.B. Zhitenev, R.C. Ashoori, and M. Shayegan, Phys. Rev. Lett. 82, 161 (1999).
38. Y.Y. Wei, J. Weis, K. von Klitzing, and K. Eberl, Appl. Phys. Lett. 71, 2514 (1997).
39. M. Koltonyuk, D. Berman, N.B. Zhitenev, R.C. Ashoori, L.N. Pfeiffer, and K.W. West, Appl. Phys. Lett. 74, 555 (1999).
40. E. Scheer, N. Agrait, J.C. Cuevas, A.L. Yeyati, B. Ludoph, A. Martin-Roderos, G.R. Bollinger, J.M. van Ruitenbeek, C. Urbina, Nature 394, 154 (1998).
41. E. Scheer, H. v. Lohneysen, A.D. Mirlin, P. Wolfle, H. Hein, Phys. Rev. Lett. 78, 3362 (1997).
42. W. Schindler, Th. Koop, A. Kazimirov, G. Scherb, J. Zegenhagen, Th. Schultz, R. Feidenhans'l, J. Kirschner, Surf. Sci. Lett. 465, L783 (2000).
43. E. Hahn, E. Kampshoff, N. Wälchli, and K. Kern, Phys. Rev. Lett. 74, 1803 (1995).
44. M. Dietterle, T. Will, D.M. Kolb, Surf. Sci. 396, 189 (1998).
45. B. Müller, B. Fischer, L. Nedelmann, A. Fricke, and K. Kern, Phys. Rev. Lett. 76, 2358 (1996).
46. E. Korutcheva, A.M. Turiel, I. Markov, Phys. Rev. B 61, 16890 (2000).
47. S. Meltzer and D. Mandler, J. Electrochem. Soc. 142, L82 (1995).
48. F. Forouzan and A.J. Bard, J. Phys. Chem. B , 10876 (1997).
49. K. Borgwarth, N. Rohde, C.Ricken, M.L. Hallensleben, D. Mandler, J. Heinze, Advanced Materials 11, 1221 (1999).
50. A.D. Muller, F. Muller, and M. Hietschold, Thin Solid Films 366, 32 (2000).
51. M. Hugelmann, P. Hugelmann, W. Schindler, submitted (2001).
52. L. Santinacci, T. Djenizian, and P. Schmuki, Abstract No. 412, 198th Meeting of The Electrochemical Society, Phoenix, October 22 - 27, 2000.
53. H.W. Schumacher, U.F. Keyser, U. Zeitler, R.J. Haug, K. Eberl, Physica E 6, 860 (2000).
54. T.P. Moffat and H. Yang, J. Electrochem. Soc. 142, L220 (1995).
55. G.P. Heydon, S.R. Hoon, A.N. Farley, S.L. Tomlinson, M.S. Valera, K. Attenborough, and W. Schwarzacher, J. Phys. D: Appl. Phys. 30, 1083 (1997).
56. A. Blondel, B. Doudin, and J.P. Ansermet, J. Magn. Magn. Mater. 165, 34 (1997).
57. J.L. Maurice, D. Imhoff, P. Etienne, O. Durand, S. Dubois, L. Piraux, J.M. George, P. Galtier, and A. Fert, J. Magn. Magn. Mat. 184, 1 (1998).
58. R. Randler, M. Dietterle, and D. M. Kolb, in: Electrolytic Metal Deposition: Fundamental Aspects and Application, eds.: A. S. Dakkouri and D. M. Kolb, Oldenburg, Munich (1999).
59. E. Leva, M. del Popolo, W. Schmickler, Surf. Sci., in press (2001).
60. S. Maupai, University of Erlangen-Nürnberg, priv. commun. (2001).
61. D.M. Kolb, R. Ullmann, and J.C. Ziegler, Electrochim. Acta 43, 2751 (1998).
62. U. Stimming, R. Vogel, D. M. Kolb und T. Will, J. Power Sources 43-44, 169 (1993).
63. D.M. Kolb, G.E. Engelmann, and J.C. Ziegler, Angew. Chem. Int. Ed. 39, 1123 (2000).
64. W. J. Lorenz, W. Wiesbeck, and G. Staikov, Abstract No. 561, ECS Fall Meeting, Honolulu October 17-22, 1999.
65. E. Budevski, G. Staikov, and W.J. Lorenz, Electrochimica Acta 45, 2559 (2000).
66. D. W. Suggs and A. J. Bard, J. Amer. Chem. Soc. 116, 10725 (1994).
67. D. W. Suggs and A. J. Bard, J. Phys. Chem. 99, 8349 (1995).
68. M. R. Vogt, F. A. Möller, C. M. Schilz, O. M. Magnussen, and R. J. Behm, Surf. Sci. 367, L33 (1996).
69. M. R. Vogt, A. Lachenwitzer, O. M. Magnussen, and R. J. Behm, Surf. Sci. 399, 49 (1998).
70. M. R. Vogt, W. Polewska, O. M. Magnussen, and R. J. Behm, J. Electrochem. Soc. 144, L113 (1997).

71. J. Scherer, M. R. Vogt, O. M. Magnussen, and R. J. Behm, Langmuir 13, 7045 (1997).
72. M. Kruft, C. Stuhlmann, and K. Wandelt, Surf. Sci. 377-379, 601 (1997).
73. M. Kruft, M. Wilms, P. Broekmann, B. Wohlmann, Z. Park, C. Stuhlmann, and K. Wandelt, Abstract 826, p. 957, The Electrochemical Society Meeting Abstracts, Vol. 97-2, Pennington, NJ (1997).
74. T. P. Moffat, in: Electrochemical Nanotechnology, eds.: W. J. Lorenz and W. Plieth, p.171, Wiley -VCH (1998).
75. B. J. Cruickshank, D. D. Sneddon, and A. A. Gewirth, Surf. Sci. 281, L308 (1993).
76. J. R. LaGraff and A. A. Gewirth, Surf. Sci. 326, L461 (1995).
77. R. C. Bardwaj, A. Gonzales-Msartin, and J. O?M. Bockris, J. Electroanal. Chem. 307, 195 (1991).
78. R. C. Bardwaj, A. Gonzales-Martin, and J. O?M. Bockris, J. Electrochem. Soc. 138, 1901 (1991).
79. M. P. Ryan, R. C. Newman, and G. E. Thompson, J. Electrochem. Soc. 142, L177 (1995).
80. V. Maurice, H. Talah, and P. Marcus, Surf. Sci. 304, 98 (1994).
81. S.-L. Yau, F.-R. Fan, T. P. Moffat, and A. J. Bard, J. Phys. Chem. 98, 5493 (1994).
82. V. Maurice, W. Yang, and P. Marcus, J. Electrochem. Soc. 141, 3016 (1994).
83. T. P. Moffat F.-R. Fan, S.-L. Yau, and A. J. Bard, in Proceedings of the H. H. Uhlig Memorial Symposium, F. Mansfeld, A. Asphahani, H. Böhni, and R. Latanison, Editors, PV 94-26, p.76, The Electrochemical Society Proceedings Series, Pennington NJ (1994).
84. M. P. Ryan, R. C. Newman, and G. E. Thompson, Philos. Mag. B 70, 241 (1994).
85. M. P. Ryan, R. C. Newman, and G. E. Thompson, J. Electrochem. Soc. 141, L164 (1994).
86. V. Maurice, W. Yang, and P. Marcus, J. Electrochem. Soc. 143, 1182 (1996).
87. P. Marcus, W. Yang, and V. Maurice, in Surface Oxide Films, J. A. Bardwell, Editor, PV 96-18, p. 1, The Electrochemical Society Proceedings Series, Pennington NJ (1996).
88. V. Maurice, H.-H. Strehblow, and P. Marcus, J. Electrochem. Soc. 146, 524 (1999).
89. P. Marcus and V. Maurice, in: Electrochemical Nanotechnology - In-situ Local Probe Techniques at Electrochemical Interfaces, eds.: W. J. Lorenz and W. Plieth, p.185, Wiley-VCH, Weinheim 1998.
90. O. M. Magnussen and R. J. Behm, Abstract No. 562, 196th Meeting of The Electrochemical Society, Honolulu, October 17 - 22, 1999.
91. O. M. Magnussen, L. Zitzler, B. Gleich, and R. J. Behm, Abstract No. 566, 196th Meeting of The Electrochemical Society, Honolulu, October 17 - 22, 1999.

Nanoscience & Nanotechnology
eds. E. Balabanova, I. Dragieva, Heron Press, Sofia, 2001

ELECTROCHEMICAL BIOSENSORS

A. Kaisheva

Central Laboratory of Electrochemical Power Sources, Bulgarian Academy of Sciences, 1113 Sofia, Bulgaria

Abstract. Amperometric biosensors with enzymes-oxidases are reported. A new type enzyme/gas-diffusion electrode is developed in which oxygen needed for the enzymatic reaction is supplied in gas phase through a porous hydrophobic gas-layer of the electrode. Gas biosensors on the basis of an enzyme/gas-diffusion electrode are investigated.

Keywords: biosensor, enzyme, enzyme electrode

1. Introduction

The interest in the research and development of electrochemical biosenors is mainly due to their potential application in the medical practice, environmental control, food industry etc.

The electrochemical biosensor is a complicated device (Figure 1) which converts a biological recognition process of an analyte into an electrical signal. The biological component is usually an immobilized enzyme. It catalyses the biochemical reaction of a definite analyte. The transducer is usually a conventional electrode on which an electrochemical reaction is taking place, by which an electrical signal is obtained, proportional to the analyte concentration.

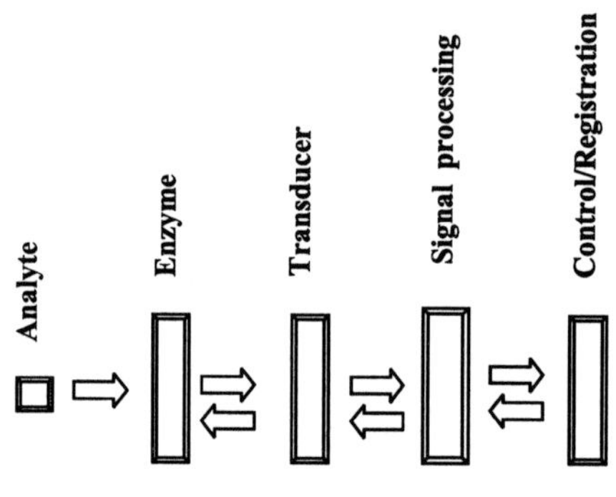

Figure 1

The obtained signal is further amplified, stored, etc. In some cases a feedback between the components of the biosensor is introduced in order to improve its performances.

The system of immobilized enzyme/conventional electrode is usually known as an enzyme electrode.

A great advantage of the electrochemical biosensor is the use of biologically recognition proteins, such as enzymes, which possesse a high biocatalytic activity, selectivity and specificity. On the other side the electrical signal of the biosensor can be easily registrated, amplified, memorized, etc.

Electrochemical biosensors using enzymes-oxidases have been extensively studied for a range of medical and ecological applications.

In principle the enzymes are nano-particles consisting of a long polypeptide chain, formed into a globule. The catalytic activity of the enzymes is due to the presence of some co-factors of a non-protein nature, such as inorganic ions or organic groups, attached to the enzyme protein. Recently is was shown that the zones of the contact protein/co-factor form grooves with particular geometry so that they are accessible only for the molecules of the analyte specific to the given enzyme [1].

The enzyme "active sites" are essentially insulated electrically within the enzyme molecule, so that electron transfer between them and the conventional electrode surface does practically not occur.

The enzymatic reaction in an amperometric biosensor with enzyme oxidase is given as an example. During this reaction the active center of the enzyme is reduced. It is reoxidized to it active form by the molecular oxygen.

$$\text{Analyte} + O_2 \xrightarrow{enzyme^{ox}} \text{Product} + H_2O_2 \qquad (1)$$

$$\text{Enzyme}^{red} \xleftrightarrow{O_2} \text{Enzyme}^{ox} \qquad (2)$$

To obtain an electrical signal of the electrochemical biosensor it is needed electrons from the active center to be transferred to the conventional electrode surface and electric current to be generated. However in most cases direct electron transfer between the enzyme and the electrode surface is not observed.

That's why an indirect ways are used to obtain electrical signal of the biosensor. In many cases an electrochemical signal of the biosensor is obtained by the electrochemical oxidation of the H_2O_2 generated in the enzymatic reaction (1).

$$H_2O_2 \xrightarrow{catalyticelectrode} O_2 + 2H^+ + 2e^- \qquad (3)$$

Various types of electrochemical biosensors in which the signal is obtained by the electrochemical reaction of H_2O_2 are investigated.

2. Membrane Biosensors

In the membrane-type of biosensor [2] the enzyme is immobilized on a porous membrane, placed in a close proximity to the electrode surface. Platinum electrodes are usually used as a catalytic electrodes for the H_2O_2 oxidation.

A new type of non-platinum catalytic electrode [3] was developed for use in membrane enzyme electrode. The catalytic electrode consists of a carbon support and a thin layer of non-platinum catalyst, prepared by pyrolysis in argon of CoTMPP impregnated on a dispersed carbon. This catalyst possesses some important advantages: it possesses a high activity for the electrochemical H_2O_2 oxidation, it is not sensitive to poisons, its price is low.

A schematic cross section of enzyme electrode with sandwich-type membrane is presented in Figure 2.

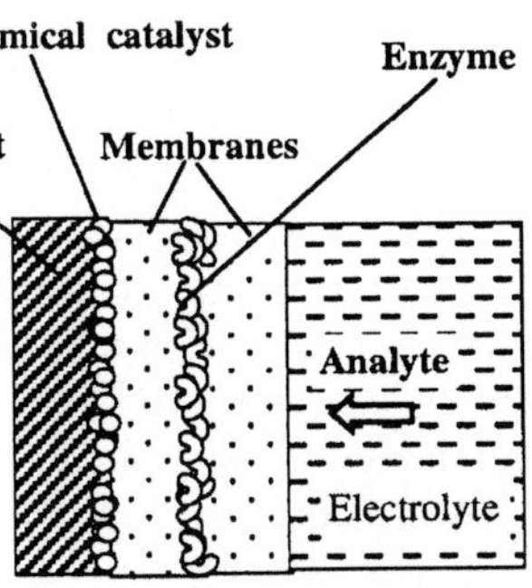

Figure 2

The membrane consists of two diffusion membranes with an enzyme layer between them.

The corresponding biosensor is an electrochemical cell with an enzyme electrode, a Ag/AgCl reference/counter electrode and electrolyte KCl. Constant electrode potential is maintained. In the presence of analyte in the electrolyte an electrical signal is obtained proportional to the analyte concentration [4].

The concentration of glucose in blood serum samples is determined with the this biosensor using glucose oxidase. The comparison (Table 1) of the results with those measured by a routine procedure with BACKMAN analyzer show a very good coincidence.

Table 1

Sample No	Glucose concentration, mM	
	Glucose biosensor	Backman-2 Analyzer
1	3.4	3.3
2	3.6	3.7
3	4.8	4.7
4	3.5	3.5
5	3.9	3.9
6	3.3	3.4
7	11.5	11.5

A theoretical model of a glucose biosensor with asymmetrical sandwich-type membrane is studied [5].

It was theoretically shown that when the enzyme layer is near to the catalytic electrode the linear part of the calibration curve of the biosensor is larger, so the biosensors can operate at high analyte concentrations. On the other side the response time of a biosensor is shorter when the enzyme layer is near to the bulk solution. It can be concluded that needed performance of the biosensor can be obtained by changing the position of the enzyme layer in the sandwich membrane.

3. Enzyme/Gas-Diffusion Electrode

The enzymatic reaction of enzymes-oxidases involves oxygen, so that the rate of the enzymatic reaction will depend strongly on the concentration of oxygen in the reaction zone of the enzyme electrode.

Usually the enzyme electrodes described in the literature represent a compact electrode with an enzyme immobilized on its surface. The electrode is immersed into the electrolyte, so that both the analyte and the oxygen are supplied to the enzyme from the electrolyte. It is considered that the concentration of the dissolved oxygen is high enough to ensure a high rate of the enzymatic reaction [6].

We use our experience in the field of air gas-diffusion electrodes to design a new type enzyme/gas-diffusion electrode.

The air-breathing gas-diffusion electrode for fuel cells is a tablet which separates the electrolyte from the surrounding air. The tablet contains two layers: a porous hydrophobic layer with pores free from electrolyte through which oxygen from the air is transported in gas phase and a catalytic layer soaked with the electrolyte where the electrochemical reaction is taking place. The enzyme/gas-diffusion electrode possesses the same structure, but an enzyme is immobilized into the catalytic surface of the electrode.

During operation the analyte is supplied to the enzyme layer from the electrolyte, but oxygen needed for the enzymatic reaction is transported mainly in gas phase through the free from the electrolyte pores of the hydrophobic gas layer. In this way an effective supply of oxygen to the reaction zone of the enzyme/gas-diffusion electrode is ensured [7].

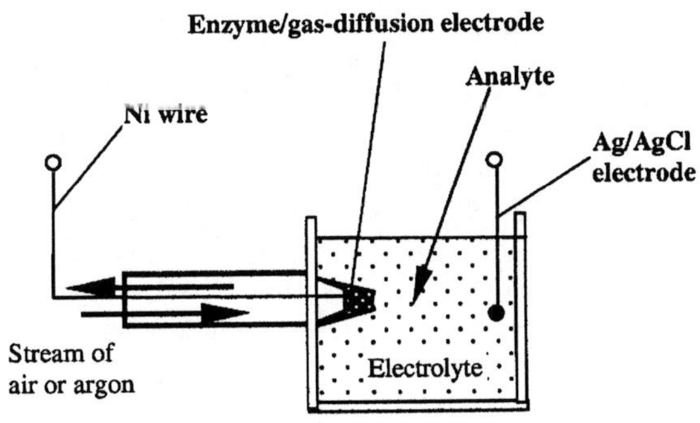

Figure 3

The enzyme/gas-diffusion electrode is mounted on the electrochemical cell (Figure 3) in such a way that its enzyme/catalytic layer is in contact with the electrolyte and the porous hydrophobic layer is in contact with the open air. The performance of the investigated enzyme/gas-diffusion electrode were studied at two different modes of oxygen-supply:

— supply in gas phase (from the air) through the porous carbon gas layer and from oxygen dissolved in the electrolyte.

— supply from the dissolved O_2 in the electrolyte only. (In this case the of oxygen through the porous carbon electrode is prevented by blowing of a stream of argon to the gas-side of the electrode).

The influence of the mode of oxygen supply on the performance of the glucose enzyme/gas-diffusion electrode with enzyme glucose oxidase is demonstrated in Figure 4.

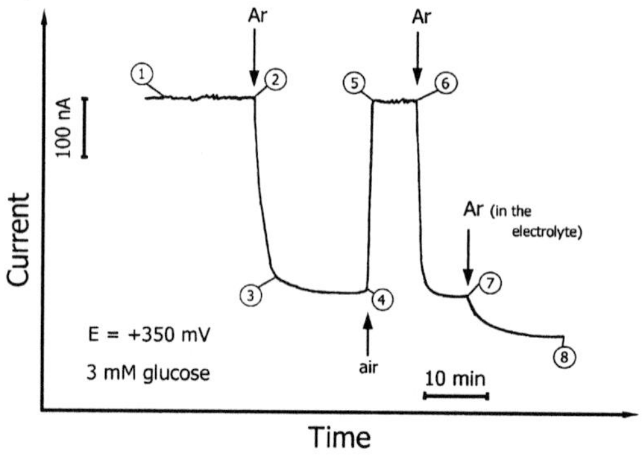

Figure 4

The electrode is operating at a constant potential at a constant glucose concentration in the electrolyte. The oxygen is supplied in gas phase from the air through the porous hydrophobic gas layer. The obtained amperometric signal is shown.

At the beginning of the experiment air is supplied in gas phase from the gas side of the electrode. The steady-state current (①to ②) corresponds to the presence of 3 mM glucose in the electrolyte. In the moment ②the supply of oxygen from the air is prevented by blowing argon to the gas-side of the electrode. Now oxygen needed for the enzymatic reaction is supplied to the working zone of the electrode from the dissolved oxygen only. This results in a rapid decrease of the signal and establishment of a lower steady-state current (③to ④) at the same concentration of glucose. When the supply of oxygen to the gas side of the electrode is restored (④), its steady-state current increases rapidly to its initial value (⑤to ⑥). The presented curve is a result of a consecutive change in the type of gas (air or argon) blown to the gas side of the electrode at one and the same glucose concentration.

If argon is blown to the gas-side of the electrode and simultaneously the dissolved oxygen is purged from the electrolyte by blowing with argon, a further decrease of the steady-state current is observed (⑦to ⑧) due to a lack of oxygen. When all dissolved oxygen will be purged from the electrolyte, no signal from the biosensor will be obtained.

From this experiment it can be concluded that the mode of oxygen supply influences significantly the performance of the enzyme electrode.

The calibration curves (the steady-state current as a function of the glucose concentration in the electrolyte) of enzyme/gas-diffusion electrodes with enzyme glucose oxidase were studied at different modes of oxygen-supply: when oxygen is supplied from the dissolved oxygen only; when oxygen is supplied in gas phase from the surrounding air through the gas-side of the electrode; when pure oxygen is supplied through the gas-side of the electrode.

On Figure 5a are presented calibration curves of one and the same enzyme/gas-diffusion electrode operating at two different oxygen concentrations in the reaction zone of the electrode.

At low analyte concentrations the diffusion control of the enzymatic reaction is predominating and the calibration curves are linear. In this region of analyte concentrations the current is given by:

$$I = \frac{nFAC_s}{d} \tag{4}$$

where d is the "thickness" of the diffusion layer; F, A, n, and C_s are the Faraday constant, the electrode surface area, the number of electrons involved in the electrochemical reaction, and the concentration of the analyte.

At high analyte concentration a kinetic control of the enzymatic reaction is predominating. In this case the catalytic current of the electro-enzymatic reactions (1–3) is given by the equation of Michaellis-Menten:

$$I = \frac{I_{\max}}{1 + \frac{k_{O_2}}{C_{O_2}} + \frac{k_s}{C_s}}. \tag{5}$$

where the maximum current $I_{\max}$ and the Michaellis constants k_s and k_{O_2} are related to the rate constants in reactions (1–3). C_s and C_{O_2} are the concentrations of the analyte and the oxygen on the electrode surface.

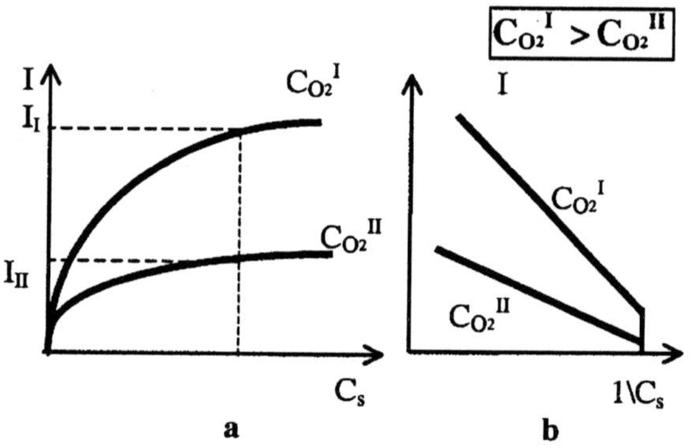

Figure 5

The values of the steady-state current (Figure 5a) at one and the same analyte concentration are much more higher at high oxygen concentration (CO_2^{I}) in the reaction zone. At low oxygen concentration (CO_2^{II}) in the reaction zone the generated current is much lower due to the lack of oxygen.

Equation (5) can be presented in the following form, known as electrochemical Eadie-Hofstee plot:

$$I = I_{\max}^{app} - K_M^{app}\left(\frac{I}{C_s}\right) \tag{6}$$

where:

$$K_M^{app} = \frac{k_s}{1 + \frac{k_{O_2}}{C_{O_2}}} \tag{7}$$

is the apparent Michaellis constant, and

$$I_{\max}^{app} = \frac{I_{\max}}{1 + \dfrac{k_{O_2}}{C_{O_2}}}. \qquad (8)$$

The calibration curves from Figure 5a are presented in Figure 5b in coordinates $I - I/C_s$ where the ratio I/C_s expresses the sensitivity of the electrode. It is seen that the sensitivity of the electrode is constant in the region of analyte concentrations in which the steady-state current depends linearly on the analyte concentration (vertical part of the curve). It is visible that this vertical part of the curve is larger in the case of higher oxygen concentration in the reaction zone.

At higher analyte concentrations, kinetic control predominates in the enzymatic reaction (slanted part of the $I - I/C_s$ curve). From the slope of the curve in this region, the value of the apparent Michaellis constant of the process can be estimated. The apparent Michaellis constant can be used as a measure of the rate of the enzymatic reaction.

From the slopes of the experimental calibration curves, presented in coordinates $I - I/C_s$, the values of the apparent Michaellis constants were determined for one and the same enzyme/gas-diffusion electrode with enzyme glucose oxidase operating at different conditions of oxygen supply.

It was found that the obtained values of the Michaellis constants are strongly influenced by the mode of the oxygen supply to the reaction zone of the electrode. For the case when air is supplied in gas phase to the gas-side of the electrode $K_M^{app} = 20.6\,\text{mM}$. The value of the Michaellis constant is more than 10 times lower ($K_M^{app} = 2.0\,\text{mM}$) when the oxygen supply is achieved from the dissolved oxygen only.

It can be concluded that the enzyme electrodes operate at conditions of oxygen deficiency when only the dissolved oxygen is taking part in the enzymatic reaction.

4. Electrochemical Gas Biosensors

Environmental pollution is one of the most substantial problems of present-day society. Often soil and surface waters around chemical plants are contaminated. A number of volatile compounds are emitted into the atmosphere. In this connection research and development of methods for the assessment of pollution levels in all types of waters and in the atmospheric air is of prime importance.

The progress in the development of gas biosensors nowadays is connected mainly with the large prospects for their practical applications for an efficient and effective detection and on-line monitoring of hazardous gases in the ambient air. On the other hand the monitoring of volatile components evolved from food can be used for the characterization of food freshness. The analysis of some volatile components in respiratory gas would provide a convenient and noninvasive methods of diagnosis of disease state.

Gas biosensors for detection of vapours [8] of some organic compounds were designed on the basis of the above described enzyme/gas-diffusion electrode: phenol gas biosensor (using an enzyme tyrosinase), alcohol gas biosensor (using the enzyme alcohol oxidase).

A schematic cross section of a gas biosensor based on an enzyme/gas-diffusion electrode is presented in Figure 6. In this case, both, the analyte and oxygen needed for the enzymatic reaction are transported to the reaction zone from the gas-side of the electrode through the hydrophobic, free from the electrolyte pores of the gas layer.

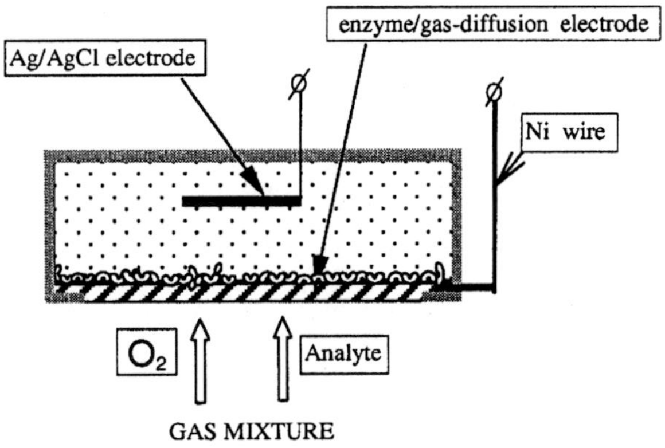

Figure 6

The investigations have shown that the amperometric signals of phenol and alcohol gas biosensors with the described structure are practically independent on the humidity of the gas mixture.

References

1. F. Scheller, F. Schubert, Biosensoren, Akademie Verlag, Berlin, 1989
2. H.Gunasingam, T. Ruth, Biosensors, 4 (1989) 349
3. I. Iliev, et al., J. Power Sources, 17 (1986) 345
4. A. Kaisheva et al., Sensors & Actuators, 8 (1992) 53
5. I. Iliev at al., Sensors & Actuators, 8 (1992) 65
6. J. Roshpon, Ext. Abstr. 117 Electrochem. Mtg, 1990, Montreal, Canada, p. 1023
7. I. Iliev at al., Electroanalysis, 7 (1995) 542
8. I. Iliev, A. Kaisheva, in Frontiers in Biosensoric II, Birkhauser Verlag, Basel-Boston-Berlin, 1997

Nanoscience & Nanotechnology
eds. E. Balabanova, I. Dragieva, Heron Press, Sofia, 2001

SPRAY PYROLYSIS IN FRONTIERS OF NANOTECHNOLOGY

V. Kozhukharov[1], M. Machkova[1], N. Brashkova[1], M. Ivanova[1] and V. Blaskov[2]

[1]University of Chemical Technology and Metallurgy, Sofia- 1756, Bulgaria
[2]Institute of General and Inorganic Chemistry at Bulgarian Academy of Sciences , Sofia 1113 , Bulgaria

Abstract. The Spray Pyrolysis (SP) process is based on rapid thermolysis of precursors in liquid state (aerosol) and a following sintering stage at high temperature. SP technology can be applied for preparation of ultra- fine, nano-crystalline metal oxide and oxyhydroxide particles. In the present work advances in the SP technology are demonstrated describing the main stages of the SP method. An analysis is presented regarding the atomizers used, selected SP equipment and the morphology of the final product. The review shows that the SP method is useful for the production of nanophase materials with application in the most important advanced and strategically functions as microsystems.

Keywords: spray pyrolysis, nanomaterials, application

1. Introduction

Nowadays the modern materials science announces a significant contribution to enhance of the quality of life. Undoubtedly the new materials and processes for their production are leading for this impact. It is well known, that the main object of materials science is the investigation of the relationships between *structure, properties, processing*, and *performance*. The methods of processing can be divided into three main directions: synthesis from (i) gas mixture, (ii) liquid and (iii) solid state phase. We would like to mark only that the synthesis from gas mixture phase is a progressive in the field of nanotechnology. Evidently advantages exist at laser beam assisted synthesis during the steps of the technological processes, as deposition, etching, doping, oxidation and reduction, starting from gas mixture phase [1]. In the present work our attention will be focused on aerosol state (dispersed liquid into gas phase) as useful one in the field of nanotechnology and as a key factor in the spray pyrolysis (SP) method.

As it is known, "*spray*" – is a system of small drops of water or other liquid highly dispersed, usually in air, based on action of an atomizer; "*pyrolysis*" – is a chemical decomposition initiated by heat. The large flexibility of the SP processes offers us wide possibilities of advanced powder synthesis on a submicron level and film deposition as well. For a SP process significantly higher temperatures (e.g. $> 300°C$) are applied to form the desired inorganic phases by *thermolysis*, i.e. solution decomposition by the action of heat. We intend to use SP as a term to describe the whole processes of ultra-fine powder synthesis in which a solution is atomized and thermolyzed at a relatively higher temperature to attain the requisite phase.

The aim of this review paper is to illustrate the progress that has been accomplished in the SP process during the last years, applying examples taken from different fields of application, including that of nanomaterials.

2. SP Nature and Stages

As a technology based on solution drops highly dispersed in a gas medium (i.e. aerosol state) the following advantages of the SP technology can be underlined: (i) high variety of the precursors which can be used; (ii) both aqueous and non- aqueous solutions can be applied; (iii) solvents or surfactants: polar or non- polar, protic or aprotic constituents can be used; (iv) easy preparation of powders with single- or complex composition at a high level of homogenization; (v) suitable for hollow or porous powder synthesis with a spherical morphology; (vi) easy preparation of ultrafine and nanosized particles and (vii) thin films deposition.

However, the SP process posseses some limitations which can be marked as follows: (i) can not be accomplished exact control of the precursor parameters; (ii) undefinite control regarding the drops coagulation process; (iii) production rates are limited out of the need to produce drops of diameter $< 5\,\mu m$; (iv) limitations exist in reduction-based aerosol thermal decomposition processes; (v) lot of technological parameters for control exist; (vi) exact evaluation of the parameters of the synthesis regarding monocrystalline and multicrystalline, fragmentation of particles and their morphology.

The stages of a SP process can be described as follows:

- Stage of preparation of spray medium
- Atomization stage
- Thermolysis (including evaporation, precipitation, drying and decomposition steps)
- Sintering stage.

Each of the mentioned stages are characterized with technological parameters as Me- salt solubility, concentration, viscosity, density, volume, temperature and pressure, surface tension as well as rate of nucleation and crystallization. Thermodynamic parameters also exist but there are not an object of discussion in the present study.

3. Preparation and Characterzation of the Spray Medium

A priori can be accepted that the nature of the liquid phase influences the next SP technological parameters. When we start from solution, characteristics such as viscosity (influence on flow rate), density, reological characteristics, solids content, miscibility effect and surface tension play an important role for the aerosol formation. In practice, both protic (containing a removable proton) and aprotic, inorganic and organic constituents in a SP process as precursors can be used. The exact selection is in close dependence of the technological parameters of the SP process, properties of the solvent utilised and the characteristics of the final product as powder or film, amorphous or crystalline, single- or multiphase.The evolution of the chemical properties of the precursors used (monomerchemistry), via controlled polymerization (polymer chemistry) to form the desired final product in solid state (e.g. film, or nanoparticles) after sintering process is very important.

For solutions, the nature and behaviour of the Me- salt is very important component for the preparation of workable precursor solution. The thermal properties of the precipitated precursor is also important during the thermolysis stage. For suspensions (dispersed a solid phase in liquid), there are three important factors for the next atomization step. These are the level of the solid phase concentration, granules and their morphology, the relationship behaviour between the solid phase and the liquid medium. The most important and critical parameter is the particle size and their morphology.

4. Atomization

Atomization stage has a dominant effect on the particle size because there exists one-to-one correlation between a drop and the final particle. Namely, the atomization stage is a process for aerosol formation. The characteristics for an "ideal" aerosol atomizer is to generate a high monodisperse media, stable, with uniform drops-size and concentration in volume as well as with a spherical morphology of the aerosol drops.

A nebulizer is a type of atomizer capable to produce a high monodisperse aerosol outcome by removing of the larger spray drops by impaction within the device. The most air jet nebulizers are both hydraulic and pneumatic pressure type nozzles. They are cappable for drop size formation from 10 to 100 μm at a velocity value $\sim$ 10 m/s. The size of aerosol particle (d_p) depends of the concentration of the solvent into precursor solution (F_c) and the drop diameter (d_d), according to the equation $d_p = d_d(F_c)^{1/3}$. Details regarding nozzles for aerosol generation and equipment are presented in [2].

5. Thermolysis

The thermolysis is a complex process of solution decomposition by the action of heat. The most important steps are *evaporation, precipitation, drying*, and *decomposition*.

The evaporation period is a complex one including phenomena as follows: i) evaporation of the solvent from the drop surface; ii) diffusion of the solvent vapours away from the drop in the gas phase; iii) shrinkage of the droplet; iv) change the temperature in the drop and v) diffusion of solute toward the center of the drop. The nature of the precursor and all primary technological parameters contribute to mass and heat transport into and on the surface of a drop. The size and number of the crystals formed is a function of supersaturation fenomena. When the degree of supersaturation is low, the nucleation rate is small relative to the growth rate. As a result only few crystals are formed [3]. The size, number and distribution of the crystals determine the size and size distribution of the pores. Both have a significant effect on the next stage of drying.

The drying of drops on precipitated salts is an important stage in the SP. Size of the drops and crystal distribution, diffusivity of the vapour, size and distribution of the pores, mass concentration of the solvent vapours, thickness and the porosity of the crust have influence on the drying step. The term "decomposition" can be treated as synonymous of the term "calcination", i.e. "calcine" – reduce, oxidize or desiccate by heat.

6. Sintering

Sintering of the drops is the final step in a SP process. Usually this stage is accomplished from 850°C to 1220°C. The selection of the exact sintering temperature is in close dependence of the droplet size, the drop thermal evolution during the previous thermolysis process and the nature of the materials used. As a result particles with spherical morphology can be produced, i.e. this confirms the main concept of a SP process to produce one particle per drop.

7. SP Set-up Arrangements

According to [4] a schematic system is presented in Figure 1 where first atomization and drying process take place in a standard spray dryer unit. Solid particle/gas separation is executed in a cyclone separator where the gas stream and the dried particle stream are separated.

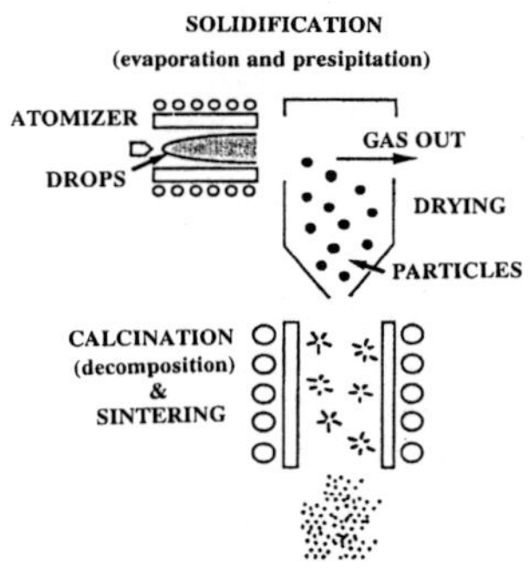

Figure 1. Set- up arrangement of a SP process after ref. [4]

The next step of thermolysis and sintering occurs through gravitation of the dried particles in a tube furnace.

Advantages of this configuration of the units are the simplicity of operation combined with the availability of commercial equipment.

8. Particles Morphology

Output products by SP method determine production of uniform high dispersed particles with spherical geometry. According to [5] spherical TiO_2 particles are successfuly prepared from $TiCl_4$ aqueous solution by SP method and after sintering at 600°C. Figure 2 shows a spherical morphology of the TiO_2 powders with almost the same normalized particle size distribution in the range of 0.2–1.4 μm.

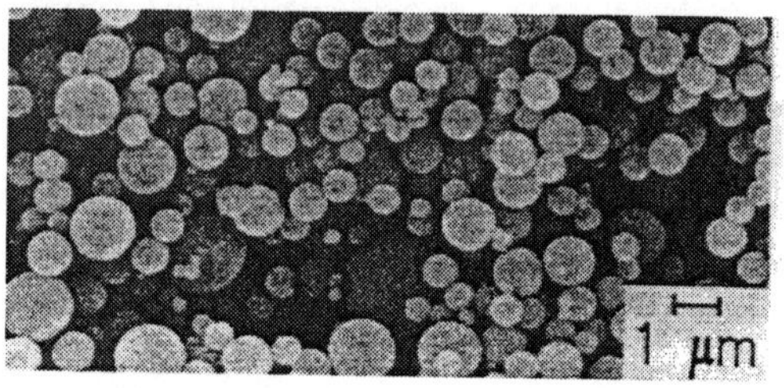

Figure 2. An illustration of spherical morphology of TiO2 powders prepared by SP method according to [5]

9. Application

It is clear from the discussion carried out that SP method can be successfully applied for synthesis of powder, single- and multi- crystalline particles, composite materials and films. The method is useful for production of solid particles, e.g. for application in thermobarier coatings, hollow and porous powder synthesis, non- oxide and metal particle synthesis, mixed nitro/carbon/oxygen multicomponent particles and advanced pigments. The useful information regarding the engineering application of the SP method towards advanced material can be found in [6,7]. Nowadays actual applications of the products outcome by SP method can be illustrated as: high Tc superconductors [4,8]; mixed conductive ceramic membranes [9]; electrode material in fuel cells (SOFCs) [10]; ultrafine, sub-mm and nanoparticles [11–13]; solid state chemical sensors [14]; thin films for diferent kind of application [15]; in the high temperature catalytic systems; in advanced environmental technology and etc.

10. Conclusions

1. During the last 20 years range of SP techniques have been developed for processing of ceramic powders and films on nanosize level.

2. SP method offers us a production of wide range of submicron powders with spherical morphology, narrow particle size distribution, single and complex compositions of a high level of homogeneity;

3. In dependence of the atomizer used, uniform drops from 0.1 to 100 mm can be produced;

4. SP process can be easily modified during the main steps of the process as: evaporation, precipitation, drying, decomposition and sintering;

5. There is a gap regarding synthesis of mix oxicarbide and oxinitride high temperature corrosive ceramic materials, nanosized powders for membrane science and technology; in the field of ceramic pigments and biomaterials, as well.

6. Limiting information exists regarding fundamental behavior of the precursors during the SP steps, the rate of nucleation in the course of the crystallization, the thermodynamic calculations and the study of the kinetic of the chemical reactions in non- equilibrium states.

Acknowledgments

The support from the Commission of the European Communities in the frame of 5- th RTD programme, ICA2-CT-2000-10030 contract is greatly acknowledged.

References

1. V. Kozhukharov and S. Vassilev, Bol. Soc. Esp. Cer. y Vidrio 208 (1994) 38
2. E. Matijevic, Amer. Geophys. Union 26 (1982) 44
3. K. Leong , J. Aerosol. Sci. 18 (1987) 525
4. C. Martin, R. Kurosky, G. Maupin, C. Han, J. Javadpour and A. Aksay, Ceramic Transactions vol. 12, Ceramic Powder Sci. III, Amer. Ceram. Soc. Westerville, OH (1990) 99
5. J. Lee, H. Cho and S. Park , Ceramic Transactions vol. 22, Ceramic Powder Sci. III, Amer. Ceram. Soc.Westerville, OH (1991) 39
6. G. Messing, S. Zhang and G. Jayanthi , J. Amer. Ceram. Soc. 76 (1994) 38
7. I. Lengoro, Ph. D. Thesis (1998) Hiroshima, Japan
8. Tonge, M. Tatsumisago, T. Minami, K.Okuyama, M. Adachi and Y. Kousaka Jap. J. Appl. Phys. 27 (1988) L1086
9. Proceed. 6th Intern. Conf. on Inorganic Membranes, June 26- 30, 2000, Montpellier, France, Host Inst. European des membranes - Montpellier, France.
10. V. Kozhukharov, M. Machkova, M. Ivanova and N. Brashkova, 7-th International Symposium on Solid Oxide Fuel Cells June 3-8, 2001 / EPOCHAL TSUKUBA, Tsukuba International Congress Center, Japan, paper accepted for publication – XII, 2000.
11. Y. Kang, S. Park and Y. Kang, Nanostruct. Mater. 5 (1995) 777
12. J. Eastnman, Y. Liad, A. Narayanasamy and R. Siegel,Mater. Res. Soc. Sym. Proc. 155 (1989) 255
13. H. Ishizawa, O. Sakurai, N. Mizutani and M. Kato, Amer. Ceram. Soc. Bull. 65 (1986) 1399
14. B. Kim, J. Lee and H. Kim, Solid State Ionics 86–88 (1996) 1088
15. Okuya, N. Prokudina, K. Mushika and S. Kaneko, J. European Ceram. Soc.19 (1999) 903

Nanoscience & Nanotechnology
eds. E. Balabanova, I. Dragieva, Heron Press, Sofia, 2001

MODIFIED CATHODES FOR HIGH-RATE Li/SO_2 PRIMARY CELLS

M. Mladenov

Central Laboratory of Electrochemical Power Sources, Bulgarian Academy of Sciences, 1113 Sofia, Bulgaria

Abstract. High rate discharge of Li/SO_2 cells is usually limited by the large activation polarisation of SO_2 discharge at the carbon surface. Heat treated carbons and catalysts were found to improve remarkably the working voltage and discharge capacity at high rate discharge (15 mA/cm^2) and low temperature (-50°C).

Keywords: lithium/SO_2 primary battery, high rate discharge

1. Introduction

Acetylene black (AB) PTFE composites on an aluminum current collector have been widely employed as cathodes for Li/SO_2 primary cells. The cathode reaction in this type of cell is the following [1]:

$$2SO_2 + 2e^- \rightarrow S_2O_4^{2-}{}_{ads} \quad (1)$$

$$S_2O_4^{2-}{}_{ads} + 2Li^+ = Li_2S_2O_4 \downarrow \quad (2)$$

Sulfur dioxide, dissolved in the electrolyte, is reduced at the porous cathode producing an insoluble precipitate of lithium dithionite. Under conditions of high-rate and/or low temperature discharge, utilization of the pore structure of the cathode becomes inefficient and limits the capacity of the cell.

The porous Li/SO_2 cathode must perform several functions efficiently. It must be a good conductor of electrons to carry the current from the positive cell terminal to the electrode /solution interface. The surface activity and surface area of the cathode are two important properties, which may serve to lower the polarization in the cell [2]. It is generally recognized that incorporation of a transition metal macrocyclic complex such as Co-phthalocyanine into the carbon cathode increases both the voltage and the capacity of Li/$SOCl_2$ at high rates of discharge [3]. The cathode must also act as a matrix to accommodate the precipitate formed on discharge.

The aim of this study was to enhance the performance of the cathode by adding materials to the acetylene black PTFE mixture, which would improve one or more of these properties.

2. Experimental

Experimental cathodes were evaluated in a laboratory electrolyte flooded glass cell consisting of demountable cathode top lead, two lithium counter electrodes and one Li reference electrode. The cell was filled with 1M LiBr in acetonytrile and then saturated with SO_2 (12%). A constant current was passed between the lithium anode and the carbon cathode. The cathode potential was monitored until 2.0 V vs. Li was reached. The specific capacity for each cathode was determined at four current densities 1.2, 3, 7.6 and 15 mA/cm^2, corresponding to 0.2, 0.5, 1.2 and 2.5 A high-rate discharge of a C cell.

Cathodes were fabricated from acetylene black plus 5% PTFE (BET surface area 60 m^2/g, true density 1.95 g/cm^3, and resistivity 0.135 ohm.cm at 1 kN/cm^2) with addition of high-conductivity carbon (G) and catalyst. The catalyst Co-phthalocyanine was incorporated into the carbon from concentrated H_2SO_4 as described in ref. [3]. Typical cathodes contained 5 wt.% of the catalyst. The acetylene black carbon was heat-treated at different temperatures 300, 600, 800°C, at which the carbon/metal complex mixture was also heated, and at 1200°C in Argon atmosphere for 1 h.

The capacities of the C-size cells containing the catalysed cathodes were also determined.

3. Results

The test results for cathodes containing two high-conductivity additives (in 12.5% amount) and for the reference cathodes are given in Table 1.

The use of the graphitic additive G-1 (NG-7, surface area 9.6 m^2/g, resistivity 0.0014 ohm.cm) leads to a decrease in the capacity but also to an increase in the operating potential at 1.2 A rate of discharge. Carbon G-2 (acetylene black treated at 1200°C, surface area 50 m^2/g, resistivity 0.062 ohm.cm at 1 kN/cm^2) yields higher capacity per gram of carbon than the reference electrode at current densities of 7.6 and 15 mA/cm^2, but the cathode potential is substantially higher (by about 100 mV at these discharge rates) than the average value. Both additives improve the particle-to-particle contact, reducing the IR (potential) drop in the cathode and consequently increasing the cathode potential. Carbon G-1 is less effective in terms of capacity performance, because of the large size of its particles (5–20 μm) and hence the reduced macropore volume between them.

The next step in our study was to examine the effect of high-temperature treatment of carbon or carbon/Co-Pc mixture on the performance of the Li/SO_2 cathodes.

Discharge behaviour of high-temperature treated non-catalysed Li/SO_2 cathodes. The discharge capacity for Li/SO_2 cathodes, including the standard ones (Table 1) and those prepared from carbon (non-catalysed) treated at 300, 600, and 800°C for 1 h (presented in Table 2), are compa-

Table 1. Evaluation of experimental carbons for cathodes in laboratory cells: baseline data, graphitic material and acetylene black treated at 1200°C

Carbon type	Concentration (AB+G) (w./o)	Current Density (mA/cm^{-2})	Capacity Cc carbon Ah/cc	Capacity Wt carbon Ah/g	Average cathode potential V vs. Li	Comments
Ac. Black(AB-1)	95+0	1.2	0.687	1.82	2.820	baseline data
Ac. Black(AB-1)	95+0	3.0	0.680	1.80	2.783	baseline data
Ac.black(AB-1)	95+0	7.6	0.607	1.61	2.553	baseline data
Ac. Black(AB-1)	95+0	15	0.483	1.28	2.431	baseline data
Carbon G-1	82.5+12.5	1.2	0.660	1.75	2.682	Graphitic material
Carbon G-1	82.5+12.5	3.0	0.653	1.73	2.660	Graphitic material
Carbon G-1	82.5+12.5	7.6	0.577	1.53	2.650	Graphitic material
Carbon G-1	82.5+12.5	15	0.415	1.10	2.559	Graphitic material
Carbon G-2	82.5+12.5	1.2	0.694	1.84	2.802	TreatedAB1200°C
Carbon G-2	82.5+12.5	3.0	0.675	1.79	2.730	TreatedAB1200°C
Carbon G-2	82.5+12.5	7.6	0.626	1.66	2.644	TreatedAB1200°C
Carbon G-2	82.5+12.5	15	0.538	1.42	2.542	TreatedAB1200°C

red at 7.6 and 15 mA/cm^2. The capacity of each cathode type was reproducible within a 2–3% deviation range. The data appear to show that sintering (high-temperature treatment) of non-catalysed cathodes has but a minor effect on their performance characteristics. Practically, identical capacity and voltage values were obtained for cathodes discharged at 7.6 mA/cm^2. At 15 mA/cm^2, the potentials of the cathodes sintered at 800°C were about 110 mV higher than those of standard baseline cathodes. However, the capacity spread of the sintered electrodes at this current density is of the same order as that found at 7.6 mA/cm^2, which implies that the thermal treatment of the carbon substrate up to 800°C does not contribute much to increasing the discharge capacity of Li/SO_2 cathodes.

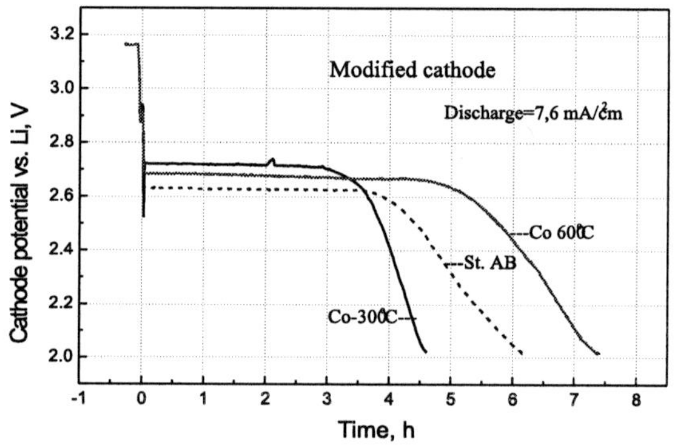

Figure 1. Discharge curves for Li/SO_2 cathodes as a function of sintering temperature at 7.6 mA/cm^2

Discharge of catalysed Li/SO_2 cathodes. The results we have obtained with Co-Pc catalysed electrodes, sintered at 300, 600 and 800°C are summarized in Table 2. Typical discharge curves at 7.6 mA/cm^2 for catalysed cathodes are presented in Figure 1.

The capacity data for catalysed cathodes given in Table 2 are compared with those for non-catalysed (standard) ones (cf. Table 1) at the respective current densities.

The data from Table 2 (and Figure 1) show that at 7.6 mA/cm^2, the catalysed cathodes exhibiting the highest activity are the ones heat-treated at 600°C. The catalyst improves the capacity by 20% along with ca. 130 mV increase in cathode potential. The carbon/Co-Pc cathodes sintered at 300°C have a slightly inferior performance as compared to the standard baseline electrodes, although the average voltage remains the same. The electrodes sintered at 800°C yield the same capacities as the standard baseline cathode, however, their voltage performance is about 200 mV higher

Table 2. Effect of sintering on the room temperature performance of cathodes doped with Co-Pc

Carbon type	Cathode sintering temp. °C	Current density mA/cm^2	Capacity cc carbon Ah/cc	Capacity wt carbon Ah/g	Average cathode potential V vs. Li.	Comments
Acet. black(AB-2)	300	7.6	0.596	1.59	2.530	Non-catalysed cath.
Acet. black(AB-2)	300	15	0.680	1.26	2.483	Non-catalysed cath.
Acet. black(AB-2)	600	7.6	0.610	1.63	2.564	Non-catalysed cath.
Acet. black(AB-2)	600	15	0.483	1.50	2.431	Non-catalysed cath.
Acet. black(AB-2)	800	7.6	0.601	1.64	2.582	Non-catalysed cath.
Acet. black(AB-2)	800	15	0.550	1.46	2.560	Non-catalysed cath.
Acet. black(AB-3)	300	7.6	0.543	1.44	2.540	Catalysed cath.
Acet. black(AB-3)	300	15	0.415	0.95	2.451	Catalysed cath.
Acet. black(AB-3)	600	7.6	0.739	1.96	2.682	Catalysed cath.
Acet. black(AB-3)	600	15	0.637	1.69	2.560	Catalysed cath.
Acet. black(AB-3)	800	7.6	0.633	1.68	2.682	Catalysed cath.

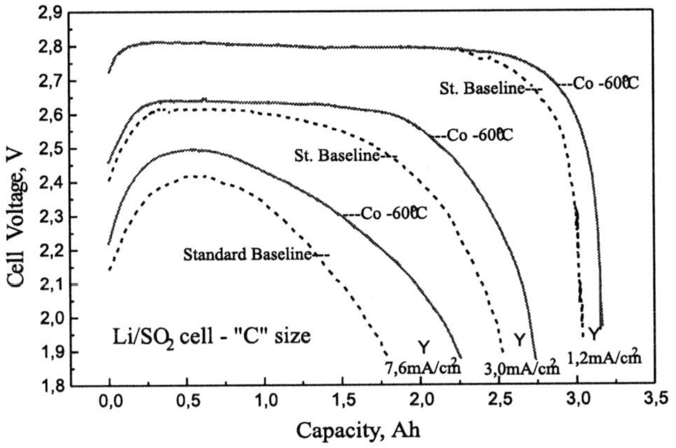

Figure 2. Discharge curves for Li/SO$_2$ in C-cells at 1.2, 3.0 and 7.6 mA/cm^2, corresponding to 0.2, 0.6 and 1.2 A battery discharge rate

than that of the standard cells. This behaviour is typically observed with a transition metal complex catalysed Li/SOCl$_2$ cell [4]. The catalyst may play a dual role.

Firstly, it may lower the activation energy for electron transfer from the carbon substrate to SO$_2$. Secondly, the catalysed carbon may have a higher adsorption capacity for SO$_2$ and thus increase the concentration of the SO$_2$ that is reduced on the cathode.

The data in Table 2 for cathodes sintered at 300°C indicate that Co-Pc on a porous carbon electrode has a poisoning effect with respect to SO$_2$ reduction, unless it is activated at about 600°C or higher temperatures. Increasing the sintering temperature to 800°C leads to decomposition of the Co-Pc complex and chemical bonding of the metal to the carbon [3], and results in a catalyst that improves mainly the cell voltage with only minor effects on improving the cell capacity.

The results obtained with catalysed Li/SO$_2$ cathodes encouraged us to conduct experiments aimed at improving the performance characteristics of real C-sized cells at high discharge rates and negative temperatures.

Figure 2 illustrates the improvement in performance of a Li/SO$_2$ cell "C" size, using cathode modified with AB/Co-Pc (treated at 600°C), at three different discharge rates. It can be seen that at slow discharge modes of 0.2 A, the effect of the catalyst is negligible. Under these conditions, the process of SO$_2$ reduction proceeds until the porous matrix is filled with the precipitation product, due to the low polarization limitations observed with non-catalysed electrodes. At medium discharge rates of 3 mA/cm^2, the use of a catalyst improves the efficiency of the cathode. At high discharge rates (7.6 mA/cm^2), the catalytic effect of Co-phthalocyanine is most pronounced. The capacity of cells with calatysed cathodes is about 15% higher than that of their non-catalysed counterparts.

The effect of modified cathodes on the discharge capacity at different temperatures is presented in Figure 3 for Li/SO$_2$ cell "C" size (CLEPS), compared with those of "Mallory" – USA, and "Duracell" – USA.

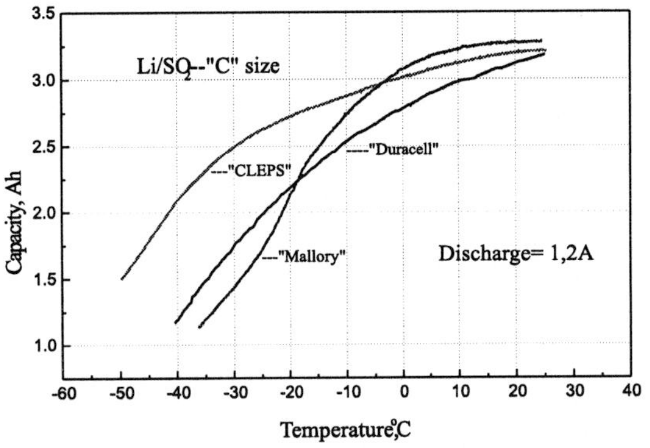

Figure 3. Comparison of the discharge capacities as a function of temperature for "C"-size cells produced by "Mallory" – USA, "Duracell" – USA and "CLEPS" – BG

The cell with modified cathodes (CLEPS) exhibits the best capacity at 0.2 A rate of discharge and temperatures as low as -50°C.

Conclusion

Some carbonaceous materials have been tested as additives to improve cathode capacity or operating voltage in laboratory tests. The graphitic additive imparts lower capacity to the cathodes, but leads to an increase in operating potential in Li/SO$_2$ cells at discharge rates higher than 7.6 mA/cm^2. Treated acetylene black G2 is more effective in terms of Ah/g at 7.6 mA/cm^2 rate of discharge compared to G1, because of its small particles' size (30 nm), which does not reduce the pore volume of the cathode and facilitates the reduction process in the electrode interior.

Both G-1 and G-2 improve the particle-to-particle contact, reducing the IR drop in the cathode and thus increasing the cathode potential. The sintering temperature of the carbon/metal complex mixture has been identified as an important factor that affects the catalytic activity of cobalt-phthalocyanine (Co-Pc). Data obtained with thermally treated acetylene black catalysed with Co-Pc (sintered at about 600°C) evidence that this catalyst improves the cathode performance in terms of both higher load voltage and capacity up to a current density of 15 mA/cm^2. Increasing the sintering temperature to 800°C leads to decomposition of the Co-Pc complex and chemical bonding of the metal to the carbon, and results in a catalyst, which mainly improves the cell voltage with only minor beneficial effects on cell capacity.

References

1. R. Moshtev et al., J. Electroanal. Chem., 1980, vol.8, N:3, p 335.
2. T. Reddy et al., Optmization of Ultra High Rate Lithium-Sulfur-Dioxide Cells, 4-th Monhly Latter Report, Contract DAAR-20-81-c-0410 Power Convertion, Incorporated, Mount Vernon, New York, 1981.
3. K. M. Abraham et al., Metal Phtalocyaninr-Catalysed Li-SOCl$_2$ Cells, Electrochim. Acta, Vol.37, N:3, p. 531, 1992.
4. R. Nowak et al., Electrochim. Acta, 33, 1331, 1988.

Nanoscience & Nanotechnology
eds. E. Balabanova, I. Dragieva, Heron Press, Sofia, 2001

NANO PARTICLE SIZE CARBON AS INTERCALATING ANODE MATERIAL FOR LITHIUM BATTERIES

S. Uzunova, B. Banov, B. Puresheva, A. Momchilov

Central Laboratory of Electrochemical Power Sources, Bulgarian Academy of Sciences, 1113 Sofia, Bulgaria

Abstract. The conductive binders play an important role in the preparation methods and the electrochemical behaviour of the test electrode. A new class of materials Ultra Fine Carbon, (UFC) became in the last time, a real "top of the art". These materials are proposed as liquid suspensions in which the total amount of the dry residue does not exceed 10%. On the other hand, the very well known TAB2 present absolutely dry mixture of 70% acetylene black and 30% PTFE. The advantages of the both conductive binders reflect in the new proposed dry Nano particle size Carbon material (NC).

Keywords: alternative to lithium anode materials, lithium intercalation anode compounds, intercalating carbon materials, nano particle size carbon materials

1. Introduction

Many different carbon materials were investigated as an active material for the anodes in lithium batteries. These materials have either ordered structure as graphite or disordered one as cokes. Carbon fibres with isotropic or anisotropic properties were investigated too [1]. Recently a new class conductive additives, Ultrafine carbon (UFC) suspension was proposed, described in the literature and patents [5,6,7]. The UFC conductive additives are carbon-polymer aqueous or nonaqueous suspensions in which the total amount of dry substance is 10% [6]. The dry substance consists of lightly branched chains of acetylene black or carbon black and polymer (PVA or PVP) particles with small size, which keep the carbon chains in a suspended state.

We have combined the advantages of the new conductive additives, small particle size allowing perfect contact between the active particles of the electrode material and the advantages of the classical conductive additives, 100% dry mixture with binder. One Nano particle size Carbon (NC) material was used for the preparation of the teflonized dry mixture.

This work illustrates the electrochemical properties of those obtained active electrode material as anode in lithium battery.

2. Experiment

One Nano Carbon (NC) material with median particle size of about 100–300 nm was used as it is supplied by the producer. The teflonized mixture contents 5 weight per cent of PTFE. The specific surface area (SSA) of the pellet prepared of NC was measured by the BET method by using a Strohlein & Co. AREA Meter.

The test electrodes were prepared by pressing an appropriated amount of NC on a copper foil, or by spreading on the same support a paste of NC in alcohol. After that the those obtained electrodes were sintered at 300°C for 10 min.

The XRD measurements were carried out on the dry residue and on the electrodes before and after cycling on a Philips APD diffractometer with Cu K_α radiation.

The slow voltammetry and cycling tests were performed in three-electrode laboratory button cells with a Li reference electrode. The electrolyte solution was 1M $LiClO_4$ in EC + DEC (1:1) with water content lower than 20 ppm. All cycling tests were carried out galvanosatically with a current density of 0.4 mA/cm^2 in the voltage range 0.05 to 1.50 V on a computer cycling device.

Results and Discussion

The specific surface areas of the NC teflonized mixture is given in Table 1. These results are compared with the specific surface area of the TAB2 conductive binder. The specific surface area of the NC mixture is very close to that of TAB2.

Table 1

Conductive material	NC	0.9NC +0.1TAB2	TAB2
BET [m^2/g]	37.7	36.9	34.9

The XRD pattern of the NC powder is shown on Figure 1. A large halo from 16° to 30° 2θ is present in the pattern of the NC sample.

This halo confirms the nano particle size of the used carbon material and the preparation method of the mixture (absence of agglomeration of the carbon particles). These patterns are changed significantly after applying the mixture onto the copper current collector. The diffractogram of the NC anodes before cycling is shown in Figure 2.

A representative slow voltammogram with a scan rate 15 mV/s of the NC material is shown in Figure 3. The shape of NC voltammogram is similar to those of carbon fibres [1]. Two peaks are observed in the smooth curve during the first lithiation of the NC carbon material. These peaks most

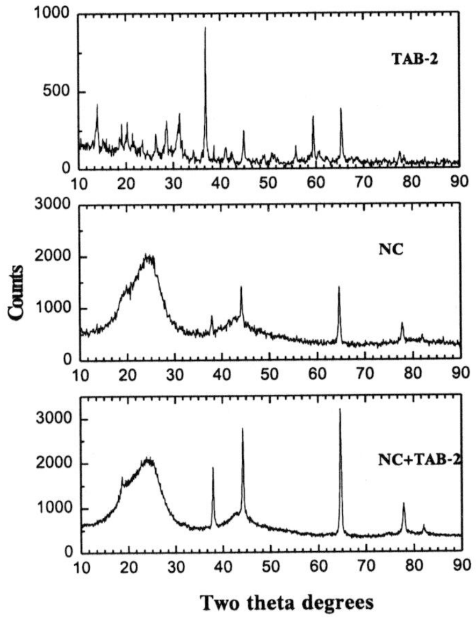

Figure 1. XRD of TAB-2, NC, and NC+TAB-2 mixture

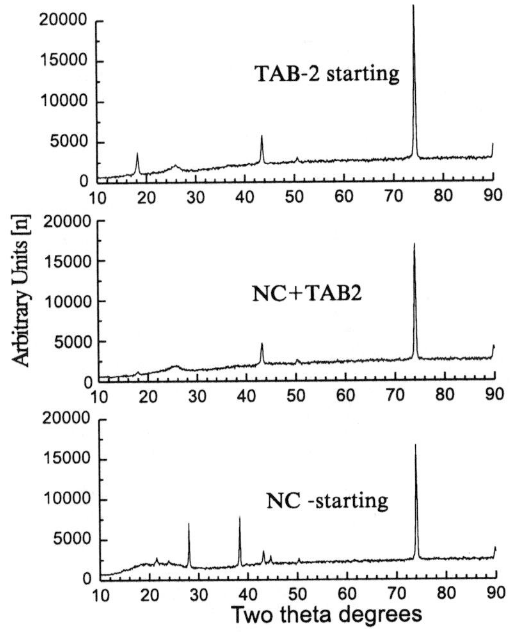

Figure 2 XRD of test electrodes prepared from TAB-2, NC+TAB-2 and NC

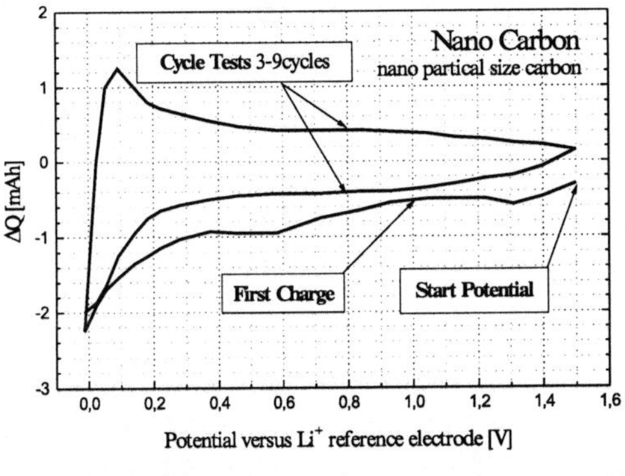

Figure 3 Cyclic voltammetry of Nano particle size Carbon NC sample, formation cycle (lithiation) and the following cycling test 3-9 charge discharge curves

probably are due to the formation of a passive film on the surface of the electrode [1].

The data from the electrochemical cycling tests of anodes based on NC materials, or mixture with TAB2 are illustrated in Table 2. The results obtained with anodes with TAB2 (containing only acetylene black) are also shown in the same Table 2.

Table 2. Specific reversible capacity test of NC, NC+TAB2 mixture and pure TAB2

Cycles [n]	1	3	5	9
Specific cap. of NC [mAh/g]	273	244	230	212
Specific cap. of 0.9NC+0.1TAB2 [mAh/g]	239	218	203	185
Specific cap. of TAB 2 [mAh/g]	180	177	165	142

Comparing the specific capacity exhibited by TAB2 and NC and the XRD patterns it can be concluded that NC materials defer from the acetylene black, but it is close to some kind of hard carbon. The specific capacity of that NC material is not so large – from 200 to 240 mAh/g at the fifth cycle. The work is in progress and we expect an improvement of the specific characteristics of that Nano Carbon material as active anode material also as conductive additive in cathode mixtures.

Conclusions

1. Combining the advantages of TAB2 material – high active carbon material content, and the miscellanies structure of PTFE we succeed to prepare nano particle size carbon NC with the same structure.

2. Thus prepared material can be successfully used as both, conductive binder with nano particle size, or as active intercalating anode material in lithium primary and secondary cells.

3. We have demonstrated that the electrochemical characteristics of thus prepared nano particle size intercalating carbon material are very close to the hard carbon materials used as anode materials in lithium ion batteries.

References

1. Roh, K. Jeong, H. Cho, H. Kang, Y. Lee, S. Ryu, B. Lee, Journal of Power Sources, 68 (1997) 271
2. T. Tran, L. Spellman, W. Goldberger, X. Song, K. Kinoshita, Journal of Power Sources, 68 (1997) 106
3. V. Manev, I. Naidenov, B. Puresheva, G. Pistoia, Journal of Power Sources 57 (1995) 133
4. F. Disma, L. Aymard, L. Dupont and J. Tarascon, Journal of Electrochemical Society, Vol.12, December 1996
5. A. Kozawa, K. Fujita, A. Sato and D. Brodd, Proceeding of the 12th Annual Battery Conference on Applications and Advances, p.277, Long Beach, Jan. 14-17, 1997
6. A. Kozawa, S. Mase, N. Suzuki and A. Sato, Progress in Batteries & Battery Matrials, Vol.16, 1997, p.350
7. Momchilov, A. Trifonova, B. Banov, B. Pourecheva, A. Kozawa, Progress in Batteries & Battery Matrials, Vol.17, 1998, p.123
8. V. Manev, A. Momchilov, K. Tagawa and A. Kozawa, Progress in Batteries & Battery Matrials, Vol.12, 1993, p.157

Nanoscience & Nanotechnology
eds. E. Balabanova, I. Dragieva, Heron Press, Sofia, 2001

SYNTHESIS AND ELECTROCHEMICAL PROPERTIES OF $LiCoO_2$ WITH CONTROLLED PARTICLE SIZE

S. Uzunova, B. Banov, B. Puresheva, A. Momchilov

Central Laboratory of Electrochemical Power Sources, Bulgarian Academy of Sciences, 1113 Sofia, Bulgaria

Abstract. It is very well know that the active electrode material particle size (high specific area of working electrodes) plays very important role for the electrochemical behaviour and design of the cell. The lithium electrochemical systems dose not defer to this basic rule. In the last time the micro batteries became one new trend in the battery development. The high current drain and the fast charge discharge rates are commonly used. All these requirements can be achieved only by new materials or by already know but prepared in different way to ensure the above requirements. We have successfully developed a new universal way for synthesis of electrode materials with controlled grain size. The $LiCoO_2$ cathode materials have been prepared using this new synthesis method, and the electrochemical characteristics are presented.

Keywords: lithium batteries, electrode materials cathode materials, intercalation compounds, metal oxides, lithiated metal oxides, cobaltate, nickelate, manganese dioxides, spinel, lithium ion cells

1. Introduction

The particle size of the active cathode materials for lithium ion batteries plays one important role for the electrochemical behaviour [8–12] of the cell. The different cathode materials possess different optimal particle size of the crystals related to the structure, crystal lattice and allowing the multiple insertion/deinsertion of lithium ion from the original structure without deformation or internal stress formation. The achievement of this principle purpose can be done by different ways: grinding, sintering at low temperature and applying multiple grinning or by applying new synthesis methods [1,3,4]. The new synthesis methods are based on sol-gel process or semiconductors technics [4]. The grinding technic is widely appalled and leads to satisfactory results. During this process the crystal particles are subjected to big stress and the initial structure is deeply affected. These effects decrease the lithium diffusion into the crystal structure of the compound and finally led to complete deterioration [5–7] of the cathode material electrochemical behaviour. Finally the process of grinding led to wide particle size distribution and it is a compromise between electrochemical and mechanical characteristics. In the last few years many research groups around the world try to find a new and effective way to synthesis of materials for lithium batteries with controlled particle size. A sol-gel technic [2,3,6], or drying in a stream of hot air are proposed as prospective synthesis way.

The aim of the present work is to demonstrate a new technique allowing the production of materials with controlled particle size. The method is based on the fast evaporation of solvent in an appropriated liquid media, thus obtaining particles with desired dimensions and spherical shapes.

2. Experimental

The new proposed method consists of fast evaporation in liquid unmixable media of aqueous solution of the desired compound. The principal schematic of the method is given at Figure 1. It consists of 4 steps. Preparation of starting aqueous compound solution with controlled parameters: concentration, density, temperature, speed of mixing. The second one is the drying in the unmixable liquid media with controlled parameter as follow: density of liquid media, temperature, speed of mixing of starting solution. The next step consists of separation of prepared starting precursor from the liquid drying media, and final drying. All these steps are drown in Figure 1. Considering these principles we have try to prepare a real cathode material with controlled grain size.

steps	Operations
1	**Solution containing Me^+** *Controlled parameters* concentration $\rightarrow C$ [mol/l] density $\rightarrow \rho$ [g/ml] temperature $\rightarrow t$ [°C] speed of mixing $\rightarrow V$
2	**Drying in liquid media** *Controlled parameters* density $\rightarrow \rho$ [g/ml] temperature $\rightarrow t$ [°C] speed of mixing $\rightarrow V$
3	**Washing from the drying liquid**
4	**Drying**
5	**Starting precursor**

Figure 1. Schematic of the new method for preparation of active electrode materials with controlled grain size.

We have choose the $LiCoO_2$ as material that can be prepared by the new technique due to the fact that this material is easy for synthesis by solid state reaction and the advantages or the disadvantages of the method can be detected easier. A lithium and cobalt nitrate at the appropriated ratio 1:1 were mixed. The thus obtained solution was dropped in

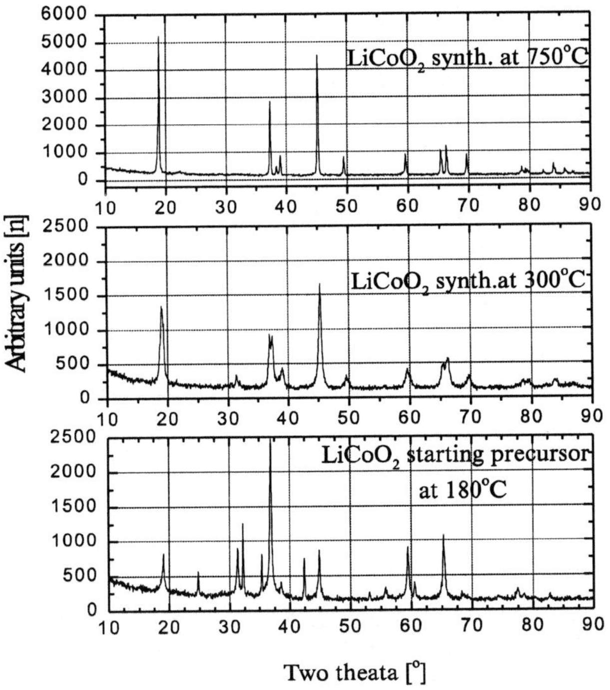

Figure 2. XRD patterns of LiCoO2 pretreated at different temperatures, starting precursor at 180°C, fired at 300°C and final product synthesized at 750°C

a liquid reagent media and the starting precursor of stochiometric $LiNO_3.Co(NO_3)_2$(hydrate) ratio was formed. After that the suspension was filtered and dried at temperature of 100–150°C for one hour. A DTA and DTG test have shown that there are three important temperatures that must be observed vigorously. The first one is 300°C, the temperature of dehydration of the starting precursor and the nitrate decomposition. The second one is 570°C the starting of the real synthesis and the last one is the final firing temperature of 730°C.

The test electrodes were prepared as previously described in [8,9] and subjected to electrochemical tests. A three electrode test cell with lithium reference electrode and limited electrolyte volume was used. The electrolyte in use was 1M $LiClO_4$ in EC:PC:DMC, 2:1:1 by volume.

3. Results and Discussion

As described above a $LiCoO_2$ was successfully synthesised and XRD characterised. The XRD patterns of the lithiated cobaltate are presented at Figure 2. It can be seen the final product prepared at 750°C and bellows it the XRD of the same product fired at 300°C and the starting precursor prepared at 180°C. It must be noted that the XRD at 180°C is reported for the first time in the literature. It is very important to mention the very high crystalinity of this starting precursor. The XR diagram presents almost all picks as the XRD at 750°C. The XRD at 300°C is absolutely pure $LiCoO_2$ without any extra picks.

This is very important because even the sol-gel process doses not give clear XRD at temperature bellow 350°C.

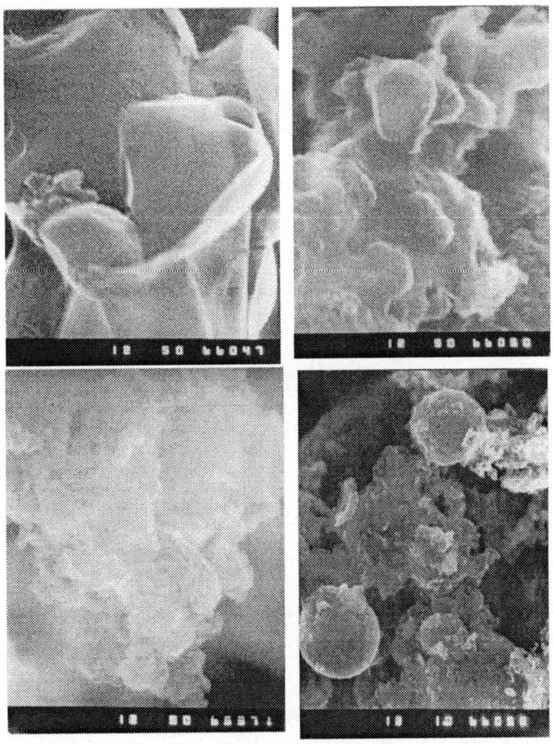

Figure 3. SEM picture tree samples synthesized according scheme 1 from different concentrations. The median particle size is ranged from 600–1000 nm, for the first picture 200–400 nm for the second, and 50–100 nm for the third one. The fourth picture present the starting precursor with median size of 20 μm

Even at this temperature the XRD is mixture from many compounds (J.P. Ramos, Synthesis by soft chemistry...). This fact can be attributed to the small particle size of the starting precursor that diminish the synthesis temperature. The checking of the influence of the mentioned above aqueous solution parameters we have carried out as a series of experiments with different concentrations. The results of these experiments are presented at four SEM pictures of the starting precursor obtained at 180°C. The Figure 3 demonstrates the dependence of the type of the grain size from the starting concentration. It can be seen that is possible to obtain a variety of grain size changing only the concentration of the starting solution. The obtained particles are: at the first photo with median size of about 0.5–1.5 μm, at the second one 200–400 nm, at the third one 50–200 nm. The four one present at low magnification the particles formed in the liquid drying media: they have spherical shape of about 20 μm. All these pictures confirm our starting ideas that are possible to form electrode material with controlled grain size.

The electrochemical tests of the thus obtained materials are presented at Figure 4. The first discharge curve of the $LiCoO_2$ synthesised by the new technique exhibit very large discharge capacity – 178 mAh/g. This result is proud for publication, but it must be noted that the reversibility of the material is not so good.

Our main purpose for the instance was to develop and to optimise the new way for synthesis of electrode materials for lithium batteries with controlled particle size. We have

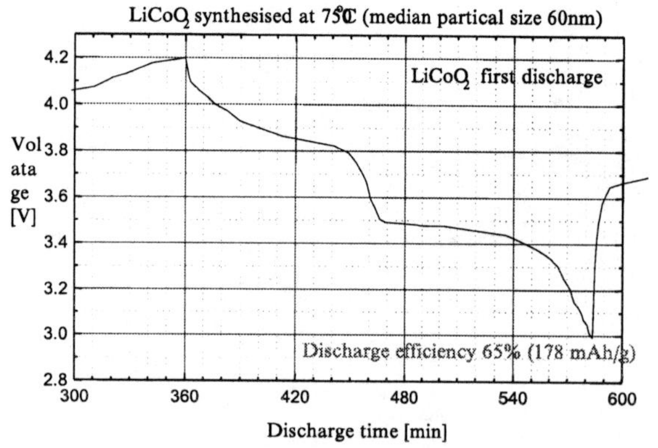

Figure 4 First discharge curve of $LiCoO_2$ prepared with median grain size of about 60 nm prepared at 750°C for 4 hours

demonstrated that the method gives adequate results and we can prepare the desired active electrode material with the calculated particle size distribution. The work is on progress.

4. Conclusions

A new method of synthesis was successfully developed allowing the synthesis of wide materials range with controlled grain size.

Applying the new method of synthesis lithium cobaltate was synthesised successfully by drying the starting precursor in a liquid agent.

The new way of synthesis allows us to obtain a prospective active cathode material with controlled particle size starting from few micro meters to tenth nano meters.

Due to the small (60–100 nm) particle size of the synthesised material, the high homogeneity of the starting compound (mixture of lithium and cobalt salts at ionic level) and the small reacting volume of the particles in contrast of the used generally solid state reactions a layered structure was successfully formed even at 180°C.

The electrochemical test carries out a high initial discharge capacity of 187 mAh/g, but the cyclability of the that obtained cathode material must be improved.

References

1. J.M. Tarascon and D. Guyomard, Eelectrochemica Acta 38(9) (1993) 1221
2. W. Ebner, D. Fouchard and L. Xie, Solid State Ionics 69 (1994) 238
3. T. Ohzuku, Lithium Batteries: in: New Materials, Developments and Perspectives (Ind. Chem. Library, 1993) p.224-245
4. M. Broussely, F. Perton, Ph. Biensan, J.M. Bodet, J. Labat, A. Lecerf, C. Delmas, A Rougier and P.P. Peres, Journal of Power Sources, 54 (1995) 1862
5. B. Banov, J. Bourilkov and M. Mladenov, Journal of Power Sources, 54 (1995) 268
6. Zhetcheva and R. Stoyanova, Solid State Ionics 66 (1993) 143
7. Gummov and M.M. Tackeray, Solid State Ionics 53-56 (1992) 681
8. N. Ilchev and B. Banov, J. Power Sources, 35 (1991) 175
9. N. Ilchev, B. Banov, Progress in Batt. & Batt. Materials, JEC Press, Vol.10 (1991) 232
10. N. Ilchev, B. Banov, Y. Bourilkov, 6IMLB Ext. Abstr. 329-333
11. N. Ilchev, B. Banov, R. Kvachkov, 6IMLB Ext. Abstr. 416-420
12. B. Banov, A. Momchilov, A. Trofonova, B. Puresheva, Journal of Power Sources 81-82 (1999) 562-565

Nanoscience & Nanotechnology
eds. E. Balabanova, I. Dragieva, Heron Press, Sofia, 2001

ELECTROCHEMICHAL AND XRD STUDY OF COPPER NANOPARTICLES SYNTHESIZED BY BOROHYDRIDE REDUCTION

S. Bliznakov, M. Mitov[1], E. Lefterova, A. Popov, G. Ivanova

Central Laboratory of Electrochemical Power Sources, Bulgarian Academy of Sciences, 1113 Sofia, Bulgaria
[1]Department of Chemistry, N. Rilski South – West Univ., Blagoevgrad, Bulgaria

Abstract. Copper nanoparticles are synthesized by borohydride reduction method and tested by means of X-ray diffraction (XRD) and cyclic voltametry in alkaline solutions. An electrochemical, assigned to hydrogen absorption/desorption reaction, takes place on the prepared from Cu nanoparticles electrodes in the potential range between -1.15 and -0.55 V (vs.Hg/HgO). Three crystalline phases (Cu, Cu_2O and CuO) are found to exist in copper nanoparticles produced from aqua solutions. After charge/discharge cycles both oxides are reduced to elemental Cu, which is accompanied with the increasing of the crystalline size of the same phase.

Keywords: Copper nanoparticles, borohydride reduction, XRD, electrochemical hydrogen absorption

1. Introduction

In previous articles a reaction, associated with an electrochemical hydrogen absorption/desorption, was observed to take place on electrodes, prepared from $Co_xB_yH_z$ nanoparticles, produced by borohydride reduction in the potential range between -1.2 and -0.55 V (vs.Hg/HgO) [1–3]. In the range of those potentials a reaction with the participation of Co also takes place:

$$Co(OH)_2 + 2e^- \leftrightarrow Co + 2OH^-$$
$$E_0 = -0.828\,\text{V (vs. Hg/HgO)}$$

The electrochemical oxy-reduction reactions of cations of noble metals as Cu, Ag, Au, Pt, Pd etc., take place at more positive potentials than the potential range studied [1–3]. For this reason, the present study aims a synthesis of cooper containing nanoparticles by borohydride reduction and tests of electrodes prepared from them by using electrochemical methods and XRD.

2. Experimental

The synthesis was carried out into a reactor with hydrodynamic regime of full mixing at mechanical stirring [4–6]. Three kinds of samples of copper nanoparticles were prepared using reduction of $CuSO_4$ aqua solution with sodium borohydride solution.The conditions of syntheses are present in Table 1. The powder obtained was washed and dried under vacuum for 24 h.

A pre-determined amount of Cu nanoparticles is mixed with Teflonized carbon Vulcan XC-72 (ratio 5:3). The mixture is pressed onto a nickel mesh in cylindrical matrix with diameter 15 mm, used as a current collector, at 100 MPa [1].

Table 1. Conditions of the syntheses

Samples #	$CuSO_4.5H_2O$ M	$NaBH_4$ (in water solution of 0.6 M NaOH) M	pH of the filtrate
1	0.017	0.41	7.9
2*	0.017	0.41	7.7
3	0.017	0.83	9.4

* – with hydrophobic protection (substance – xylene)

The electrochemical studies are carried out in a conventional three-electrode glass cell. The as prepared electrodes are connected as a working electrode and a large area nickel foil is used as a counter electrode. All potentials are measured against Hg/HgO reference electrode. As a working electrolyte is used 6M KOH solution [1–3]. Electrodes are tested by means of cyclic voltametry using a Potentiostat-galvanostat PJT-30-2 (Radiometer-Taccusel) with electrochemical special interface IMT-101.

The samples are analyzed by means of X-ray diffraction (XRD) before and after electrochemical studies using Philips-APD 15 diffractometer with Cu K_α radiation.

3. Results and Discussion

Three crystalline phases – Cu, Cu_2O and CuO, in different proportion are registered in samples of as-prepared nanoparticles by means of XRD (Figure 1). In Figure 1, for sample 1 from Table 1 the three crystal phases are observed, but the CuO phase prevailing, which value of 2θ (according to ASTM-data) are presented by the strip diagram. The use of the hydrophobic protection during the synthesis of the nanoparticles (Figure 1 – sample 2*) leads to increase of the Cu_2O content (shown for these sample as a strip diagram) in respect to the other two phases – CuO and Cu. The sample 3 contains the largest amount of elemental copper according to the X-ray spectrum, represented on Figure 1. A presence of both oxides (Cu_2O and CuO) is also observed into this probe (# 3 sample from Table 1) and the wider peaks indicate smaller crystalline size and strong compositional heterogeneity of the probe. The same sample is obtained by using higher concentration of $NaBH_4$ solution. It is well known that the higher the rate of the reduction pro-

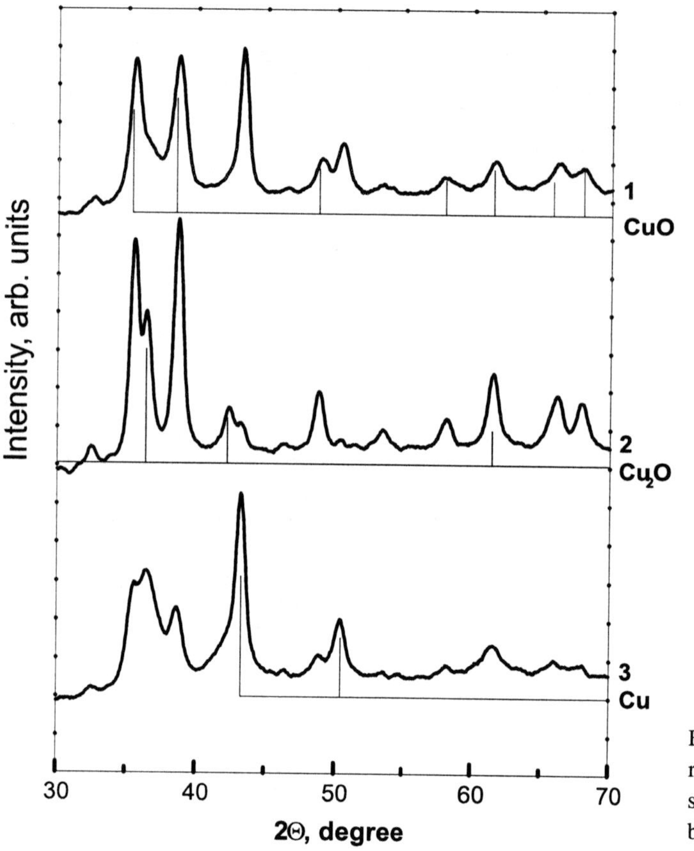

Figure 1. XRD-spectra of samples 1, sample 2 and sample 3 from Table 1

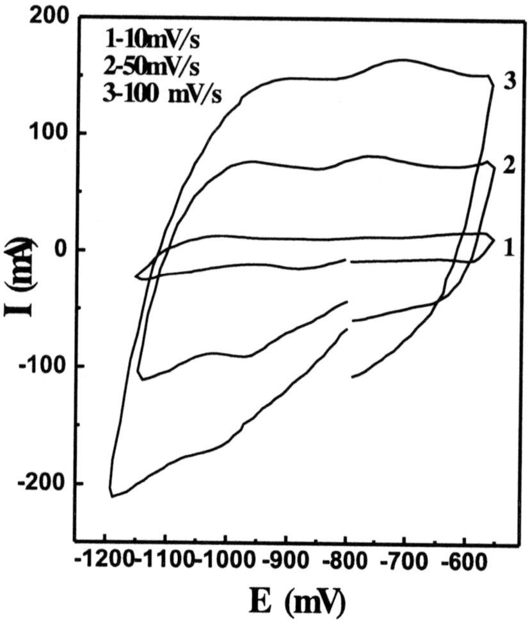

Figure 2. Cyclic voltamograms from sample 3 with different sweep rates: 10, 50 and 100 mV/s

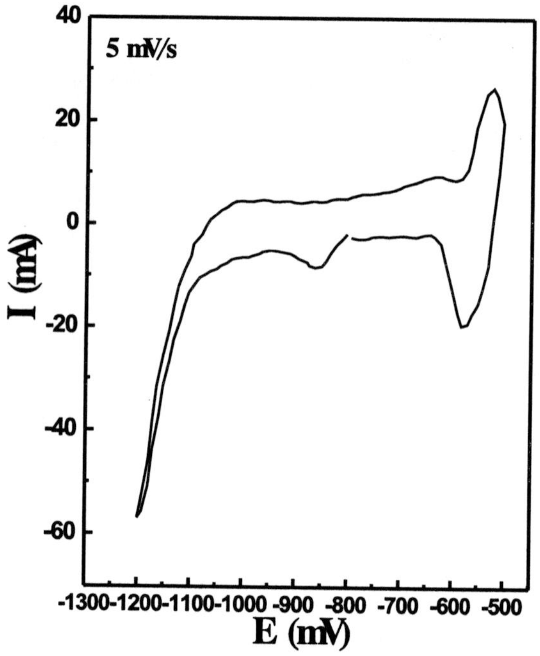

Figure 3. Cyclic voltamograms of the sample 3 – the potential range is enlarged in more positive direction. same time, the intensity of the second anodic peak increases, while that of the first one becomes smaller

cess, the smaller are crystalline size of Cu nanoparticles produced in our case.

Two anodic and one cathodic peaks are observed, when the potential is swept between -1.15 and - 0.55 V (Hg/HgO) in Figure 2. The cathodic and the anodic peaks, possessing more positive potential value, have close potentials and the same magnitude with those, obtained with $Co_xB_yH_z$ [3]. At small sweep rates, both anodic peaks have almost equal magnitudes. First of them (at more negative potentials) may be attributed to oxidation of adsorbed hydrogen atoms, while the second one can be related to electrooxidation of stronger bonded hydrogen. The cathodic peak can be associated with the reaction of electrochemical reduction of water to adsorbed hydrogen atoms. The potential difference between cathodic and anodic peaks grows up when the potential sweep rate increases which indicates that the electrochemical reaction is quasi-reversible [3].

When the potential range is enlarged into the positive direction, a large second cathodic peak at about -0.53 V (Hg/HgO) appears (Figure 3). It indicates another reaction, which even by the visual change of electrode color from black to reddish can be attributed to reduction of copper oxides to elemental copper.

XRD-spectra of electrodes as probes, prepared from copper nanoparticles, produced at the conditions and called as a sample 3 from Table 1, are shown in Figure 4.

In Figure 4 (a) the XRD spectra of the electrode as an initial probe for electrochemical test procedure is shown. After the first charge in Figure 4(b) an increase of the Cu and Cu_2O crystalline phases content and disappearance of the CuO phase is observed. At the same time, the peaks become narrower, which corresponds to increasing of the crystalline size. For Cu phase this increase is from 13 nm (before charge) to 28 nm (after charge cycling). During the next discharge (Figure 4c) and charge cycling (Figure 4d) procedures the Cu_2O phase gradually disappears and transforms to the phase of elemental copper (strip diagrams of Figure 4c). That is especially obvious from Figure 4d in which the diffraction spectra of the oxide phases are lacking.

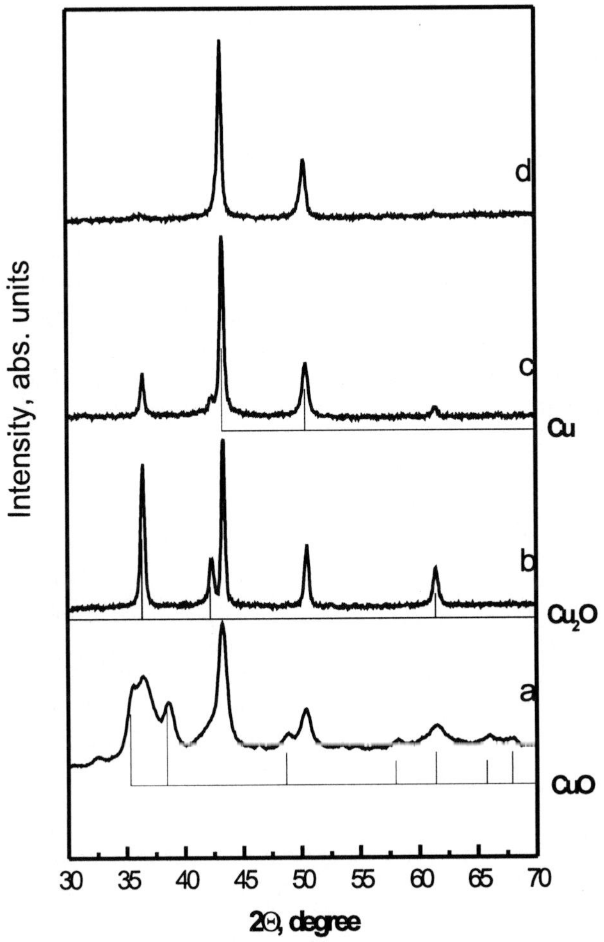

Figure 4. XRD-spectra of the electrode prepared from sample 3: a) before cycling procedure; b) after first charge (10 mA, 14 h, KOH); c) after discharge (3 mA, KOH); d) after second charge (10 mA, 14 h, KOH)

4. Conclusion

Three crystalline phases (of Cu, Cu_2O and CuO) are found to exist in copper-boron-hydrogen nanoparticles, synthesized by borohydride reduction in aqua solutions.

After the charge/discharge processes in an alkaline medium (6M KOH) the transformation of the copper oxide phases into a crystalline phase of elementary copper is determined by X-ray method, joined by an increase in the size of the crystallites in the nanosize range – from 13 to 28 nm diameter.

In the potential range between -1.15 and -0.55 V (vs.Hg/HgO) occurs electrochemical hydrogen absorption/desorption reaction into electrode probes, prepared from thus obtained Cu nanoparticles. This is the same electrochemical reaction, which occurs on the electrodes prepared from $Co_xB_yH_z$ nanoparticles [1–3]. In the same potential range is impossible electrochemical reaction with copper participation because the reduction of copper oxides to elemental copper takes place at more positive potentials. Hence, in these potentials on the electrodes prepared from nanoparticles produced by borohydride reduction, takes place only electrochemical hydrogen absorption/desorption reaction.

The investigation on the reasons for an occurring of the observed reduction of the copper oxides to an elementary copper during the charge/discharge cycles and whether it is due to the boron content of the initial nanoparticles will be the focus of our another works.

References

1. M. Mitov, A. Popov and I. Dragieva, "Nanoparticles produced by borohydride reduction as precursors of metal hydride electrodes", J. Appl. Electrochem, 29 (1999) 59.
2. M. Mitov, A. Popov and I. Dragieva, "Possibilities for battery application of colloid CoxByHz particles ", Colloids and Surfaces A: Physicochem.and Engineer.Aspects, 149 (1999) 413.
3. M. Mitov, Ph.D. Thesis "Electrochemical behavior of nanoparticles, obtained by borohydride reduction", South-West University, Blagoevgrad, 2000.
4. I. Dragieva, M. Slavcheva and D. Buchkov, "Structural xharacteristics of fine boron-containing metal particles" J. Less-Common Metals, 117(1986)311.
5. G. Glavee, K. Klabunde, C. Sorensen and G. Hadjipanayis, Langmuir, 10(1994) 4726.
6. I. Dragieva, G. Ivanova, S. Bliznakov, E. Lefterova, Z. Stoynov, I. Markova and M. Pankova, Phys. Chem. Glasses, 41 (2000) 264.

Nanoscience & Nanotechnology
eds. E. Balabanova, I. Dragieva, Heron Press, Sofia, 2001

ENZYME/GAS-DIFFUSION ELECTRODES

R. Kazareva-Boukoureshtlieva, S. Christov, I. Iliev, A. Kaisheva, N. Tzonkov

Central Laboratory of Electrochemical Power Sources, Bulgarian Academy of Sciences, 1113 Sofia, Bulgaria

Abstract. An amperometric enzyme/gas-diffusion electrode, which combines a porous air-breathing gas-diffusion electrode with an immobilized enzyme has been developed. The behavior of such electrodes is investigated when oxygen needed for the enzymatic reaction is supplied in gas phase through the porous structure of the gas-diffusion electrode. It is shown that the fast oxygen supply in gas-phase results in an increase of the amperometric signal and an enlargement of the linear part of the calibration curve.

Keywords: enzyme/gas-diffusion electrode; glucose; phenol

1. Introduction

The development of electrochemical biosensors which combine the specificity and selectivity of the enzyme catalyst whit appropriate electrochemical techniques is of great interest in health care, environmental control, agricultural, food and other industries [1,2].

Enzyme/gas-diffusion electrodes are investigated for the determination of glucose, lactate and phenol.

The reaction in the enzyme electrode can be expressed as follows:

Enzymatic reaction:

$$\text{substrate} + O_2 \xrightarrow{\text{enzyme}} \text{product} + H_2O_2;$$

Electrochemical reaction:

$$H_2O_2 \xrightarrow{\text{catalyst}} O_2 + 2H^+ + 2e^-.$$

Amperometric enzyme electrodes for determination of glucose and lactate are based on the enzymatic reaction of the substrate catalyzed by enzyme-oxidase. H_2O_2 is produced by the enzymatic reaction. The electric signal from the enzyme electrode can be obtained by the electrochemical oxidation of H_2O_2 on a catalytic electrode.

There is a difference in both reactions for the tyrosinase enzyme electrodes.

The tyrosinase enzyme (EC 1.14.18.1) catalyzes the oxidation of phenol to catechol and the subsequent dehydrogenation to *o*-quinone. The electrochemical reduction of *o*-quinone is used as a detection reaction by which an electrical signal proportional to the phenol concentration is obtained.

Enzymatic reaction:

$$\text{Phenol} + O_2 \xrightarrow{\text{tyrosinase}} \text{Catechol};$$

Electrochemical reaction:

$$o - \text{Quinone} + H^+ + 2e^- \longrightarrow \text{Catechol}.$$

It should be mentioned that the catechol produced in the electrochemical reaction is also taking place in the enzymatic reaction, so that an "enhancing effect" is observed [3].

In our investigations we use enzymes oxidases-glucose oxidase (GOD), lactate oxidase (LOD) and polyphenol oxidase (PPO).

It is typical for that class of enzymes that oxygen is taking part in the enzymatic reaction. Oxygen concentration should influence the reaction rate and the performance of the electrode. Usually it is considered that the concentration of oxygen, supplied to the reaction zone from dissolved O_2 in the electrolyte is enough to ensure sufficient rate of the enzymatic reaction. Our investigations have shown that the increase of oxygen concentration in the reaction zone through the gas phase O_2 supply results in a significant increase of the signal of the enzyme electrodes.

2. Experimental

The enzyme/gas-diffusion electrode combines an air gas-diffusion electrode with an enzyme immobilized into its catalytic surface. The electrode comprises 2 layers: a porous, highly hydrophobic, electrically conductive gas-supplying layer, made from a special PTFE-containing carbon material and a thin catalytic layer from CoTMPP pyrolyzed at 700°C in Ar [4]. This catalyst is very active towards the electrochemical oxidation of H_2O_2. Then enzyme is immobilized into the catalytic layer from a solution containing Nafion 117 (Figure 1).

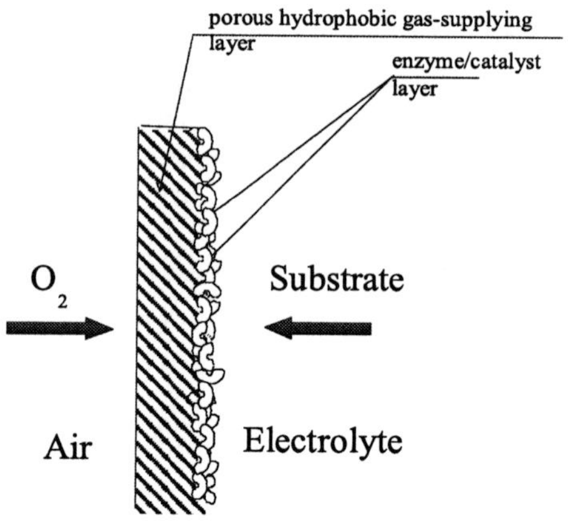

Figure 1. Schematic cross-section of an enzyme gas-diffusion electrode

The enzyme/gas-diffusion electrode is placed in a conic plastic tube in such a way that the enzyme/catalytic layer is on the top of the tube and is in contact with the electrolyte. The opposite side of the plastic tube is in contact with the open air. During operation, the dissolved substrate is transported to the immobilized enzyme from the electrolyte, but the oxygen needed for the enzymatic reaction is transported to the immobilized enzyme in gas phase from the atmospheric air through the free from electrolyte hydrophobic pores. In such a way an improved supply of oxygen to the zone of the enzymatic reaction is ensured.

In some cases, Ar or pure O_2 can be supplied through the plastic tube to the gas-supplied layer of the electrode. The design makes it possible to study the behaviour of the enzyme electrode for different ways of O_2 supply to the zone where the enzymatic reaction is taking place.

3. Results and Discussion

In Figure 2a is presented the steady-state current of the investigated GOD/gas-diffusion electrode as a function of time for consecutive additions of portions of glucose at a constant potential of +0.350 V vs. Ag/AgCl.

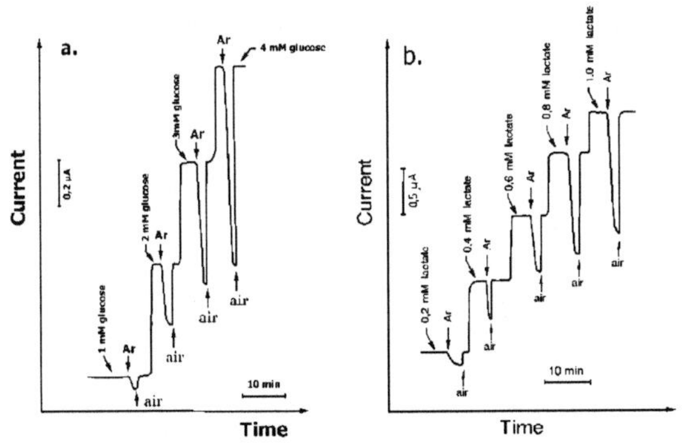

Figure 2. Amperometric response of an enzyme/gas-diffusion electrode to additions of the substrate at a constant potential E. After reaching the steady-state current, argon and air ar blown to theelectrode consecutively: (a) GOD/gas-diffusion electrode, $E = +350\,\text{mV}$ vs.Ag/AgCl; (b) LOD/gas-diffusioelectrode, $E = +400\,\text{mV}$ vs.Ag/AgCl

In the beginning of the experiment, air is available to the electrode both from the gas-side and in the electrolyte. After the addition to the cell of a certain amount of glucose the current increases and reaches its steady-state value. If now Ar is blown to the gas-side of the electrode the current decreases rapidly to a new lower value. The decrease in the value of the steady-state current in this case is due to a decreased O_2 concentration in the zone of the enzymatic reaction. After restoring the air supply to the gas-side of the electrode the current increases to its initial steady-state value. In such a way the steady-state current of the GOD/gas-diffusion electrode is obtained for several glucose concentrations both when air or Ar is blown to the gas-side of the electrode. If pure O_2 is blown to the gas-side of the electrode the steady-state current corresponding to a given glucose concentration is higher than that obtained when air is blown to the gas-side of the electrode. The experiments show that in the presence of the substrate when air or pure O_2 is available from the gas-side of the electrode no change in the steady-state current is observed when the dissolved O_2 is purged from the electrolyte by argon.

A similar current-time curve is presented in Figure 2b with a LOD/gas-diffusion electrode by consecutive additions of portions of Li-lactate to the electrolyte in the cell both when air or Ar is blown to the gas-side of the electrode.

In Figure 3a are presented calibration curves of the GOD/gas-diffusion electrode for the case when air is supplied from the gas-side of the electrode, when pure O_2 is blown to the gas-side and when Ar is blown to the gas-side of the electrode.

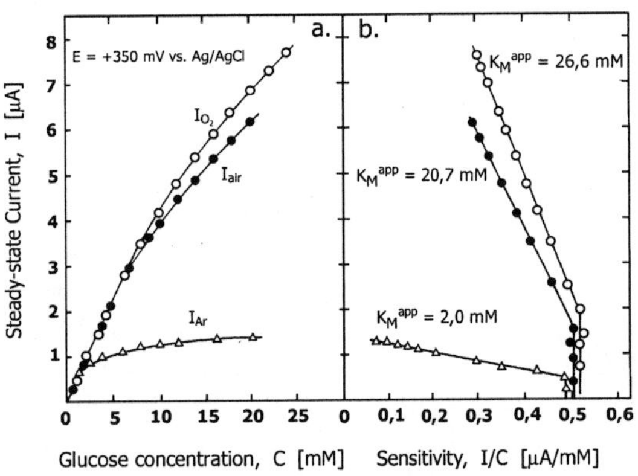

Figure 3. (a) Steady-state anodic current of a GOD/gas-diffusion electrode as a function of the glucose concentration when oxygen (I_{O_2}), air (I_{ar}) is blown to the gas-side of the electrode. (b) Corresponding electrochemical Eadie-Hofstee plot

It can be seen that the values of the steady-state current are higher when pure O_2 is blown to the gas- side of the electrode in comparison with the case when air is available from the gas-side. The steady-state current is up to 5 times lower when Ar is blown to the gas-side of the electrode. The values of the steady-state current as a function of the ratio of the steady-state current and the glucose concentration at which it is obtained (Eadie-Hofstee plot) are presented in the next part of Figure 3b. From the slope of the curves the apparent Michaelis constant K_m^{app} of the process is estimated. It can be seen that the obtained values of K_m^{app} are strongly influenced by the type of O_2 supply to the zone of the enzymatic reaction. When air is available from the gas-side K_m^{app} is 20.7 mM, when pure O_2 is blown to the gas-side K_m^{app} is higher (26.7 mM), and when Ar is blown the value of K_m^{app} is about 10 times lower (2.0 mM). It can be concluded that the enzymatic reaction is strongly influenced by the concentration of the O_2 in the reaction zone. From the presented calibration curves it is clear that the linear part is larger when air (or pure O_2) is available to the gas-side of the electrode.

Figure 4 presents a typical amperometric response of phenol enzyme/gas-diffusion electrode to consecutive additions of phenol solution to the electrolyte at a constant potential, for the two O_2 supply modes. It is seen that the amperometric signal of the enzyme electrode at a certain

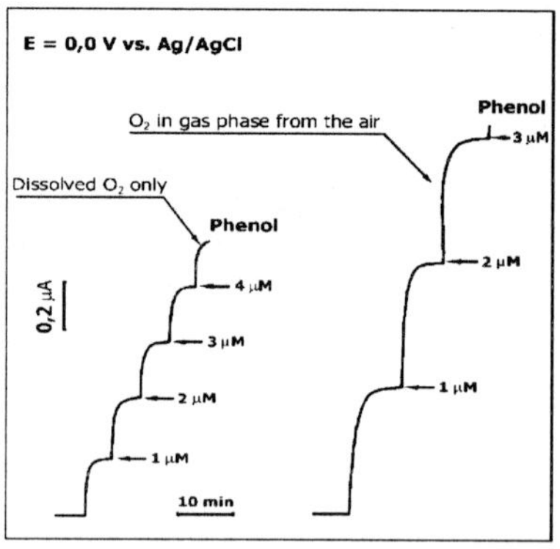

Figure 4. Amperometric response of an enzyme/gas-diffusion electrode to consecutive addition of phenol solution to the electrolyte for two modes of oxygen supply

phenol concentration is higher when O_2 is supplied in gas phase than that obtained when O_2 is supplied from the dissolved one only.

Figure 5 presents the calibration curves of phenol electrode for the different O_2 supply modes. It can be seen that the amperometric signal of the electrode strongly depends on the O_2 concentration. The enzyme electrode operates at conditions of O_2 deficiency when only the dissolved O_2 is taking part in the enzymatic reaction. The rapid O_2 supply in gas-phase from the air practically eliminates this O_2 deficiency.

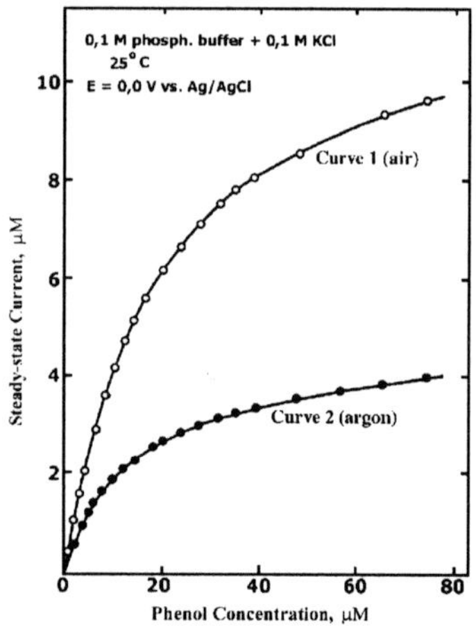

Figure 5. Steady-state current versus phenol concentration: curve 1, O_2 supplied both in gas phase from the air and from the dissolved oxygen; curve 2, O_2 supplied from the dissolved oxygen only

It should be noted that the effective diffusion coefficient of O_2 in gas-phase through the hydrophobic pores of the carbon electrode is several orders of magnitude higher [5] than that of the dissolved O_2. The sufficient O_2 supply results both in increase of the amperometric signal and in enlargement of the linear part of its calibration curve.

The influence of the O_2 concentration on the performance of the alcohol enzyme/gas-diffusion electrode will be investigated in our further studies.

4. Conclusions

Biosensors for determination of glucose, lactate and phenol concentration in water solution are designed, based on a enzyme/gas-diffusion electrode with different enzymes. The influence of the O_2 concentration on the performance of the enzyme/gas-diffusion electrode was investigated. Using this experience, enzyme/gas-diffusion electrodes with appropriate enzymes can probably be a base of designing biosensors for detection of various substrates in water solution.

References

1. Biosensors: Fundamentals and Applications, Turner A.P.F., I. Karube, G.S. Wilson (eds.), Oxford University Press, Oxford, 1987.
2. Biosensors: A Practical Approach, edited by A.E. Cass, Oxford University Press, 1990
3. M. Dennison, J. Hall, A. Turner, Anal. Chem., 67 (1995) 67
4. A. Kaisheva, I. Iliev, S. Gamburtzev, J. Power Sources 13 (1984) 181
5. I. Iliev, Bull. Soc. Chim. Beograd, 48 (Suppl.) (1983) S317–S338

Nanoscience & Nanotechnology
eds. E. Balabanova, I. Dragieva, Heron Press, Sofia, 2001

MATHEMATICAL MODELING OF ENZYME GAS-DIFFUSION ELECTRODES

Ilia Iliev, Nikola Tzonkov

Central Laboratory of Electrochemical Power Sources, Bulgarian Academy of Sciences, 1113 Sofia, Bulgaria

Abstract. A mathematical model of an enzyme-gas diffusion electrode for detection of phenol in solution is described. The system of differential equations is solved numerically. Spatial distributions of the concentrations of the species considered, as well as the electric current generated by the electrode are obtained for discrete moments of time until steady state is reached. A calibration curve is built, which shows good correlation with experimental data. It is shown that the product of the indicator reaction, reinvolved in the enzyme reaction, results in amplification of the signal.

Keywords: biosensor, enzyme/gas-diffusion electrode, tyrosinase, mathematical modeling

1. Introduction

Since the development of the first enzyme electrode by Clark (1962) much has been done to improve the biosensor reliability and to widen the range of measured concentrations. A vast variety of electrodes have been developed for monitoring of different chemical compounds. In recent years much attention is paid to mathematical modeling of the biosensors.

Many reports on models for enzyme electrodes have been published. A detailed theoretical review of various models and the means of solution was published by Schulmeister (1990). Various publications are reviewed by E. Hall, J. Gooding, and C. Hall (1995).

In most cases enzyme electrodes are based on oxydase enzymes. The reactions with these enzymes involve oxygen as co-factor. Jobst, G., Moser, I., Urban, G. (1996) published a model of a bi-enzymed electrode for detection of glucose in solution. Their results show that the response of the electrode highly depends of the concentration of oxygen in the solution.

Attempts have been made to increase the oxygen concentration in the solution, such as bubbling oxygen. A principally different approach is the application of enzyme-gas diffusion electrodes (Iliev et al., 1994, Kaisheva et al., 1995). An enzyme-gas diffusion electrode is in contact with both the bulk solution and the atmosphere and the oxygen is imported directly from the air to the place of the enzymatic reaction.

2. The Biosensor

The biosensor for determination of phenol concentration in water solution is an electrochemical cell. The working electrode (Kaisheva et al., 1996) is an enzyme gas-diffusion electrode with enzyme tyrosinase. The electrolyte is 0.1 M phosphate buffer + 0.1 M KCl. The working electrode consists of three subsequent layers: a porous carbon material, covered by an enzyme layer and a diffusion membrane. diffusion membrane is in contact with the solution and the carbon material is in contact with the atmosphere. During the operation of the electrode, the substrate diffuses through the membrane and through the enzyme layer, where enzymatic reaction occurs along the diffusion path:

$$\text{Phenol} + O_2 \xrightarrow{\text{tyrosinase}} \text{Catechol} \qquad (1)$$

$$\text{Catechol} + O_2 \xrightarrow{\text{tyrosinase}} o\text{-Quinone} \qquad (2)$$

The change of the local concentration of the substances in the enzyme layer is due both to diffusion and to local consumption/generation, resulting from the enzymatic reaction. Part of the *o*-quinone produced diffuses towards the surface of the porous carbon material, where an electrochemical reaction occurs:

$$o\text{-Quinone} + H^+ + 2e^- \longrightarrow \text{Catechol}. \qquad (3)$$

In this reaction *o*-quinone turns to catechol and two electrons are transferred. The electrochemically produced catechol diffuses backwards through the enzyme layer and is involved again in the enzymatic reaction (2). The amperometric signal is produced by the indicator reaction (3).

3. The Model

The model electrode (Figure 1) consists of a diffusion membrane (in contact with the solution), carbon electrode, and an enzyme layer between them. It is accepted that the membrane and the enzyme layer are homogeneous and have flat parallel surfaces. A one-dimensional coordinate

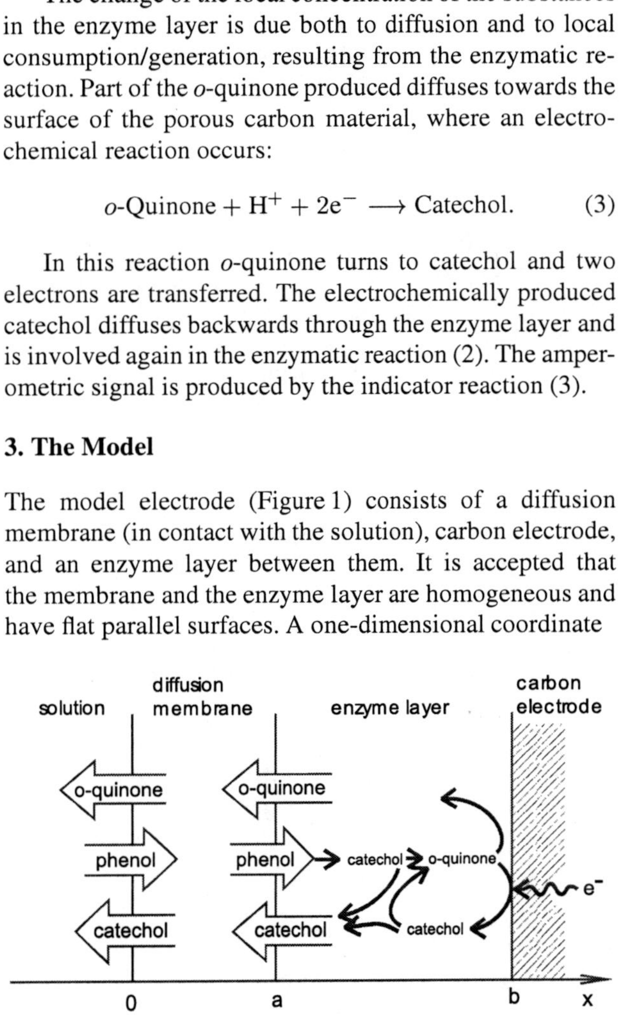

Figure 1. Model of enzyme/gas-diffusion electrode for determination of phenol solution

system with x-axis perpendicular to the surfaces of the layers is introduced. The beginning ($x = 0$) is at the boundary between the solution and the membrane. The boundary between the membrane and the enzyme layer is at $x = a$, and the carbon electrode surface is at $x = b$. Three processes are taken into account: transport of substances; enzym reaction (1) – (2) and electrochemical reaction (3).

Assumptions

1. The transport mechanism of the substances is diffusion, described by the Fick's law.

2. The effective diffusion coefficients of phenol, catechol and *o*-quinone (correspondingly D_p^M, D_c^M, and D_q^M in the membrane and D_p^E, D_c^E, and D_q^E in the enzyme layer) are constant within a given layer.

3. The kinetics of the enzymatic reaction is described by the Michaelis-Menten equation – the local consumption of phenol (R_p) and catechol (R_c) depend only on the local concentration of phenol (C_p) and catechol (C_c) respectively and are given by:

$$R_p = V_{\max}^p/(1 + K_p/C_p) \qquad (4)$$

$$R_c = V_{\max}^c/(1 + K_c/C_c) \qquad (5)$$

4. The carbon electrode potential is sufficiently high, so that the electrochemical reaction (3) is diffusion limited – the concentration of *o*-quinone is practically zero at the electrode surface;

5. The surface of the carbon electrode is impermeable for phenol, i.e. there is no flow of phenol through it.

6. At the beginning the concentrations of the species considered in the solution and in both layers are zero.

7. The concentration of phenol in the bulk solution increases stepwise from zero to a constant value C_p^0 at the moment $t = 0$.

8. The stirring of the bulk solution is strong enough so that for $t \geq 0$ the concentration of phenol is constant and equal to C_p^0 throughout the bulk solution. The concentration of catechol and *o*-quinone in the bulk is constant and equal to zero.

9. The boundary diffusion layer between the solution and the membrane is neglected, so that the concentration of phenol, catechol and *o*-quinone at $x = 0$ are equal to those in the bulk solution (correspondingly C_p^0, 0, 0).

Equations

With the above assumptions, the transient behaviour of the model is described by the following systems of partial equations ((5) – (7) and (8) – (10)). In the diffusion membrane ($0 \leq x \leq a$) where no enzymatic reaction occurs, the diffusion of phenol, catechol, and *o*-quinone is described by homogeneous equations:

$$\partial/\partial t\, C_p(x,t) = D_p^M \partial^2/\partial x^2\, C_p(x,t) \qquad (6)$$

$$\partial/\partial t\, Cc(x,t) = D_c^M \partial^2/\partial x^2\, C_c(x,t) \qquad (7)$$

$$\partial/\partial t\, Cq(x,t) = D_q^M \partial^2/\partial x^2\, C_q(x,t) \qquad (8)$$

In the enzyme layer ($a \leq x \leq b$), the diffusion is described by the equations:

$$\partial/\partial t\, C_p(x,t) = D_p^E \partial^2/\partial x^2\, C_p(x,t) - R_p \qquad (9)$$

$$\partial/\partial t\, C_c(x,t) = D_c^E \partial^2/\partial x^2\, C_c(x,t) + (R_p - R_c) \qquad (10)$$

$$\partial/\partial t\, C_q(x,t) = D_q^E \partial^2/\partial x^2\, C_q(x,t) + R_c \qquad (11)$$

At the first stage of the enzymatic reaction (1), phenol is turned to catechol, so in equation (9) contains an expression, describing the local consumption of phenol. In equation (10), the term $(R_p - R_c)$ is included to describe the local generation of catechol. It is taken into account that catechol is generated from phenol, hence R_p (4), but also that catechol is consumed to produce *o*-quinone, hence -R_c (5). In equation (11), the term that describes the generation of *o*-quinone is introduced. In fact it is equal to the term R_c (5) that describes the consumption of catechol with sign "+".

The terms that describe the local consumption of phenol (generation of catechol) and the local consumption of catechol (generation of *o*-quinone) are:

$$R_p(x,t) = V_p^{\max} C_p(x,t)/(K_p^M + C_p(x,t)); \qquad (4')$$

$$R_c(x,t) = V_c^{\max} C_c(x,t)/(K_c^M + C_c(x,t)). \qquad (5')$$

Initial and Boundary Conditions

At the initial moment $t = 0$:

At $x = 0$ (solution/membrane boundary):

$$C_p(0,0) = C_p^0 = \text{const}; \qquad (12)$$

$$C_c(0,0) = 0; \qquad (13)$$

$$C_q(0,0) = 0. \qquad (14)$$

For $0 \leq x \leq b$:

$$C_p(x,0) = 0; \qquad (15)$$

$$C_c(x,0) = 0; \qquad (16)$$

$$C_q(x,0) = 0. \qquad (17)$$

For $t > 0$:

At $x = 0$ (solution/membrane boundary):

$$C_p(0,t) = C_p^0 = \text{const}; \qquad (18)$$

$$C_c(0,t) = 0; \qquad (19)$$

$$C_q(0,t) = 0. \qquad (20)$$

At $x = a$ (membrane/enzyme layer boundary):

$$D_p^M \{\partial/\partial x\, C_p(x,t)\}_{x=a} = D_p^E \{\partial/\partial x\, C_p(x,t)\}_{x=a} \qquad (21)$$

$$D_c^M \{\partial/\partial x\, C_c(x,t)\}_{x=a} = D_c^E \{\partial/\partial x\, C_c(x,t)\}_{x=a} \qquad (22)$$

$$D_q^M \{\partial/\partial x\, C_q(x,t)\}_{x=a} = D_q^E \{\partial/\partial x\, C_q(x,t)\}_{x=a} \qquad (23)$$

The above conditions (21–23) reflect the assumption that the amount of substance leaving a layer is equal to that entering the subsequent layer.

At $x = b$ (surface of the carbon electrode):

$$C_q(b,t) = 0; \qquad (24)$$

$$\{\partial/\partial x\, C_p(x,t)\}_{x=b} = 0; \qquad (25)$$

$$D_c^E\{\partial/\partial x\, C_c(x,t)\}_{x=b} = -D_q^E\{\partial/\partial x\, C_q(x,t)\}_{x=b}. \qquad (26)$$

Expression (24) reflects the diffusion limitation, (25) reflects the impermeability, and (26) describes the fact that the amount of *o*-quinone that reaches the carbon electrode surface is equal to the amount of the electrochemically produced catechol.

An enzyme electrode can operate in *diffusion-limited* or *reaction-limited* mode. In the first case the transport of substrate to the place of the enzymatic reaction is impeded and the enzymatic reaction can be considered as infinitely fast. Thus the formation of electrical signal is diffusion limited only. In the second case the enzymatic reaction is impeded; the concentration of the reagents is high enough and the kinetics of the reaction completely determines the speed of the process as a whole. In the present model we consider the case when both diffusion and reaction kinetics are taken into account simultaneously in the mathematical formulation.

4. Computer Simulation

Since the analytical solution of the systems under these initial and boundary conditions is not known, we have used a fully explicit scheme. As shown in Smith (1991), the correctness of this method depends of the ratio between the steps of discretization of space (h) and time (τ). The solution is correct if the following condition is fulfilled: $0 < D\tau/h^2 \leq 0.5$ (D is the effective diffusion coefficient). In our case, for all values of the effective diffusion coefficients used, this relation is $0.01 < D\tau/h^2 \leq 0.03$.

5. Results and Discussion

The numerical solution of the equations was implemented with a Delphi 3.0 program, which receives the following parameters: effective diffusion coefficients of the species involved; apparent Michaelis constants and maximal reaction rates for the two stages of the enzymatic reaction (1), (2); steps of discretization in time (τ) and distance (h); area of the working surface of the enzyme electrode, and thickness of the layers. As a result the program renders: 1. Concentration profiles, given as matrices $C(i,j)$, where i is the location and j is the moment of time; 2. Spatial distributions of the rate of phenol and catechol consumption, given as matrices $R(i,j)$; 3. Current through the electrode $I(j)$ for every moment of time j; 4. Flow of phenol at $x = 0$; 5. Flow of *o*-quinone at $x = b$.

For different values of the bulk phenol concentration the following data is obtained: concentration profiles of phenol, catechol, and *o*-quinone; profiles of the reaction rates for the first and the second stage of the enzymatic reaction; the steady-state current. On this data is based the calibration curve shown on Figure 2. The curve is linear for a

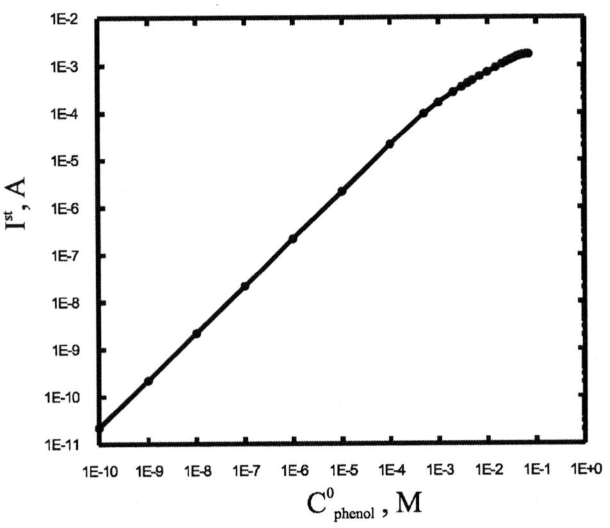

Figure 2. Calibration curve obtained from the calculated values of the stationary current

wide range of initial phenol concentrations. It is similar to the experimentally derived calibration curve of the real electrode (Kaisheva et al., 1996).

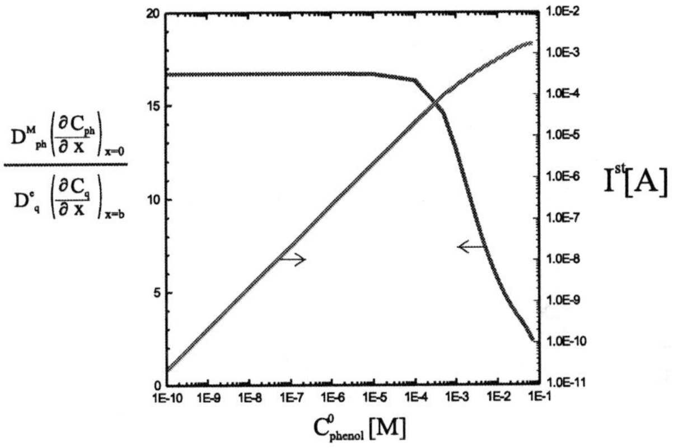

Figure 3. The dependency of the gain from the initial phenol concentration

As it was mentioned above, a typical feature of this electrode is that each one particle phenol may turn to catechol, which in turn may turn to *o*-quinone. This *o*-quinone is reduced to catechol, and may be looped again in the second stage of the enzymatic reaction. This cycle may be repeated several times, i.e. one single particle phenol may cause electron transfer several times. So we can talk about gain in the amperometric signal. This would make the electrode more sensitive to lower phenol concentrations in the bulk solution. In order to express this gain quantitatively, we accept as its measure the ratio of the phenol flow, which enters the membrane from the solution, to the flow of quinone on the carbon surface, in the steady state. Figure 3 shows the dependency of the gain from the initial phenol concentration. We may see that for low concentrations the gain is almost unchangeable; it decreases significantly for the higher values of the initial concentration. If we juxtapose this curve with the calibration curve, we may see that the gain decreases for these values of the initial concentration, for which the calibration curve is not linear.

6. Conclusion

This paper illustrates just preliminary results of the model calculation. The presented model is a subject of further investigation. The parameters of the model, such as thickness of the layers will be varied in order to optimize the sensor performance. Further, the properties of the materials also affect the response of the electrode and could be considered in future work on that matter. A vast research work in that field can be performed, aiming to developing of a serially manufacturable device, which could be applied for environmental monitoring.

References

1. Clark, L.C. Jr., Lyons, C., Ann. N.Y. Acad. Sci.1962, 102, 29
2. Hall, E. A. H., Gooding, J., Hall, C., Redox Enzyme Linked Electrochemical Sensors: Theory Meets Pratice. Microchim. Acta 121, 119-145, 1995
3. Iliev, I., Kaisheva, A., Scheller, F., Pfeiffer, D., Amperometric Gas-Diffusion/Enzyme Electrode, Electroanalysis 1995, 7, No.6
4. Jobst, G., Moser, I., Urban, G., Numerical simulation of multilayered enzyme sensors. Biosensors & Bioelectronics, 11, No.1/2, pp. 111-117, 1996
5. Kaisheva, A., Iliev, I., Kazareva, R., Christov, S., Wollenberger, U., Scheller, F., Enzyme/gas-diffusion electrodes for determination of phenol. Sensors and Actuators B 33 (1996) p.39
6. Schulmeister, T., Mathematical Modelling of the Dynamic Behaviour of Amperometric Enzyme Electrodes. Selective Electrode Rev.,1990, Vol. 12, pp. 203-260
7. Smith, G. D., Numerical Solution of Partial Differential Equations, Oxford University Press, 1969

Nanoscience & Nanotechnology
eds. E. Balabanova, I. Dragieva, Heron Press, Sofia, 2001

AMPEROMETRIC GAS BIOSENSORS

S. Christov, R. Kazareva-Boukoureshtlieva, I. Iliev, A. Kaisheva, N. Tzonkov

Central Laboratory of Electrochemical Power Sources, Bulgarian Academy of Sciences, 1113 Sofia, Bulgaria

Abstract. Electrochemical gas biosensors are developed, based on a new type of enzyme/gas-diffusion electrode [1]. A biosensor for detection of phenol vapours in the air with an enzyme tyrosinase is described. Calibration curves of the biosensor are investigated in the presence of vapours of phenol, 4-chlorphenol and p-cresol. It is shown that phenol vapour concentration in the ppb range are detectable. Biosensor for determination of alcoholic vapour in the atmospheric air with an enzyme alcohol oxidase is also described. The transient response and the calibration curve of the biosensor are investigated in the presence of methanol or ethanol in the air.

Keywords: electrochemical gas biosensor, enzyme/gas-diffusion electrode, tyrosinase, alcohol oxidase

1. Introduction

Phenolic compounds are important materials in a broad range of chemical manufacturing processes such as coal gasification, plastic industry, petrochemical production, etc. A number of volatile phenol compounds are emitted in the atmosphere. Because of the environmental and toxicological risk strict limits on phenol concentration are generally prescribed by law. In this connection simple and fast methods are desired for determination of phenolic compounds in the atmospheric air.

Recently an amperometric biosensor for measuring of phenol vapour in the gas phase was reported [2]. The biosensor detects phenol vapour concentrations in the ppm range. Its normal operation requires a strict control of the humidity of the air. A significant decrease of the biosensor's signal with the decrease of the air humidity is observed.

A new type of biosensor for determination of the concentration of vapours of phenolic compounds in the atmospheric air is designed by the use of an enzyme/gas-diffusion electrode [3]. Investigations of this phenol-vapours biosensor are described in the present paper.

The development of electrochemical biosensors for the determination of the concentration of alcohols is of prime importance in clinical laboratories in food and fermentation industries. Measurement of the alcohol blood levels are becoming prevalent in order to enforce drinking and driving laws. It is known that the alcohol concentration in the alveolar air is proportional to that in the blood. The analysis of respiratory gases with an appropriate device can allow the police to perform the measurements of the alcohol concentration directly, without the need of medically qualified personnel. The electrochemical biosensors is particularly attractive in this respect. The biosensors couple the inherent specificity of the biocatalytic reaction with the remarkable sensitivity of amperometric transduction and can be easily accomplished by light weight and low cost instrumentation.

Electrochemical gas biosensor for alcohol based on an interdigitated gold microelectrode system is reported in [4]. A gel layer containing the electrolyte in which the enzyme (alcohol oxidase or alcohol dehydrogenase) is immobilized is applied to the working area of the microelectrode system. Another type of ethanol gas biosensor is reported [5] consisting of a two compartment cell separated by a porous PTFE membrane. The biosensor shows sensitivity in the concentration range 40–1200 ppm ethanol vapours. A short life-time of the biosensor is observed due to the deactivation of the enzyme.

2. Experimental

The investigated phenol-vapour biosensor comprises an tyrosinase/gas-diffusion electrode, an Ag/AgCl electrode used as a reference and counter electrode; both electrodes placed in an electrolyte space filled with 0.1 M phosphate buffer containing 0.1 M KCl. The enzyme/gas-diffusion electrode is mounted in the biosensor in such a way that its enzyme-containing layer contacts the electrolyte. The porous hydrophobic gas layer of the electrode is exposed to the air.

A mushroom tyrosinase (EC 1.14.18.1-Sigma) is used. It catalyzes the oxidation of the phenolic compounds via hydroxylation with molecular oxygen to catechols and subsequent dehydrogenation to the corresponding quinone compounds:

$$\text{Phenol} + O_2 \xrightarrow{\text{tyrosinase}} \text{Catechol}, \qquad (1)$$

$$\text{Catechol} + O_2 \xrightarrow{\text{tyrosinase}} o\text{-Quinone}. \qquad (2)$$

The strong oxidizing power of oxygen makes the overall reaction irreversible.

The electrochemical reduction of the produced in the above enzymatic reaction *o*-quinone is used as a detection reaction by which an electrical signal proportional to the phenol concentration is obtained:

$$o\text{-Quinone} + H^+ + 2e^- \xrightarrow{\text{tyrosinase}} \text{Catechol}. \qquad (3)$$

The catechol produced in the electrochemical reaction (3) is also participating in the enzymatic reaction (2) so that an "amplifying effect" is observed [2].

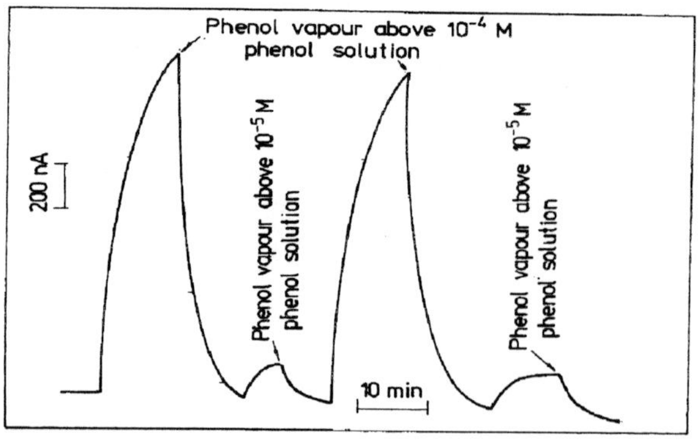

Figure 1. Transient response of a gas biosensor when consecutively placed in the air above 10^{-4} and 10^{-5} M phenol solutions

It should be noted that beside the phenol vapours also oxygen from the atmospheric air is transported in gas phase through the porous hydrophobic gas layer of the enzyme/gas-diffusion electrode to its enzyme layer. This is important since oxygen is taking part in the enzymatic reactions (1) and (2). The transport in gas phase ensures fast oxygen supply to the enzymatic reaction zone which allows a high rate of the enzymatic oxidation of phenol to be achieved.

The electrode, used in the alcohol gas biosensor, comprises a porous, highly hydrophobic electrically conductive carbon layer (made from a special PTFE-containing carbon material) and a thin catalytic layer from CoTMPP pyrolyzed at 700°C in Ar [6]. The enzyme alcohol oxidase is soaked and immobilized into the catalytic layer from a solution containing Nafion 117. The enzyme/gas-diffusion electrode is placed in a conic plastic tube in such a way that the enzyme/catalyst layer is on the top of the tube.

The biocatalytic reaction of alcohols in the presence of alcohol oxidase can be expressed as follows:

$$RCH_2OH + O_2 \xrightarrow{\text{alcochol oxidase}} RCHO + H_2O_2.$$

The amperometric signal is a result of the electrochemical oxidation of H_2O_2 catalyzed by the pyrolyzed CoTMPP-catalyst:

$$H_2O_2 \xrightarrow{\text{catalyst}} O_2 + 2H^+ + 2e^-.$$

3. Results and Discussion

Phenol vapours electrochemical biosensor. An amperometric signal is obtained when the described biosensor is placed ca. 3 cm above the level of a phenol solution in such a way that the hydrophobic gas layer of the electrode is facing the solution. In such a case the signal of the biosensor is due to the presence of the phenol vapour in the air above the solution. When the biosensor is placed above pure water no amperometric signal is observed. The amperometric signal is investigated when the biosensor is placed above phenol solution with different concentration. The working potential of the phenol vapour biosensor is 0.00 V vs. Ag/AgCl (at this potential the background current of the biosensor is practically zero). When the biosensor is placed consecutively (Figure 1) above water solution of phenol with one and the same concentration, amperometric signals with one and the same values are obtained.

The biosensor showed no response to a range of solvent vapours including chloroform and acetone.

The lack of data in the literature for the concentration of the saturated phenol vapours which are in equilibrium with phenol water solutions with low concentrations complicate the calibration of the phenol vapour biosensor. That's way in the abscissas of the curves presented in this paper are given the concentrations of the phenol solution above which the measurement is performed. In Figure 2 is presented the amperometric signal of the investigated biosensor placed above phenol solutions as a function of the phenol solution concentration. It is seen that a signal of the biosensor is obtained at very low concentration of phenol vapour (above 5×10^{-7} M phenol solution). The signal depends linearly for a wide range of phenol concentrations. The experimental points are obtained after 43 days storage of the biosensor at 4°C.

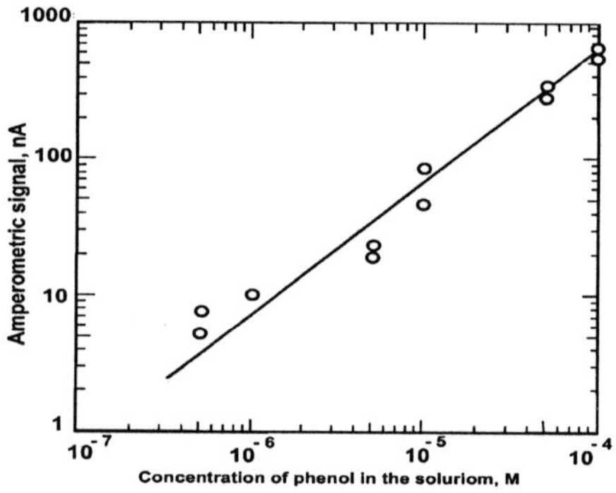

Figure 2. Calibration curve of a gas biosensor exposed to vapour above phenol solutions with different concentrations

In [7] are given data for the concentration of saturated phenol vapours in equilibrium with phenol solutions dawn to 0.00079 wt.%. It is shown that at low concentrations of the phenol water solution the ratio of the saturated phenol water solution the ratio of the saturated phenol vapour concentration and the solution concentration is ca. 2.32. In Table 1 are presented data from [7] and data from lower phenol concentrations calculated (*) under the assumption that this ratio is preserved.

Table 1

Phenol solution M	3.4×10^{-3}	7×10^{-4}	1×10^{-4}	1×10^{-5}	1×10^{-7}
Phenol vapour ppm	265	49	6.7*	0.67*	0.0067*

The calculated data obtained by extrapolation are not exact and can be used only for estimation of the order of magnitude of the phenol vapour concentration.

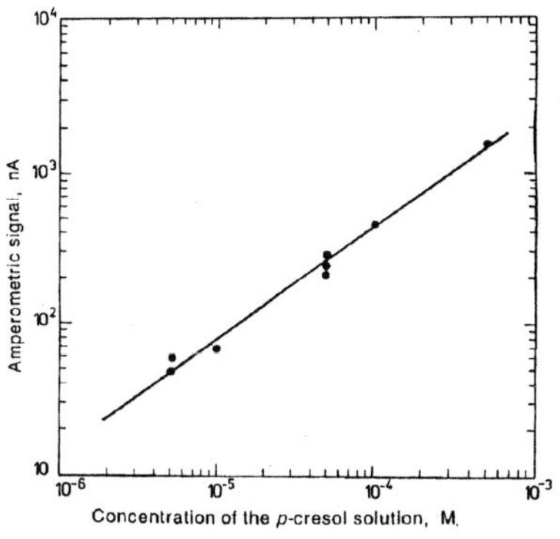

Figure 3. Calibration curve of a gas biosensor exposed to vapour above p-cresol solutions with different concentrations

The described biosensor is sensitive also to vapours of some phenolic compounds – *p*-cresol, 4-chlorphenol, etc. Calibration curve with *p*-cresol vapours is presented in Figure 3. It is seen that linear calibration curve is obtained.

Alcohol vapours electrochemical biosensor. An amperometric signal of the proposed gas biosensor for determination of alcohol vapours in the air is studied when the biosensor is placed above the level of an alcohol solution. The gas biosensor is arranged in such a way that the porous carbon membrane of the enzyme/gas-diffusion electrode is facing the solution. In such a case the obtained amperometric signal of the gas biosensor is due to the presence of the alcohol vapour above the solution.

The working potential of the alcohol vapour biosensor is 0.350 V vs. Ag/AgCl.

The transient response of the gas biosensor is investigated when the biosensor is consecutively placed above water solutions of ethanol with different concentrations, Figure 4. It is seen that amperometric signal are obtained proportional to the concentration of the ethanol. It can be concluded that the proposed gas biosensor can be used for the determination of the alcohol vapour concentration in the air.

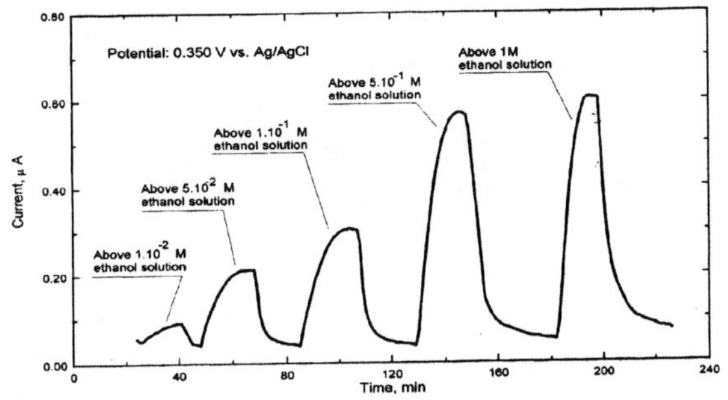

Figure 4. Transient response of a gas biosensor when consecutively placed above ethanol solutions with different concentrations

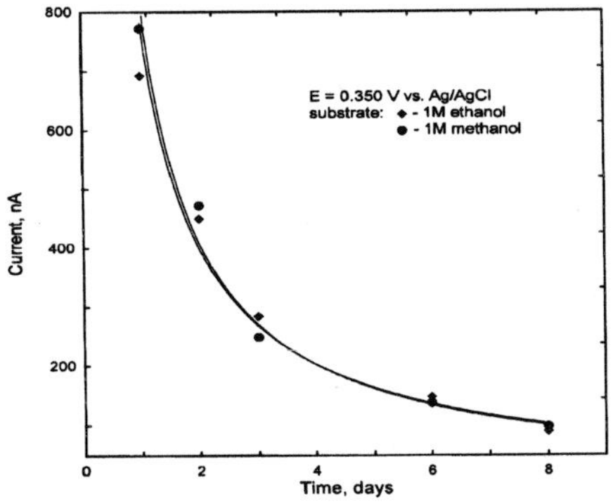

Figure 5. Amperometric signal of the alcochol gas biosensor as a function of a storage time

The amperometric signal of one the same gas biosensor obtained from both the ethanol and methanol vapours above 1 M solution of ethanol and 1 M solution of methanol is followed for more than a week. The obtained data are presented in Figure 5.

It is seem that for both cases the activity of the immobilized enzyme alcohol oxidase decreases with time. In between the measurements the biosensor was kept at 4°C.

4. Conclusions

A biosensor for the determination of phenol vapour concentration is designed, based on a tyrosinase/gas-diffusion electrode. This biosensor can be also used for the detection of other phenolic compounds, such as *p*-cresol and 4-chlorphenol.

Amperometric gas biosensor with enzyme alcohol oxidase is sensitive to vapours of methanol, ethanol and other alcohols. Amperometric gas biosensors with analogous construction with suitable enzymes can be used for the detection of various gaseous substrates.

References

1. A. Kaisheva, I. Iliev, R. Kazareva, S. Christov, Sensors and Actuators B 26-27 (1995), 425
2. M. Dennison, J. Hall, A. Turner, Anal. Chem., 67 (1995), 67
3. A. Kaisheva, I. Iliev, R. Kazareva, S. Christov, F. Scheller, U. Wollenberger, 8th Conf. EUROSENSORS IX, Stokholm, June 1995, Digest of technical papers, Vol.1, p.490-493
4. C.K.O' Sullivan, S. Lafis, M.I. Karayannis, M. Dennison, J.M. Hall, A.P.F. Turner, B. Hobbs, W.J. Aston, 4-th European Workshop on Biosensors for Enviromental Monitoring, 1995, Barcelona, Spain, Proceeding, p. 44-45
5. K. Mitsubayashi, K. Yokoyama, T. Takeuchi, I. Karube, Anal. Chem., 66 (1995), 3297
6. A. Kaisheva, I. Iliev, S. Gamburzev, J. Power Sources, 13 (1984), 181
7. N.V. Chalov, Journal of Applied Chemistry, 19 (1946), 1371 (in russian)

Nanoscience & Nanotechnology
eds. E. Balabanova, I. Dragieva, Heron Press, Sofia, 2001

ELECTROCHEMICAL CORROSION OF AMORPHOUS AND NANOCRYSTALLINE $Cu_{66}Ti_{34}$ ALLOYS

V. Zaprianova, R. Raicheff, E. Kashchieva

University of Chemical Technology and Metallurgy, 8, Kl. Ohridski blvd., Sofia 1756

Abstract. The electrochemical corrosion of $Cu_{66}Ti_{34}$ alloys obtained by rapid quenching in relation with the change of their microstructure upon heat treatment has been studied. The samples of the amorphous alloy were subjected to nanoquasicrystallization, partial nanocrystallization and full crystallization. The potential-sweep technique in conjunction with TEM, XRD amd DTA measurements have been used in order to investigate the corrosion behaviour of the alloys in model solutions (10% H_2SO_4 and 5% HNO_3). Heat treatment of the amorphous $Cu_{66}Ti_{34}$ alloy at 300°C for 1 h leads to nanoquasicrystallization. In partial nanocrystallization of the alloy (at 500°C for 0.5 h) a nanocrystalline phase of Cu_3Ti_2 is indicated in the amorphous matrix, while the full crystallization of the alloy (at 500°C for 12 h) leads to three-phase crystallization – Cu_3Ti_2, β-Cu_3Ti and Cu_2Ti. It has been found that amorphous and nanocrystalline $Cu_{66}Ti_{34}$ alloys exhibit a high and closed to pure titanium corrosion resistance in both sulfuric and nitric acid solutions. The thermal treatment leading to full crystallization decreases the corrosion resistance of the alloy.

Keywords: corrosion, amorphous, nanocrystalline, copper-titanium alloys

1. Introduction

Nanostructured materials possess unequal properties as high mechanical strenght and good soft magnetic characteristics, which cannot be obtained for conventional crystalline and amorphous alloys [1–3]. One of the methods of producing metallic materials with nanostructure is the crystallization of an amorphous phase.

As known, the electrochemical corrosion behaviour of metallic materials is sensitive to their structure. In principe, the corrosion resistance increases by transition from a multiphase to a single phase alloy. The corrosion properties of multiphase alloys are basically determined by the nature and distribution of the crystalline phases [6].

In the recent years interesting results have been obtained for metallic alloys with amorphous structure. Gutzow [7] has shown that the dissolution rates of the amorphous alloys are higher than those of their crystalline counterparts. This is attributed to the unstable (from thermodynamical point of view) state of the alloys with unordered structure and higher energy. The greater activity of the amorphous alloys with respect to dissolution is confirmed in the case of a single phase crystallization when the chemical composition of the amorphous alloy is equal to the stoichiometric composition of the formed crystalline phase (e.g. the amorphous $Ni_{67}B_{33}$ alloy [8]). The multiphase crystallization, however, leads to an increase of the dissolution rate as established for amorphous alloys of the types Fe-Ni-Cr-P-C [9], Ni-Ti [10], Ni-P [11] and Fe-Ni-P-B [12].

The effect of the structure on the corrosion behaviour of metallic alloys is an interesting problem for further investigation. Although corrosion resistance is a desirable property in nanostructured alloys, we still could find very scarce literature data concerning this aspect [13,14].

The aim of the present work is to investigate the effect of nanostructure resulting from heat treatment of $Cu_{66}Ti_{34}$ amorphous alloy on the electrochemical corrosion behaviour.

2. Experimental

The samples of the amorphous $Cu_{66}Ti_{34}$ alloy were produced in the form of ribbon (12 mm wide and 50 μm thick) by rapid quenching from a melt. In order to obtain various structures, samples of as-quenched alloy were subjected to 3 types of heat trearment in evacuated quartz tubings:

— nanoquasicrystallization at 300°C for 1 hour
— partial nanocrystallization at 500°C for 0,5 hour
— full crystallization at 500°C for 12 hours

The structure of the alloys was characterized by TEM (EM-400 Philips), XRD (DRON-UM 1 powder diffractometer) and DTA (Paulik apparatus). The XRD measurements were made using CuK_α radiation.

Potentiodynamic polarization technique was applied in the electrochemical measurements. Potentioscan (Radelkis-OH-405) with a logarithmic converter of the current, an conventional three-compartments electrochemical cell and a specially designed Teflon holder for the sample-electrode were used [15]. The polarization potential-current curves of the alloy in the model corrosive solutions (10% H_2SO_4 and 5% HNO_3) were automatically recorded (potential sweep rate 1 $mV.s^{-1}$). The potentials are given in respect to Saturated Calomel Electrode (SCE).

3. Results and Discussion

In Figure 1 are shown the X-ray diffraction patterns and electronmicroscopy microphotographs of $Cu_{66}Ti_{34}$ alloy.

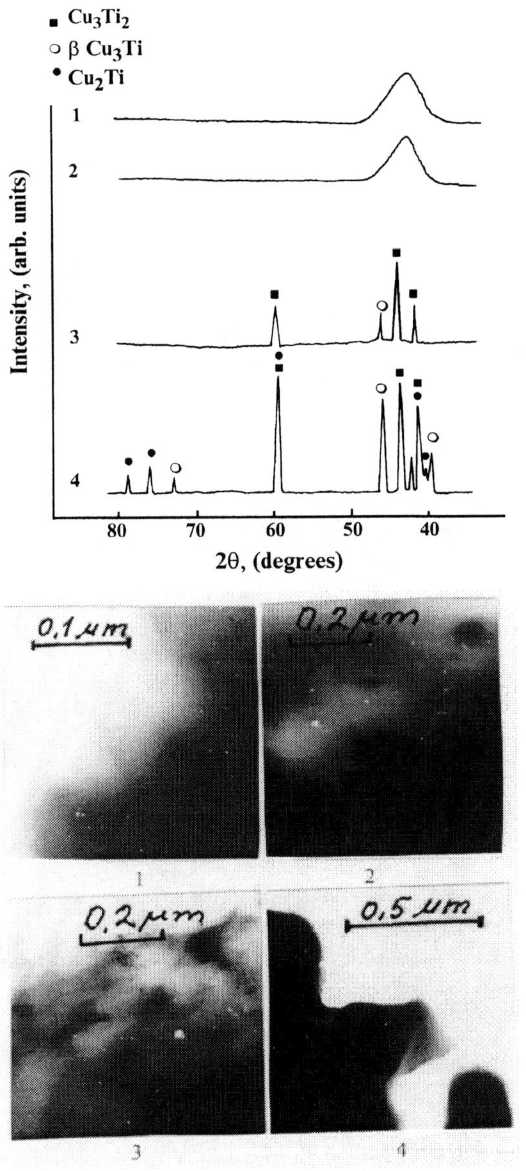

Figure 1. X-ray diffraction patterns and electronmicroscopy microphotographs of $Cu_{66}Ti_{34}$ alloy: 1 – as-quenched, 2 – heated at 300°C/1 h, 3 – heat treated at 500°C/0.5 h, 4 – heated at 500°C/12 h

The as-quenched alloy shows a homogeneous amorphous structure. After heat treatment at a temperature lower than the temperature of crystallization (300°C/1 h) nanoscale particles (the average particle size is in the range of 50 nm) precipitate from the amorphous matrix and a nanoquasicrystalline structure is observed. Heat treatment at a temperature higher than the temperature of crystallization (500°C/0.5 h) leads to the formation of crystalline regions with a size up to 100 nm embedded in an amorphous matrix. The nanocrystalline phase was identified as Cu_3Ti_2. This methastable phase changes into an equilibrium structure of Cu_3Ti_2, β-Cu_3Ti and Cu_2Ti with a size of the crystalls above 100 nm after prolonged heating at the same temperature (500°C/12 h).

The amorphous, nanoquasicrystalline and partial nanocrystalline alloys have a high corrosion resistance. However, the full crystallization decreases considerably the corrosion resistance of the alloy (Figure 2 and Figure 3).

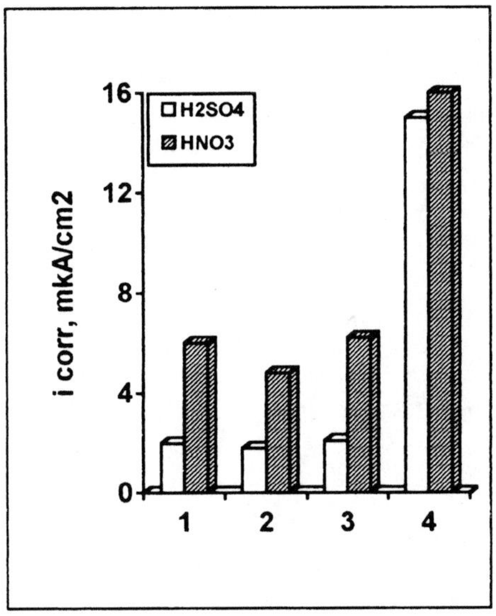

Figure 2. Corrosion current (i_{corr}) for amorphous (1), nanoquasicrystalline (2), partial nanocrystalline (3) and full crystalline (4) $Cu_{66}Ti_{34}$ alloy in 10% H_2SO_4 and 5% HNO_3

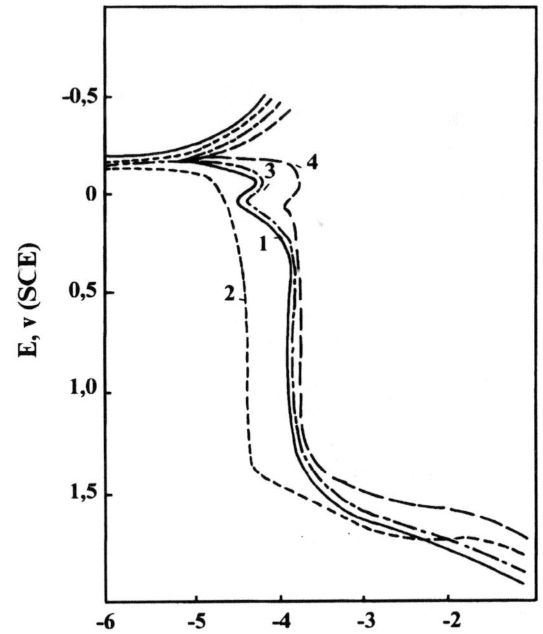

Figure 3. Potentiodynamic polarization curves of amorphous (1), nanoquasicrystalline (2), partial nanocrystalline (3) and full crystalline (4) $Cu_{66}Ti_{34}$ alloy in 10% H_2SO_4

The improved succeptibility to passivation of the nanoquasicrystalline amorphous alloy $Cu_{66}Ti_{34}$ may be attributed to the homogeneous distribution of the nanoparticles, which act as an effective barrier of the dissolution of the amorphous matrix due to the their greater thermodynamical stability.

The finding that the corrosion resistance of amorphous alloys is improved by the precipitation of nanoscale particles points to the future development of nanoquasicrystalline amorphous alloys for application.

Typical polarization curves of the nanoquasicrystalline amorphous alloy and the pure metals titanium and copper in 10% H_2SO_4 and 5% HNO_3 are shown in Figure 4.

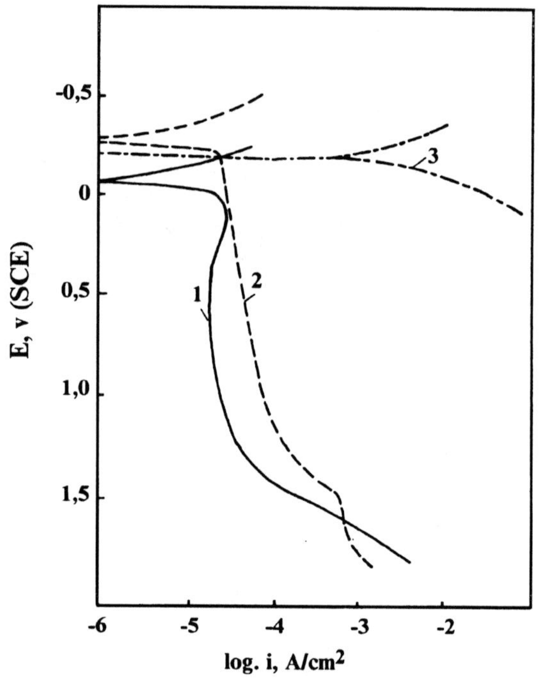

Figure 4. Potentiodynamic polarization curves of nanoquasicrystalline $Cu_{66}Ti_{34}$ alloy (1) and pure metal titanium (2) and copper (3) in 5% HNO_3

The nanoquasicrystalline amorphous alloy as well as titanium passivate in a wide potential range in both acids. The passive current of the alloy in both solutions is lower than those of titanium. This makes possible the application of the alloy as a subsitutiour of Ti (for example as material for anodes).

4. Conclusion

On the basis of the results in the present study it may be concluded: (i)The amorphous, nanoquasicrystalline and partial nanocrystalline $Cu_{66}Ti_{34}$ alloys possess a high corrosion resistance both in sulfuric and nitric acid solutions; (ii) The nanoquasicrystallization of the alloy exerts a beneficial effect on its passivity; (iii) The full crystallization results in formation of heterogeneous structure composed by Cu_3Ti_2, β-Cu_3Ti and Cu_2Ti crystalline phases and leads to a considerable decrease of its corrosion resistance.

References

1. A. Inoue, T. Masumoto, in: Current Topics in Amorphous Materials: Physics and Technology, ed. Y. Sakurai, Y. Hamakawa, T. Masumoto, K. Shirae, K. Suzuki, Elsevier Science Publ., 1993, p.177
2. H. Gleiter, Nanostr. Mater., 6(1995)3
3. A. Inoue, Mater. Jap., 34(1995)934
4. T. Masumoto, H. Kimura, A. Inoue, Y. Waseda, Mater.Sci.Eng., 23(1976)141
5. B. Jalnin, I. Kekalo, I. Skakov, E. Shelehov, Fiz. met. 79(1995)94
6. R. Raicheff, Chemical Resistance and Corrosion Protection of Materials, Technica, Sofia, 1990, p.74
7. I. Gutzov, J. Non-Cryst.Sol., 45(1981)301
8. K. Heusler, D. Huerta, J. Electrochem. Soc., 136(1988)65
9. M. Naka, K. Hashimoto, T. Masumoto, Corrosion, 36(1980)679
10. I. Skvorzova, N. Tomashov, J. Kovneristy, E. Osipov and E. Trofimova, Zashch. Met., 18(1982)323
11. V. Zaprianova, R. Raicheff, E. Kashchieva and L. Fachikov, Bulg. Chem. Commun., 27(1994)162
12. V. Zaprianova, R. Raicheff, Proc. International Symposium on Glass Problems, IGC, Istanbul, 1996, v.2, p.82
13. C. Souza, S. Kuri, F. Politti, J. May, C. Kiminami, J. Non-Cryst.Sol., 247(1999)69
14. C. Souza, M. Oliveira, J. May, W. Botta, N. Mariano, S. Kuri, C. Kiminami, J. Non-Cryst.Sol., 273(2000)282
15. V. Zaprianova and R. Raicheff, Proc. Eurocorr ,91, edited by I. Karl and M. Bod ,GTE, Budapest,1991, v.1, p.41

Nanoscience & Nanotechnology
eds. E. Balabanova, I. Dragieva, Heron Press, Sofia, 2001

NANO PARTICLES MANGANESE COBALT OXIDE – SPINEL OVER ACTIVATED CARBON

L. Ljutzkanov

Institute of Chemical Engineering, Bulgarian Academy of Sciences, 1113 Sofia, Acad. G. Bonchev St. Bl. 103, Bulgaria

Keywords: one-step process, activated carbon, catalyst, intercalating carbon , materials, nano particle size, manganese cobalt oxide – spinel

This one-step process is a promising alternative to the expensive conventional processes for production of activated carbons and catalysts. The temperature range 600–1000°C in this one-step treatment also has an important role in preparation of activated carbons [1]. The aim of the present work is to study one-step activation in a horizontal-rotating reactor [2].

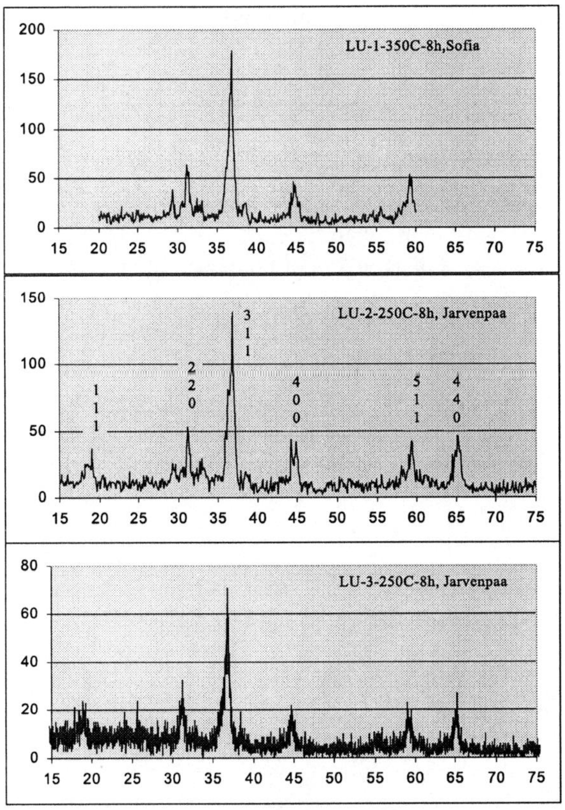

Figure 1. X-ray diffraction patterns for Mn-Co_2-O_x spinel annealed at 250–350°C

The apricot stones are impregnated with aqueous solution of the cobalt acetate and manganese (II) acetate in molecule ratio Co:Mn= 2:1. After drying the samples are activated in a one-step treatment, employing a horizontal rotating stainless steel reactor 0.7 l, situated in an electrically heated oven. The treatment is conducted at 750°C for a period of 30 to 75 min.

The samples have been prepared studied by X-ray diffraction and were found Co_2MnO (Figure 1).There was found a spinell structure after heat treatment at 250°C/8 h and a better one after 48 h. The sise of spinells are chenged from 1.46 nm after 8 h heat treatment to 138 nm after 48 h. The best activity did get the sample with sise of spinells about 60 nm.

The third one had too much spinell and did not work well. Anyway the results were very good. The second sampple contained about 10.78% and had been done by slow speed. The fist one with 8.14% salts was also okay. Their size was sample 1.46 nm after 8 h heat treatment and 138 nm after 48 h and second one was about 60 nm after fist 8 h and under 100 nm after 48 h heat treatment.

References

1. K. Gergova, N. Petrov, S. Eser, Carbon 32, 1994, 693.
2. L. Ljutzkanov, A. Atanasov, at all. Method of Treatment of carbon-containing materials, registry Nr 101705 for Bulg. pat. /30.06.1997/.

Nanoscience & Nanotechnology
eds. E. Balabanova, I. Dragieva, Heron Press, Sofia, 2001

VAPOUR-PHASE FABRICATION OF POLYIMIDE NANOCOMPOSITE FILMS USING CLUSTER GENERATION IN PLASMA

G. Danev[1], E. Spassova[1], J. Assa[1], D. Oliver[2], R. Enikov[2]

[1]Central. Lab. of Photo Processes BAS,G. Bonchev str., bl.109, 1113 Sofia, Bulgaria
[2]Institute of Electronics BAS, 72 Tzarigradsko Shaussee blvd. 1784 Sofia, Bulgaria

Abstract. A method for composite films deposition is described. Polyimide is deposited by evaporation of monomers. Carbon nanoparticles are added in the film. They are produced by plasmachemical decomposition of hydrocarbons.

Keywords: composite films, polyimide, plasma chemistry, nanoparticles

1. Introduction

The development of new materials based on a polymer matrix with embedded carbon particles has been the subject of permanent interest and intensive investigations. These materials possess an enhanced mechanical resistivity and conductivity in comparison with those of the polyimide matrix [1,2,3]. The possibility for obtaining polyimide layers through deposition of the monomers from a gas phase [4,5] allows the combination of the two processes - the simultaneous deposition of the both – the polymer matrix precursors and the carbon in vacuum [6]. The polyimides (PI) are aromatic polymers whose obtaining even from a gas phase [7,8] renders opportunities for the deposition of specific nano-composite thin layers. It is based on a polycondensation of two oligomer molecules of oxydianiline (ODA) and pyromellitic dianhydride (PMDA) deposited from the gas phase on a substrate. In view of preserving the stecheometry the temperature of the substrate should not exceed 305–310 K.

2. Methods

Different methods for the generation of nano-composite layers from polyimide containing carbon or refractive metal clusters from a gas phase are being developed. Wynn et al. [9] have been investigated the UV laser carbonisation of PI films to produce an electrically conducting polymer film. Also, preliminary experiments of ours with evaporated PI and PI-C composite films showed that their irradiation with KrF laser leads to increasing of the conductivity with 4–5 orders of magnitude. Most probably, the conductivity increase observed is in both cases due to the carbon

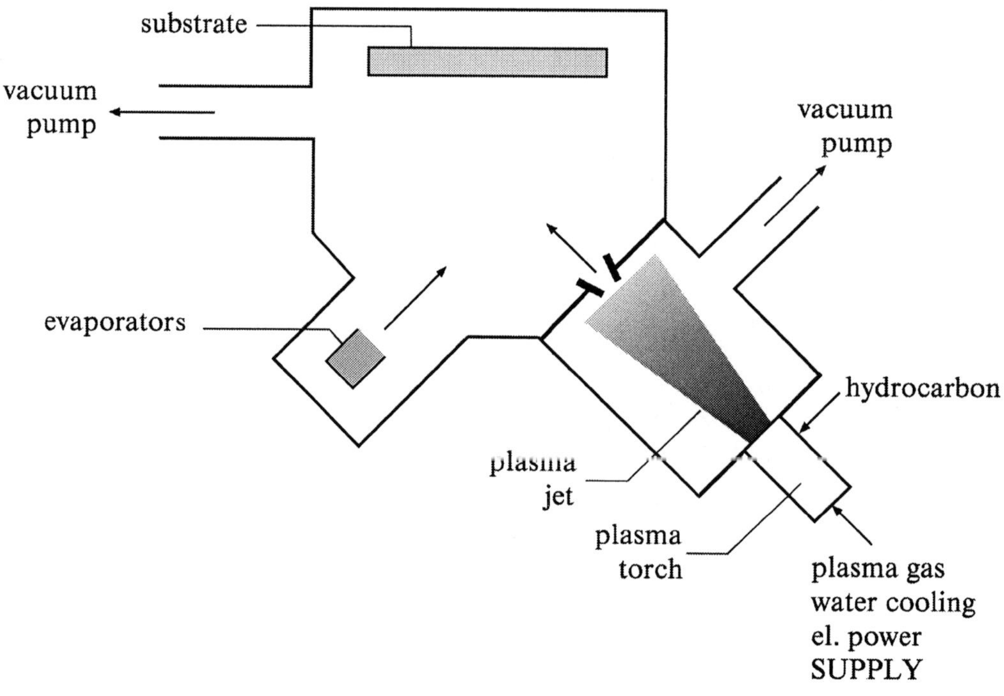

Figure 1. Experimental set-up

concentration augmentation resulting from the laser action.

In a previous work [6], the C particles were intermittently incorporated by means of quasi-arc evaporation and erosion of the graphite electrode in the course of the organic layer formation and growth. The measured conductivity of such layers is several order higher than the one of the vacuum deposited PI films. Massive eruption of eroded carbon macroparticles from the electrodes and heating of substrate (450–500 K) were observed. The polymer layers are with strongly expressed mechanical and chemical defectivity.

Another method for creation of carbon particles allow for overcoming the disadvantages of the arc evaporation from carbon electrode. Supersonic flows of expanding plasmas [10,11] were used for the generation of carbon clusters and carbon-containing active particles [12].

3. Experimental Set-up

In our investigation the plasma source represents a D.C. arc plasma torch. The values of plasma temperature and pressure are $(5–10)\times 10^3$ K and 20–50 Pa. The plasma torch is discharging into a large-volume tank at reduced pressure (0.1–1 kPa) thus forming a supersonic jet. A reagent hydrocarbon is introduced into the torch nozzle, which is decomposed under the action of the high temperature. These parameters can be controlled and changed at the time of work so that the necessary composition of active particles – radicals and clusters can be generated. The temperature of the expanding plasma jet is significantly lower ($\geq$ 1000 K), the chemical processes are retarded and thus the chemical composition is preserved. Since the conditions for the polyimide coating formation are characterized by significantly lower temperatures ($\sim$ 300 K) and pressure (10^{-3} Pa) in our experimental set-up (Figure 1) the evaporation vacuum chamber is separated from the plasma area by a water-cooling wall.

Through an aperture in it, the active particles fall onto the substrate situated opposite it. The carbon clusters are embedded in the polyimide matrix. The system thus developed allows for other substances such as metal clusters and metal compounds to be generated and introduced in the deposited layers.

Acnowledgements

This work has been financially supported by Bulgarian Science Foundation under contract F-426.

References

1. Yang, L., Osborne, D. H. and Haglund, K. F., Appl. Phys. A, 1996, 62, 403.
2. Gianneli, E. P., Adv. Mater., 1996, 8, 29..
3. Dawson, J. C. and Adkins, C. J., J. Phys. Condens. Matter., 1996, 8, 832.
4. Grunze, M. and Lamb, R. N., J. Vac. Sci. Technol. A, 1987, 5, 1685.
5. Spassova, E., Jivkov, I., Danev, G., Dimitrova, T., Koprinarova, J. and Paskaleva, A., Vacuum, 1996, 47, 1345.
6. Rangelov, N., Assa, J., Spassova, E. and Danev, G., Vacuum, 1998, 51, 193.
7. Strunskus, T.,Grunze, M. in "Polyimides - Fundamentals and Applications". Ed. M.U. Chosh and K.L.Mittel, Marcel Deker Inc.,NY, Basel, 1996, p.187.
8. Assa, J , Spassova, E., Danev, G. Vacuum, 51(2), 185, 1998
9. Wynn, G. and Fontain, A.W.,,J.Electrochemical Soc.1997, 144, 3769
10. Timmermans, C.J., Rosado, R.J. and Schramm, D.C., . Naturf. 40a (1985) 810.
11. Djakov, B.E., Enikov, R., Oliver, D. Contrib. Plasma Phys 37, 1, 57-76. (1997)
12. Beulens, J.J., Kroesen, G.M.W, Schram, D.C, Timmermans, C.J, Crouzen, P.C.N., Schuumans, H.J.A., Beijer, C.B., Werner, J. J.Appl.Polym. Sci., 46 (1990) 527.

SECTION 5:
Synthesis and Processing (Mechanical Properties, Fullerenes)

Nanoscience & Nanotechnology
eds. E. Balabanova, I. Dragieva, Heron Press, Sofia, 2001

THE "OSNET" No GTC1 – 2000 – 2820. PROJECT OF THE FIFTH FRAMEWORK PROGRAMME

S. Stavrev, J. Karadjov, L. Markov, A. Petrova, Z. Karagyozova

Space Research Institute, Bulgarian Academy of Sciences, 1000 Sofia, P. O. Box 799

1. Introduction

The "OSNET" project is of the "Development" Programme has key activities "Innovationproduct, processes and organization". In itself it's a network, and covers allimportant branches of output, working up and utilization of the decorative stones, equipment, environment and safety.

Objects:

a) To assist modernization of the production and adaptation of industrial possibilities and informative potential by means of more flexibility and possibility for reaction on required time, by stimulation the interbranch exchange and participation in small and medium size companies, taking into consideration their specific needs and part in the chain of supplies, using methods, that can create and support sufficient places of work in Europe, stop the decline in industrial work load, improve generally the quality of work..

b) Essentially to improve the common quality, decreasing the inefficiencies and expenditures, linked with products during their all cycle of life.

c) To minimize the resources consumption (materials, power, and water) aims essentially to decrease the effect from delivery, utilization and maintenance of the product during its all cycle of life.

From the "OSNET" are awaited such basic achievements as: increasing productivity and efficiency, decreasing value, decreasing refuse production, and as a whole decreasing the influence over environment in every on of its sectors, reaching good and standard products quality.

As a result the offered Network is an addressed to activities of TRA 1.9 ("infrastructure: safe and effective instruments for production, building and civic infrastructure").

The aim of TRA is to encourage projection, building and operation of steady means of production, manufacture and civil buildings, infrastructure as well as to modernize and update abandon exist installations.

The aim is to direct the attention to quality, efficiency, safety and reliability by means of innovation, integral designing and moulding, building, exploiting and support.

2. Meaning and Contribution for and to the European Union Politics

The offered Network has strong expressive European character, because includes almost all Union member countries. The Network covers a wide range of experts in investor's technologies, stones producers, equipment producers, consumers, national organizations, trading agencies and it has pointedly industrial orientations. There is significant participation of small companies.

As a result of its transnational character, the offered Network will give assistance to get over the geographical and cultural barriers, that obstruct the technical progress in this field of study and will guarantee the correctly formulation of problems in the sector, will give splendid opportunity of technological transfer within the framework of the European Union, will make easily popularization of results from scientific researches among potential users.

The benefit from Network is very important for its members, which are working on research and technological studies, because one of the aims of the 'OSNET' is identification of research priorities and organization of research plans and project by creating groups that will use their scientific and technical experience for project and development of better decisions on different problems.

In this way will avoid recovering in the research activities.

3. Economic Development, Scientific and Technological Perspectives

Policy of functioning: Basic results which will be generated by the Network:

a) Staff (working group) will achieve technologies and theory knowledge of the Network.

b) Deeper knowledge of the structure.

c) Basic technologies will be develop in the investigations, financed by the EU (EU financed investigation projects).

The provided products can be used in wide range of key area e.q. as diamond instruments, machines, equipment, etc

For instance, using them as diamond instruments will lead to a longer exploitation of the instruments as will as to an improved functioning

In addition a working plan has been developed that will be enlarged and improved during the implementation of the Network. In the middle of the implementation period we'll develop, another plan, concerning the implementation of the technologies.

The first plan, mentioned above, will lead to more efficient using of the Network proguct. It contains the following steps:

— identification of the knowledge, which will be useful for the participants;

— identification of new technologies;

— identification of present technologies;

— creation of the network user group;

— qualification of different user groups;

— conversion of the technologies in order to make them more accessible for different companies;

— translation of the literature into different languages spoken by the different user groups.

The policy of using. Leading committee-user group.

Working groups in the following sectors:

— identification of the useful knowledge;

— available innovations;

— area- different areas;

— conversion into suitable form;

— identification of specific user groups;

— user groups approach.

Policy of network – functioning. Contains:

— Distribution policy;

— Public company;

— Presentation;

— Open days of industrial technology transfer;

— Network publications;

— Thematic meetings of the staff (working group);

— Network web page.

4. Expected Results of the "OSNET" Project

Region	Network way	Results	Marketing results
Stone-croping	Elective stone-croping methods; New methods of underground stone-croping	Higher degree of stone-restoring; Higher productivity; Less waste products	Lower expenses; Lower expenses for waste products
Processing	Optimization of the processing	Shorter processing period; Higher quality; Economy of material	Lower increased expenses; Competitive products; Less waste products
Stone qualification	Standard stone taxonomy; Improved control of the quality	Standards of the stone market standards of the productions	Production of stone standard qualitative; Labels; Easier access to the market; Effective production and sale; Data collection
Instruments and equipment	Higher live of usage; Higher speed of cutting; New designs	Shorter process of cutting; Shorter cycle	Decrease of production expenses; Decrease of capital investment
Ecological aspects	Safety codes and better practice; Ecological processing	Harmless danger; More health conditions; More favorable effect over the environment	Safer working environment; Adherence to the laws of the safety environment
Technology transfer	Technology distribution; Courses; Technological consultation	Assimilation – inclusion of more appropriate technological; Qualification of the staff	Higher productivity; Power production expense; Better quality of products; Appointment of more qualified staff

5. Scientific and technological expectations

Working packets 5 (WP) and a chart of working power has been created in this project together will the working plan. There's a leading committee and the network functions are carefully planned.

Working period (the project period) – 36 mounths

General coordinator of the project: – Prof. d-r Ioanis Paspaliaris, Ntua - Athens, Greece Heads in different sectons: Prof. N. Terzopulos (NTUA), Dr. S. Midha (UWE), Prof. R. Bruno (DICMA), eng. Karozio (DAPP), eng. S. Pereira (CEVALDR), and Prof. M. Faunty (NTUA)

Total price of the project: 2355488 EURO. Beginning (starting date) 01. 01. 2001

Working group Bulgarian scientists from SRI-BAS, supervised by Prof. Dr. S. Stavrev

Under the contract and in accordance to the (working), plan the staff offers technology and new extra rigid materials. This will provide higher speed of cutting, shorter cycle processing cycle, longer exploitation of instruments.

In addition the expenses and capital investments will be reduced considerably.

Nanoscience & Nanotechnology
eds. E. Balabanova, I. Dragieva, Heron Press, Sofia, 2001

INVESTIGATION ON THE PROCESS OF α-Si_3N_4 SYNTHESIS

P. Petrov, N. Gourov[1], N. Stoichev

Institute for Metal Science, Bulgarian Academy of Sciences, 1574 Sofia, Bulgaria
[1]Department of Electrical Measurements, Technical University – Sofia, Bulgaria

Abstract. Silicon nitride (Si_3N_4) is a compound known since the end of the XIXth century, but the interest for its practical application enhanced in the last thirty years because of its high mechanical strength and thermal resistance. Thus silicon nitride is now the ceramic material with the widest application for construction elements. Properties of these details are determined by the characteristics of the Si_3N_4 powders used e.g. their phase composition, size of the particles, chemical composition, impurities, additions etc.

The process of synthesis of silicon nitride and its phase composition influenced by temperature, gas atmosphere and annealing time are studied in this report. The results are considered in connection with the chemical bonds and crystal structure of the compound.

The influence of reaction time, temperature, and gas phase content on the crystal modification ratio of the synthesized silicon nitride powders is revealed. A nano-scale dimension of α-Si_3N_4 observed confirms its higher tendency for quicker diffusion processes, than coarser β-Si_3N_4 particles. The structure of silicon nitride powders is studied with respect of reaction gas atmosphere content so to obtain either α-Si_3N_4 or its β- modification as predominant phase.

Keywords: ceramic material; silicon nitride, crystal modifications of, synthesis of

1. Introduction

Silicon nitride (Si_3N_4) is a compound known since the end of the XIXth century, but the interest for its practical application enhanced in the last thirty years because of its high mechanical strength and thermal resistance. Thus silicon nitride is now the ceramic material with the widest application for construction elements. Properties of these details are determined by the characteristics of the Si_3N_4, powders used e.g. their phase composition, size of the particles, chemical composition, impurities, additions etc.

The process of synthesis of silicon nitride and its phase composition influenced by temperature, gas atmosphere and annealing time are studied in this report.

Crystal structure and chemical bonds in silicon nitride. Atoms of the silicon and the nitrogen in the Si_3N_4 form saturated covalent bonds [1], determined so that every forth hybrid Sp^3-orbit of Si-atom overlaps with Sp^2-orbit of a nitrogen atom. Silicon nitride has two modifications α- and β-Si_3N_4, which have hexagonal crystal symmetry and appeared in structures similar to Be_2SiO_4 [2]. As in the case of silicates, the structure of Si_3N_4 is build up by silicon nitride tetragons SiN_4 [3], which can be combined in a 3-D set by a nitrogen atom and in this case they are located at angle 120° one towards the other [4].

The specific mechanism of growth determines the synthesis of different modification. α-Si_3N_4 is the crystal phase with lower thermodynamic stability compared to β-Si_3N_4, and at temperatures higher than 1400°C, the α-phase turns into β but back converting does not occur. The transformation temperature is strictly fixed and is determined by the history of the process, the previous treatment, structure defects and impurities in the silicon nitride powder.

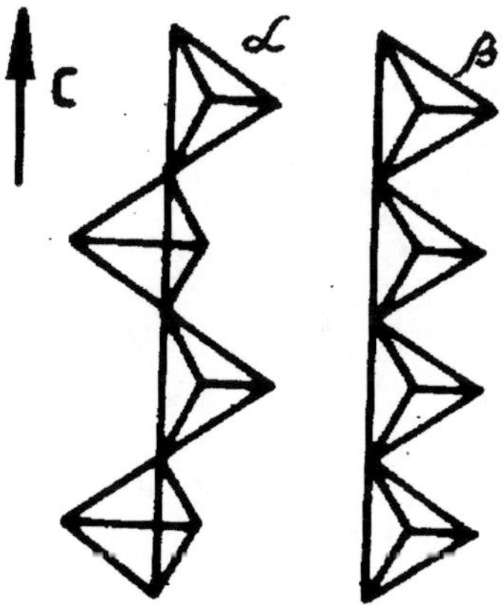

Figure 1. Crystal structure of α- and β-Si_3N_4

Crystal modifications of Si_3N_4 differ by location of the tetraheders towards the axis "C" (Figure 1). The $\alpha \rightarrow \beta$ transformation is typical for converting in the second coordinate region, in which the location of the atoms groups is changing, with preserving the structure inside the tetraheders. A monotropic $\alpha \rightarrow \beta$ transformation of Si_3N_4 determines the high thermodynamic stability of β-Si_3N_4. So the selfdiffusion in α-Si_3N_4 is considerably higher than in the β-Si_3N_4, that is a precondition for ability to anneal the α-Si_3N_4 at considerably lower temperatures and to obtain products with high mechanical properties. This tendency is enhanced in case of powders with nano-sized particles.

2. Experimental

The synthesis of Si_3N_4 was performed in electric resistance furnace at working temperature up to 1500°C, under adjustable gas atmosphere. Silicon powder with particle sizes lower than 20 μm was used, obtained by dispersing in a planet mill in tetrachlormethane. Nitrogen (of purity 4N), argon (4N) and hydrogen were used for nitrating. Gases were additionally purified from water, oxygen and other impurities.

Phase composition of silicon nitride powders obtained was determined by means of DRON-2.0 X-ray difractometer (Fe 25/10; 1×10^3; 10; 1^0; 600 mm.min^{-1}; blinds 2,4 and 0,25).

The ratio of the phases in the silicon nitride powder was determined by intensities of characteristic lines of Si, α- and β-Si_3N_4.

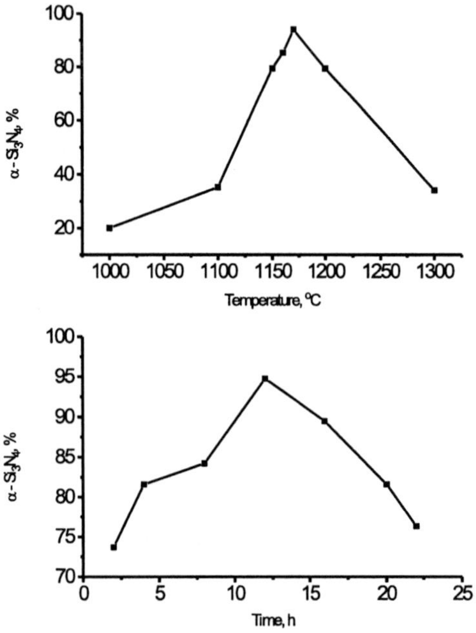

Figure 2. Yield of α-Si_3N_4 depending on: a) temperature and b) duration of the synthesis

Influence of thermal conditions on the phase composition of Si_3N_4. The crystal growth of chemical compounds can be divided in three consequent parts – forming of a nucleus, free growth of separated nuclei and diffusion controlled growth. Continuation and velocity of the process are determined by the conditions of the synthesis and the form and nature of the molecules. In result of interaction between Si and N atoms at considerably low temperature, α-Si_3N_4 has been synthesized (Figure 2).

Reaction proceeds at least 12 hours at optimum temperature of 1170°C (Figure 2a). At higher temperatures the amount of α-Si_3N_4 is decreasing, showing at 1300°C only 34,1%. Silicon nitride powders synthesized at temperatures higher than 1170°C contain predominantly β-phase.

The amount of α-Si_3N_4 depends on the reaction duration and shows a maximum at about 12 hours (Figure 2b). This indicates that $\alpha \to \beta$ transition may occur at low temperatures too. It can be supposed that up to 12 hours (up to 1200°C) the rate of α-Si_3N_4 synthesis is higher than that of β-Si_3N_4, while for reaction times longer than 12 h the forming of β-Si_3N_4 prevails.

Influence of ambient atmosphere on the content of α-Si_3N_4. It is well known that the rate of heterogeneous reactions such as Si_3N_4 synthesis depends on the composition of the gas phase. The influence of gas components ratio for two mixtures (N_2-Ar and N_2-H_2) on the yield of α-Si_3N_4 is shown on Figure 3. The conditions of that experiment are as follows: 1170°C and isothermal annealing for 4 h. Higher yield of α-Si_3N_4 is obtained for gas mixture N_2-H_2 comparing to N_2-Ar.

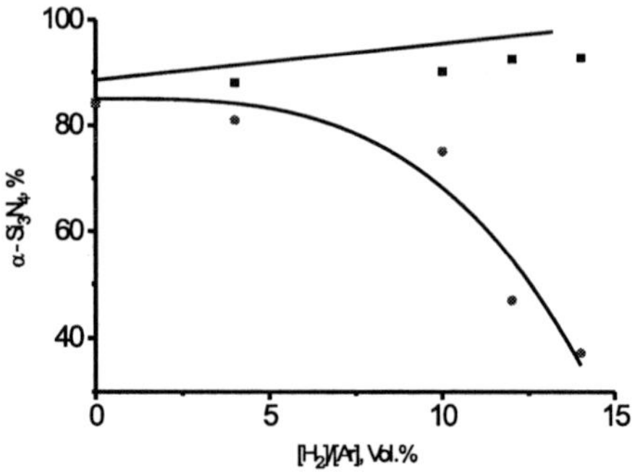

Figure 3. Influence of gas components ratio in N_2-Ar (■) and N_2-H_2 (●) mixtures on the yield of α-Si_3N_4

The α-Si_3N_4 obtained in N_2-H_2 gas atmosphere crystallizes in nano-sized needle-like particles, forming a grid structure (Figure 4).

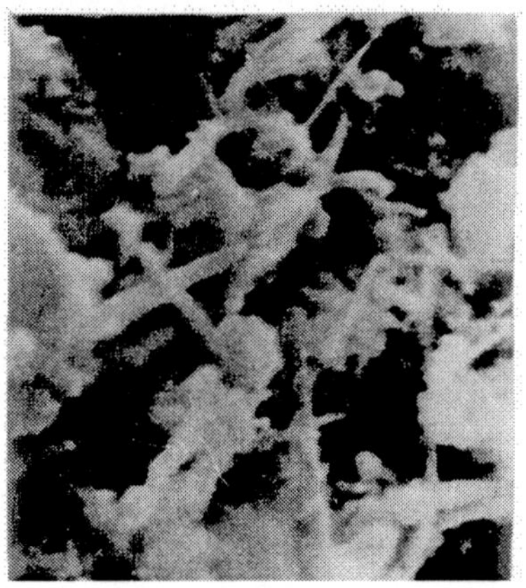

Figure 4. Typical grid structure of nano-sized needle-like α-Si_3N_4 particles; Magnification ×4000

Besides the needles, prismatic crystals of α-Si_3N_4 are observed in nearly the same nano-sized dimensions. When synthesis is realized with N_2-Ar mixture, the predominant phase is β-Si_3N_4 (Figure 5).

The sizes of α-Si_3N_4 needles vary from 7 to 14 nm in diameter and 70–78 nm in length. Particles of β-Si_3N_4 are 0.2 and 1 μm respectively. Some particles reach maximum dimensions of 9 μm in diameter and 20 μm in length. An irregular distribution of β-Si_3N_4 particles in the bulk material is observed on Figure 5 too.

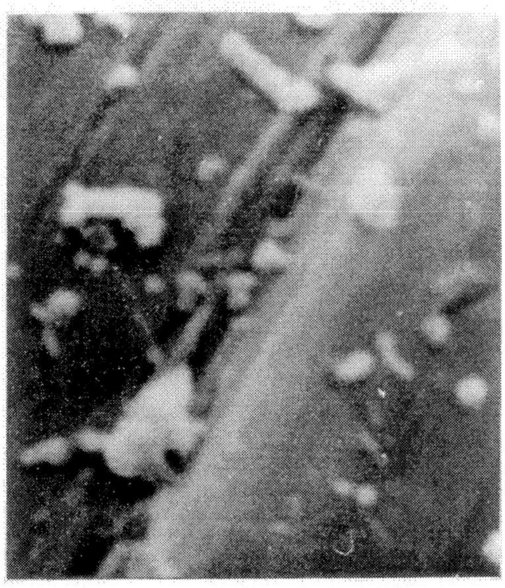

Figure 5. Electron micrograph of β-Si_3N_4 prismatic particles

3. Conclusions

1. The influence of reaction time, temperature, and gas phase content on the crystal modification ratio of the synthesized silicon nitride powders is revealed.
2. Nano-scale dimensions of α-Si_3N_4 observed confirm its ability for quicker diffusion processes, than in the case of coarser β-Si_3N_4 particles.
3. The particle sizes and their structure in silicon nitride powders are studied with respect of reactive gas composition so to obtain α-Si_3N_4 or the β- modification as predominant phase.

References

1. D. Hardie, K.H. Jack, Nature, 180 (1957) 332
2. Nitrogen Ceramics (Ed. By F.Riley-Leyten) Norrdnoff, 1977, 694
3. T.H. Miler, Izv. AN SSSR, Inorganic Materials, 15 (1979) 557-562 (in Russian)
4. R. Grinn, Acta Crystall., B35 (1979) 800-804

Nanoscience & Nanotechnology
eds. E. Balabanova, I. Dragieva, Heron Press, Sofia, 2001

PECULIARITIES OF THE ARC DISCHARGE PROCESS IN FULLERENE STRUCTURE SYNTHESIS

N. Koprinarov, G. Pchelarov, M. Marinov[1], B. Ranguelov[1], M. Konstantinova

Central Laboratory for Solar Energy and New Energy Sources, Bulgarian Academy of Sciences, 72 Tzarigradsko shose, 1784 - Sofia, Bulgaria
[1]Institute of Physical Cemistry, Bulgarian. Academy of Sciences, bl. 11 Acad. G. Bontchev str.

Abstract. The object of our investigation have been the peculiarities during sublimation of C by arc discharge. The analysis carried out shows that the high temperature obtained on the electrode surface during the arc discharge process allows us to obtain a vapor flux with a large density within the zone of synthesis. The high density increases the probability of creating fullerene type structures and accelerates their growth rate. It is demonstrated that due to this fullerene structures may be produced even at atmospheric pressure and also in an air ambient, while with other methods this can be done only at low pressure and most often in the presence of an inert gas. Attention is paid to the fact that, the multitude of physicochemical processes taking place during arc discharge are prerequisites for the conditions of synthesis to be different at different locations within the discharge space and to give rise to different types of fullerene structures. The temperature balance analysis demonstrates, that different unusual solutions are possible in electrode sublimation and hence to stimulate the growth of certain types of fullrenes.

Keywords: fullerene, arc discharge method, fullerene synthesis

Generally a fullerene structure[1,2] is only comprised of C atoms, arranged so, that they form a spatially closed network [3,4]. One of the most important conditions for building such a structure is prior to synthesis to have present wholly free carbon atoms and then to give rise to such prerequisites, that atoms are only joined to one another forming a spatially closed structure[5]. It is desirable that there are no atoms of other chemical elements around the free carbon atoms, that could attach to these and so impair fullerene structure growth. The exact conditions, that cause the carbon atoms to join only in a spatially closed structure are yet unknown, but the analysis of the data collected so far lead us to believe, that in order to create such structures in free space, there must be a very large density of carbon atoms or foreign atoms with catalytic characteristics [6] (for example Fe or Ni).

By the discharge methode [7,8] the discharge takes place mainly between the head surfaces of electrodes. Depending on the conditions created within the discharge chamber, the gas discharge region can contract and cover only part of the head surface of the electrodes or it can extend and also encompass parts of the external electrode surface. The size of the discharge region and its location on the electrode surfaces has no significance in principal as regards the nature of the physical and chemical processes taking place during synthesis.

A multitude of physical processes take place during arc discharge, as a result the local conditions at different points in space around and between the electrodes are different. The analysis of the heat emissions yields:

current heat $Q_R = \rho \frac{l}{\pi r^2} I^2$;

bulk thermal conductivity $Q_c = \pi r^2 a (T - T_1)/l$;

irradiation $Q_L = \sigma(T^4 - T_1^4) 2\pi r l$;

gas conductivity $Q_G = \int_{x=0}^{l} \lambda S \frac{T(x) - T_1}{d} dx$;

heat transported by electrons $Q_e = (K_e - W_m) I / e_0$

and shows, that the temperature at different points is different and can be calculated if the sum of all the heat flux at is determined for a given point [9]. This relation shows, that the local temperature is dependent on the way the arc discharge process is executed and on the parameters characterizing the geometry and thermal properties of the electrode material.

Using a suitable choice of construction and technologic parameters such temperature conditions can be realized, that these may lead to significant changes in discharge process utilization, this in turn can lad to a change in the conditions of synthesis. For example the cases shown in Figure 1 can be realized [10].

Whether C will sublimate on the surface of the electrodes or the carbon vapour will deposit on the respective surface, depends on the temperature. In the case of Figure 1a for example, it is observed that when a work gas is used which contains C (such as ferrocine), the growth of a deposit occurs both on the anode and cathode. The type and structure of the cathode and node deposit (Figure 2) are different [11]. Each of the other cases shown in Figure 1 has its specifics which is here only implied from the point of view of a possible application.

If we compare the arc method for C vapour generation with the thermal (electrical resistive heating) and the pyrolithic methods it can be seen, that the density of the vapour generated by the arc method is significantly higher, because generation is from the small size head surface of

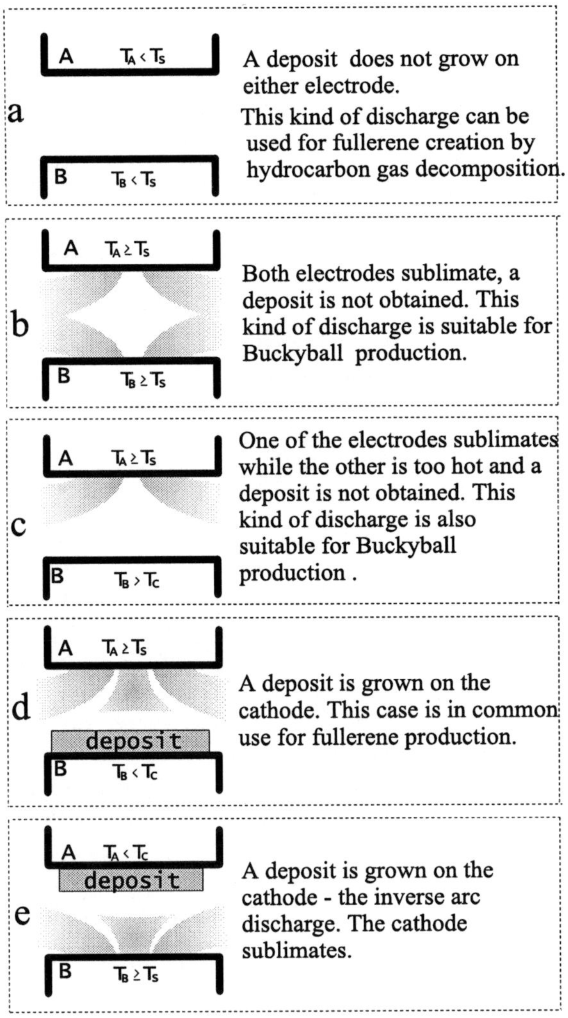

Figure 1

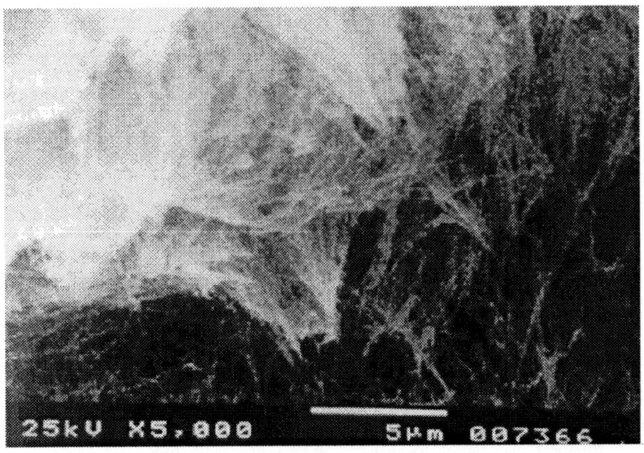

Figure 2a. A deposit on the anode

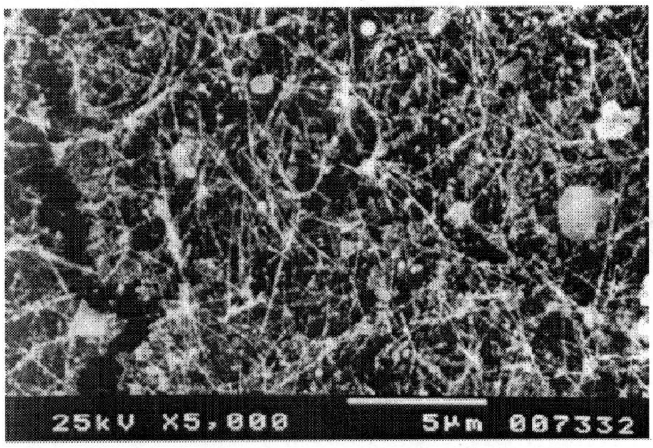

Figure 2b. A deposit on the cathode

the sublimating electrode. As a result of this a large radial and axial gradient of the density of C vapour is obtained. The flux of C vapour. can be considered as comprised of three components directed in different directions (Figure 3).

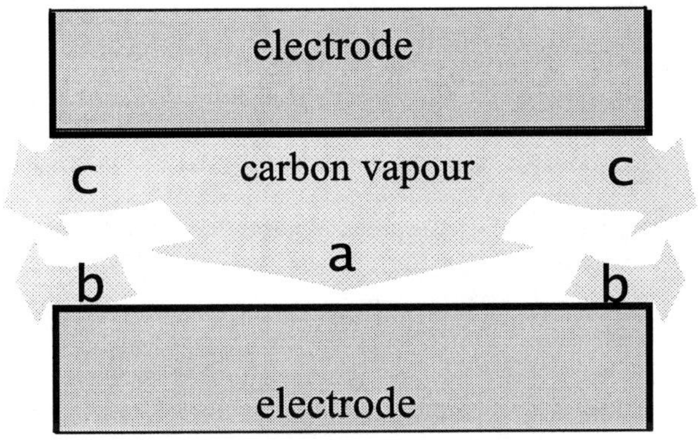

Figure 3

Component *a* is directed at the opposite electrode. It supplies material for the multitude of forms and structures arising and growing on the opposite electrode, part of these are of the fullerene type. Vapour flux *b* once it impinges on the surface of the opposite electrode is redirected tangentially along the surface, hence a large number of flat structures and graphite crystals arise. The carbon atoms from component *c* are directed internally to the chamber and when the right conditions exist supply the material in spherical or nanotube fullerene structure synthesis. The ratio between the sizes of the three components is dependent not only on the temperature gradient and the vapour density, but also on the technologic parameters and the nature of the apparatus constructed.

For pressures of about 10^4 Pa at which the arc process takes place, the length of free flight is in the range of micrometers and is much smaller than the distance maintained between the electrodes. The synthesis of a fullerene structure with the aid of C clusters can be realized only after large numbers of impacts between the atoms. All further movement of these particles is by diffusion. Therefore, the nature of the particle which will be formed following the collision between the atoms depends on the location within the discharge space between the two electrodes or further afield where the impacts between the C atoms occur.

Outside the discharge space the temperature gradient is very large and the energy of the atoms decreases. For this reason the quantities, free flight, distance between the electrodes, diameter of the discharge space, and particle diffusion length are key parameters in the process of synthesis. This is especially important when the arc process is realized for the close proximity electrode arrangement. Under a close proximity electrode arrangement we consider distances between the electrodes commensurate with or smaller than the cross section of the electrode. It can be expected that the larger the electrodesm are in size and the smaller the distance between them is the smaller the quantity of material evaporated which (component c from Figure 3) will leave the discharge region, transferred within the work chamber. This means that by decreasing the distance between the electrodes the density of flux falling on electrode which is at a lower temperature can be increased. This peculiarity may be useful when the method is used to grow an electrode (cathode) deposit that contains a large share of nanotubes. On the other hand if the aim is to introduce a greater part of the carbon vapour inside the work chamber, then component of the C flux must be increased by increasing the distance between the electrodes.

When the work gas utilized contains C (for example a hydrocarbon), a gas which dissociates pyrolithicaly in the discharge region or a gas reacting with C vapour, it is consumed within the discharge region and its concentration at the center of discharge is minimal. A gradient of its concentration appears, directed from the work chamber towards the heart of the arc. The flux of this gas is directed in opposition to components *b* and of the carbon flux. The gas enters the discharge region from the periphery of the point of discharge directed towards its center. In the case of arc discharge between close proximity electrodes, the closer the electrodes are to one another, the less the number of molecules from this gas that will reach the central region and will thus have a smaller influence on component of the carbon vapour. Their concentration at every point in space can be determined by knowing their initial concentration, their destruction within the discharge space and their diffusion. By reducing the distance between the electrodes and increasing the density of carbon vapour it is possible to avoid the detrimental effects of gases that join with C (for example O_2, N_2 or H_2) on the formation of fullerene structures if these enter the work chamber as contaminants or if their presence there is undesirable. For a head electrode surface of 10 mm^2 and a distance between them of 0.1 mm even at atmospheric pressures and given an air ambient the influence of O_2 and N_2 can be reduced so much that nanotubes can be grown in the cathode deposit. Such a deposit is shown in Figure 4.

The arc discharge method is now established as one of the more effective methods for synthesis of different fullerene structures, due to its effectiveness and simplicity.

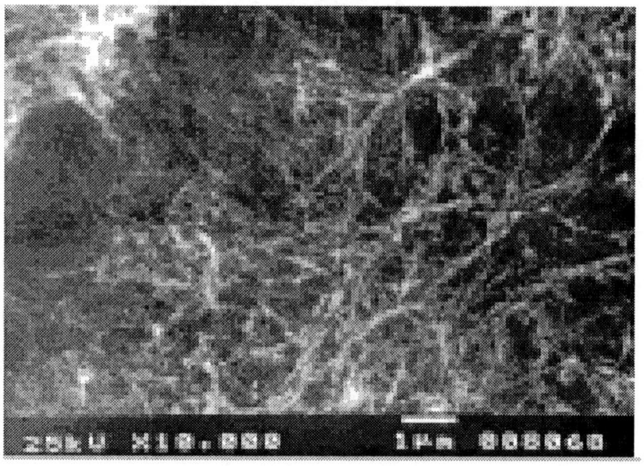

Figure 4

However, for some applications its advantages may turn to shortcomings. These originate mainly due to the fact that the generation of carbon vapour with a large density occurs only at the small electrode head surfaces. This is a major flaw if we seek to grow a deposit across large surface areas. Also the spot of arc discharge burning is very small in comparison to the work chamber volume and gives rise to large temperature and concentration gradients. For this reason the regions where it is possible to obtain uniform synthesis and uniform deposition are quite small. The process can be kept stationary by maintaining a constant distance of the electrodes in order to compensate for that part which is consumed, but that is not very easy.

References

1. Curl R F. Carbon 1992;30:1149
2. Kroto H. Science 1988;242:1139
3. Osawa E, Yoshida M, Fujita M. 1994:33
4. Ebbesen T W. Physics Today 1996:26
5. Wilson M A, Pang LSK. Carbon 1992;30:675
6. Yahachi S, Yoshikawa T, Okuda M, Fujimoto N, Sumiyama K, Suzuki K, Kasuya A, Nishina Y. J Phys Chem Solids 1993;54:1849
7. Ebbesen T W, Ajayan P M. Letters to Nature;358:220
8. Ando Y, Iijima S. Jpn Appl Phys 1993;32:L107
9. Koprinarov N, Marinov M, Pchelarov G, Konstantinova M, Stefanov R. Z Phys B 1996;99:353
10. Koprinarov N, Marinov M, Konstantinova M, Pchelarov G, Stefanov. Syntetic Metals 1996; 77:39
11. Koprinarov N, Marinov M, Konstantinova M, Rangelov B. Vacuum 2000;58:208

Nanoscience & Nanotechnology
eds. E. Balabanova, I. Dragieva, Heron Press, Sofia, 2001

COLLOIDAL DISPERSIONS OF FULLERENES IN POLAR ORGANIC SOLVENTS: INFLUENCE OF POLYMER AND SURFACTANT ADDITIVES

Rossitza G. Alargova, Shigeru Deguchi, and Kaoru Tsujii

The DEEPSTAR Group, Japan Marine Science and Technology Center (JAMSTEC); 2-15 Natsushima-cho, Yokosuka 237-0061, Japan

Abstract. The formation of colloidal dispersions of fullerene C_{70} in acetonitrile has been studied in the presence of nonionic polymer and surfactant additives preliminary dissolved in the polar medium. Stable dispersions of fullerene were readily prepared by just mixing the C_{70} solution in toluene with acetonitrile. The size of the formed particles was investigated as a function of the additive concentration using dynamic light scattering. In all cases investigated the average size of the colloidal particles was determined to be the same or larger in the presence of additives than in the case of pure acetonitrile dispersions. It is found out that the size of the fullerene particles formed when acetonitrile contained Tween 20 increases with surfactant concentration, while in the presence of polyvinyl-pyrrolidone the average size goes through a maximum and decreases again to an average value slightly larger than the one measured in the absence of polymer.

Keywords: fullerenes, C_{70}, colloidal dispersions, polar organic solvents, dynamic light scattering

1. Introduction

The unique chemical and physical properties of fullerenes and carbon nanotubes [1–3] have attracted the interest of the scientists working in various areas and have already lead to the appearance of many studies and patents suggesting various potential applications of these molecules. They exhibit an unusual behavior in many aspects and appear to be a challenging subject of study [1]. Especially important from practical and fundamental point of view is their solubility in different solvent systems. It has been found out that there is no single parameter predicting in general the solubility of C_{60} [3,4] and in addition an anomaly in its solubility as a function of temperature has been detected [5,6]. Even more interesting is the behavior of fullerenes and their derivatives in polar media including water [7–21]. In our previous studies we have shown that the fullerenes can be solubilized in polar organic solvents and water due to the formation of colloidal dispersions [22,23]. The colloid fullerenes particles turned out to be of low polidispersity and polycrystalline structure. It was shown that they remain very stable in the absence of any stabilizer for more than 10 months due to electrostatic stabilization. It was pointed out that the size of the particles depends on the type of the fullerene (C_{60} or C_{70}) and decreases with the fullerene concentration in the initial good solvent.

In the present study we examine the formation of colloidal dispersions of C_{70} in acetonitrile containing nonionic polymer or surfactant. The size distributions of the fullerene particles are studied as a function of the additive concentration using dynamic light scattering (DLS).

2. Experimental

The fullerene sample used in this study, C_{70}, (> 98%) was purchased from Tokyo Kasei Kogyo Co., Ltd., (Tokyo, Japan) and used as received. The analytical grade solvents (toluene and acetonitrile) manufactured by Wako Pure Chemicals Industries Ltd. (Osaka, Japan), were used without further purification.

The preparation procedure applied is exactly the same as that described in ref. [22]. Firstly, 0.1 wt% solution of C_{70} in toluene was prepared and stirred for at least two days in a dark place. After that it was filtered twice through 0.2 μm Teflon filter (Nihon Millipore Ltd., Japan) in order to remove the dust and remained solid fullerene particles. The exact concentration of C_{70} was determined by using UV spectroscopy. Next, 15 μl of the toluene solution was added to 5 ml acetonitrile.

The nonionic additives were preliminary dissolved in the polar medium. Their concentration was varied between 0.001 and 0.5 wt%. As a polymer additive we used polyvinylpyrrolidone (PVP) with average Mol. wt. 40 kD and as nonionic surfactant additive – polyoxyethylene sorbitan monolaurate (Tween 20) purchased from Sigma Chemical Co. Immediately after mixing the toluene solution with the acetonitrile, the color of C_{70} changed from red-brown to brown-violet indicating the formation of a colloidal dispersion. The average size and polydispersity of the particles were studied using DLS as a function of the surfactant or polymer concentration at constant concentration of fullerene (3.3×10^{-4} wt%). DLS measurements were performed on DLS-820 light scattering system (Otsuka Electronics Co., LTD., Osaka, Japan) equipped with He-Ne laser (632 nm wavelength). All measurements were carried out at 90° scattering angle. The temperature was maintained at 25°C during the experiments.

3. Results and Discussion

The formation of fullerene dispersions in the presence of Tween 20 or PVP was similar to that in the pure polar solvents [22]. The formation of particles with the size in the colloidal range and relatively low polydispersity was detected immediately after mixing the toluene solution and acetonitrile. Figure 1 shows the size distribution functions measured for colloidal dispersion of C_{70} in pure acetonitrile (1) and in the presence of 0.4 wt% Tween 20 (2) in the polar medium.

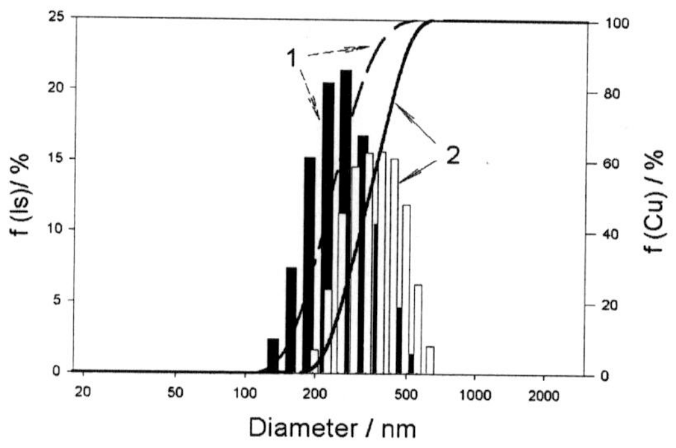

Figure 1. Size distribution functions of C_{70} colloidal dispersions formed in pure acetonitrile (1) and acetonitrile containing 0.4 wt% Tween 20 (2)

It is seen that the mean size in the presence of Tween 20 is larger, but in both cases the width of the size distribution is almost the same. The results from more detailed study of the particle size as a function of Tween 20 concentration are presented in Figure 2. One can see that the increase in the surfactant concentration does not influence the particles size up to 0.1 wt% Tween 20. Further increase in the surfactant content leads to the formation of larger particles.

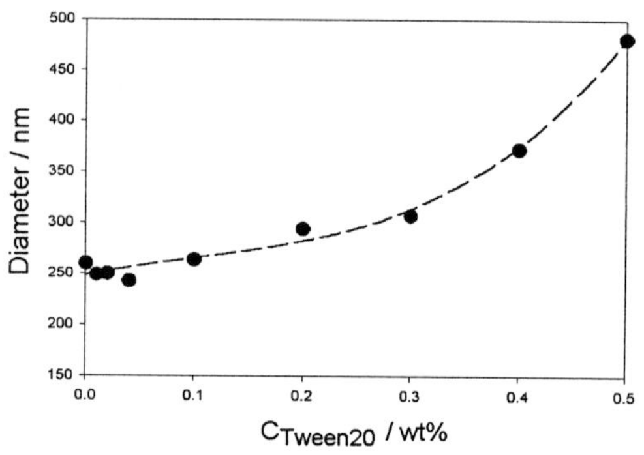

Figure 2. Average diameter of C_{70} particles formed in acetonitrile as a function of the concentration of Tween 20. The fullerene concentration is 3.3×10^{-4} wt%

The addition of surfactant was expected to decrease the size of the particles as is usually observed when dispersing a hydrophobic phase in water [24]. However, in our case the size remained unaltered or increased due to the presence of Tween 20. Of course the dispersion mechanism in water could be very different from that in acetonitrile because of the importance of the hydrophobic effect. In order to check this we prepared some samples using N-methyl pyrrolidone as a good solvent for fullerene and water as a dispersion medium. The particles were smaller than in the case of toluene/acetonitrile dispersions, but the presence of nonionic surfactant lead again to an increase in the average size. To get more insights for the possible role of Tween 20 in the formation of C_{70} colloid particles, we diluted the sample with the highest surfactant concentration (the right-most point in Figure 2) with acetonitrile. No significant change in the measured size has been detected even after 5 times dilution. This means that the size of the fullerene particles is determined almost immediately after mixing of toluene and acetonitrile and whatever the role of Tween 20 is, it is mainly important in the very first stage of the process. The electronic micrographs of fullerene particles in acetonitrile showed the regular ordering of molecules and suggest that the colloidal dispersions are formed as a result of a crystallization process [22]. Hence, the surfactant molecules could serve as a contamination at the surface of fullerene particles promoting the crystal growth.

It is shown in the literature that fullerenes can be solubilized in water in the presence of PVP [25]. In our case an interesting effect was observed when the colloidal samples were prepared with acetonitrile containing PVP. Results for the particle diameter are presented in Figure 3 as a function of polymer concentration.

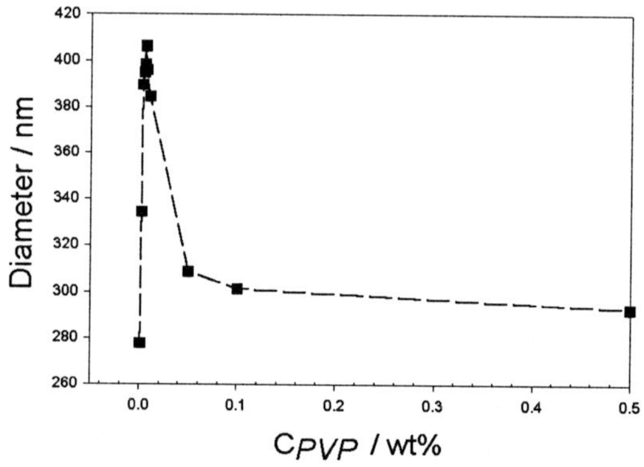

Figure 3. Average diameter of C_{70} particles formed in acetonitrile as a function of the concentration of PVP. The fullerene concentration is 3.3×10^{-4} wt%

Figure 3 shows that at low concentrations the size of the particles increases sharply and goes through a maximum at around 0.005 wt%. After that the particle diameter decreases down to 300 nm at 0.1% PVP, i.e. close to the size measured in the absence of polymer. At higher PVP concentrations, the size of C_{70} particles is slightly higher than the size in the absence of polymer and almost does not change at these conditions.

The strong influence of PVP on the size of C_{70} particles could be due to a special interaction between the fullerene and PVP molecules similar to the one reported in ref. [25] for C_{60} and PVP solubilized in water. It is also possible that at low concentrations the polymer molecules act as contamination facilitating the crystal growth, while at higher concentrations they favor the formation of more crystals smaller in size.

A comparison between Figure 2 and Figure 3 shows different behavior of the particle size formed in the presence of Tween 20, and PVP. However, in both cases, particles smaller than those formed in pure acetonitrile are not produced in the concentration range studied. Another similarity is that the size of C_{70} particles remains the same when diluting with acetonitrile. This means that the surfactant and the polymer play a role in the initial process of particles formation and the increase in the size is not a consequence of an aggregation of once formed particles provoked by the presence of additive molecules. The mechanism is most probably a complex one and more detailed study is necessary in order to find the physical explanation of the observed effects. It could include not only C_{70}/PVP interactions but solute/solvent interactions as well. In addition, it is worthy to note that the presence of polymer or surfactant changes also the properties of the dispersion medium and in this way it could affect the mixing of the components and hence the initial step of the process (see the mechanism suggested in ref. [22]).

4. Conclusions

Colloidal dispersions of C_{70} in acetonitrile containing nonionic surfactant or polymer have been prepared and studied as a function of additive concentration. The formation and the stability of the samples have been investigated with DLS. The results show that the formation of dispersions is similar to that in pure acetonitrile: colloidal fullerene particles are immediately formed when mixing a toluene solution of C_{70} with acetonitrile. The average size of the fullerene particles remains unchanged or increases when the polar solvent contains Tween 20, while in the presence of PVP the particle size exhibits a maximum at a given concentration of PVP in the low concentration range. The mechanism of the action of these additives is not clear at present and needs further investigation.

Acknowledgment

The authors would like to acknowledge Professor Koki Horikoshi, JAMSTEC, for his continuous support for this work.

References

1. Dresselhaus MS, Dresselhaus G, Eklund PC, Science of Fullerenes and Carbon Nanotubes, San Diego:Academic Press, 1996.
2. Taylor R, Walton DRM. Nature 1993;363:685.
3. Shinohara H, Sato Y, Chemistry and Physics of Fullerenes, Nagoya: Nagoya University Press, 1997 (In Japanese).
4. Ruoff RS, Tse DS, Malhotra R, Lorents DC. J.Chem.Phys. 1993; 97: 3379.
5. Ruoff RS, Malhotra R, Huestis DL, Tse DS, Lorents DC. Nature 1993; 362: 140.
6. Doome RJ, Fonseca A, Nagy JB. Colloids and Surfaces A 1999; 158: 137.
7. Friedman SH, DeCamp DL, Sijbesma RP, Srdanov G, Wudl F, Kenyon GL. J. Am. Chem. Soc. 1993; 115: 6506.
8. Sano M, Oishi K, Ishi-i T, Shinkai S. Langmuir 2000; 16: 3773.
9. Sijbesma R, Srdanov G, Wudl F, Castoro JA, Wilkins C, Friedman SH, DeCamp DL, Kenyon GL. J. Am. Chem. Soc. 1993; 115: 6510.
10. Dugan LL, Turetsky DM, Du C, Lobner D, Wheeler M, Almli CR, Shen CK-F, Luh T-Y, Choi DW, Lin T-S. Proc. Natl. Acad. Sci. USA 1997; 94: 9434.
11. Brettreich M, Hirsch A. Tetrahedron Letters 1998; 39: 2731.
12. Andersson T, Nilsson K, Sundahl M, Westman G, Wennerstrom O. J. Chem. Soc., Chem. Commun. 1992; 604.
13. Lai DT, Neumann MA, Matsumoto M, Sunamoto J. Chemistry Letters 2000; 64.
14. Scrivens WA, Tour JM, Creek KE, Pirisi L. J. Am. Chem. Soc. 1994; 116: 4517.
15. Andrievsky GV, Kosevich MV, Vovk OM, Shelkovsky VS, Vaschenko LA. J. Chem. Soc. Chem. Commun. 1995; 1281.
16. Mchedlov-Petrossyan NO, Klochkov VK, Andrievsky GV. J. Chem. Soc. Faraday Trans. 1997; 93: 4343.
17. Wang X, Goh SH, Lu ZH, Lee SY, Wu C. Macromolecules 1999; 32: 2786.
18. Jenekhe SA, Chen XL. Science 1998; 279: 1903.
19. Sun Y-P, Bunker CE. Nature 1993; 365: 398.
20. Ghosh HN, Sapre AV, Mittal JP. J. Phys. Chem. 1996; 100: 9439.
21. Mrzel A, Mertelj A, Omerzu A, Copic M, Mihailovic D. J. Phys. Chem. 1999; 103: 11256.
22. Alargova RG, Deguchi S, Tsujii K. J. Am. Chem. Soc. 2001; (submitted).
23. Deguchi S, Alargova RG, Tsujii K. J. Am. Chem. Soc. 2001; (submitted):
24. Hunter RJ, Foundations of Coloid Science, Oxford:Clarendon Press, 1995.
25. Yamakoshi YN, Yagami T, Fukuhara K, Sueyoshi S, Miyata N. J. Chem. Soc., Chem. Commun. 1994; 517.

Nanoscience & Nanotechnology
eds. E. Balabanova, I. Dragieva, Heron Press, Sofia, 2001

STABLE DISPERSIONS OF NANOCRYSTALLINE C_{60} IN WATER

Shigeru Deguchi, Rossitza G. Alargova and Kaoru Tsujii

The DEEPSTAR Group, Japan Marine Science and Technology Center (JAMSTEC);
2-15 Natsushima-cho, Yokosuka 237-0061, Japan

Abstract. Stable dispersions of nanocrystalline C_{60} in water were prepared by injecting a saturated solution of C_{60} in N-Methyl-2-pyrrolidone (NMP) into water, followed by dialysis against water to remove NMP. C_{60} was dispersed as very fine nanocrystalline clusters. The C_{60} dispersion thus prepared has very unique spectroscopic properties. These properties could be ascribed to strong interaction between NMP and C_{60}, probably through charge-transfer.

Keywords: fullerenes, C_{60}, N-methyl-2-pyrrolidone, water, dispersion, nanoparticle

1. Introduction

Interesting biological activities of water soluble fullerene derivatives have been discovered [1–4], and there has been increasing interest in preparation of aqueous solutions of fullerenes [1–8]. As fullerenes themselves are virtually insoluble in water [9], it has been achieved by chemical modification of fullerenes [1–4], formation of water-soluble host-guest complexes [5–7], or solubilization by surfactants [6–8].

While a true molecular solution of fullerenes in pure water can not be prepared, there have been some attempts to disperse fullerenes stably in water without any stabilizer [10–14]. The dispersions can be prepared with ease, and would contribute significantly to the investigation of biological activities of fullerenes. The dispersion could also be applied to other water-based applications such as those in food or cosmetics. The stable dispersion of a pure carbon allotrope, fullerene, in water without any stabilizing agents is also intriguing from viewpoint of colloid science. Besides the interest in the aqueous dispersions of fullerenes, there have been several reports regarding the formation of the aggregates of fullerenes in organic solvents or binary solvent mixtures including ours [15–19]. Thus, the cluster formation itself would be a general property of fullerenes, although the details such as the size of the cluster depend on the choice of the solvent. At present, the accumulation of experimental data in different solvent systems seems to be necessary to understand this phenomenon.

We have developed a simple preparation procedure for the aqueous dispersion of fullerene [14]. In our procedure, fullerenes are first dissolved in an organic solvent that is miscible with water such as tetrahydrofuran (THF). The solution is then injected into excess amounts of water, and the organic solvent is finally removed. It is interesting to evaluate the effect of the initial organic solvent used for the preparation, but the choice of the organic solvent is very limited because fullerenes are soluble in a limited number of solvent that are miscible with water [9]. N-Methyl-2-pyrrolidone (NMP) is one of such solvents, and dissolves unusually high amounts of C_{60} (890 μg/mL) compared with other polar solvents such as THF (10 μg/mL), ethanol (1 μg/mL), or acetone (1 μg/mL) [9,14], although the origin of the high solubility of fullerenes is unclear. Its polymeric analogue, poly(N-vinylpyrrolidone) (PVP), has also been shown to solubilize large amounts of C_{60} in water [6].

In this paper, we report the new preparation method and preliminary characterizations of a stable aqueous dispersion of nanocrystalline C_{60} clusters, prepared using NMP. Special attention has been paid to the unique effect of NMP on the UV-visible absorption spectra of the C60 dispersions.

2. Materials and Methods

Materials. C_{60} ($>$ 99.9% pure) was purchased from Tokyo Kasei Kogyo Co., Ltd. (Tokyo, Japan), and used as received. NMP was purchased from Wako Pure Chemical Industries, Ltd. (Osaka, Japan), and used without further purification. Millipore water was used throughout the work. Care was taken to minimize exposure of the samples to light during the experiments.

Sample Preparation. A small quantity of solid C_{60} and NMP were placed in a glass vial with a screw cap, and the mixture was stirred overnight at room temperature under argon atmosphere in dark. Excess solid C_{60} was filtered off with a PTFE-membrane filter (nominal pore size 0.45 μm, Acrodisk, Gelman, Ann Arbor, Michigan) to give a saturated solution of C_{60} in NMP. The solution (0.1 mL) was injected into water (10 mL), and the sample was transferred into a dialysis bag (Spectra/Por 1, cut-off molecular weight 6-8000 Daltons, Spectrum Laboratories, Inc., USA), and was dialyzed against water for one month with daily water change. A small quantity of dark orange precipitate formed after dialysis for one week, but there was no visible change in color of the supernatant or the amount of precipitate thereafter. Filtration of the suspension with a membrane filter (nominal pore size 5 μm, Millex-SV, Millipore, Bedford, Massachusetts) gave a brownish and visually clear solution of C_{60}.

Characterization. UV-visible absorption spectra were recorded on a Shimadzu UV-2400PC UV-VIS recording spectrophotometer (Shimadzu, Kyoto, Japan). All the measurements were done at room temperature. Specimens for TEM observations were prepared by depositing 4 μL of the sample solution on a 150-mesh copper grid, coated with a collodion film and deposited by carbon. The specimen was air dried in dark, and was examined on a JEOL JEM-1210 (JEOL, Japan) operating at 120 kV incident beam energy.

3. Results and Discussion

Considering the cut-off molecular weight of the dialysis membrane and the size of the C_{60} molecule, it is likely that C_{60} forms a cluster in water and remain in the dialysis bag. Attempts to measure the cluster size by dynamic light scattering failed because of the low scattered light intensity. TEM observation revealed fine clusters which look like crystals in shape (Figure 1). Size of the clusters is less than 5 nm, and is significantly smaller than those reported previously [10,12,14]. As the mean atom-to-atom diameter of the C_{60} molecule is 0.71 nm [20], simple calculation shows a few hundreds of C_{60} molecules make one cluster.

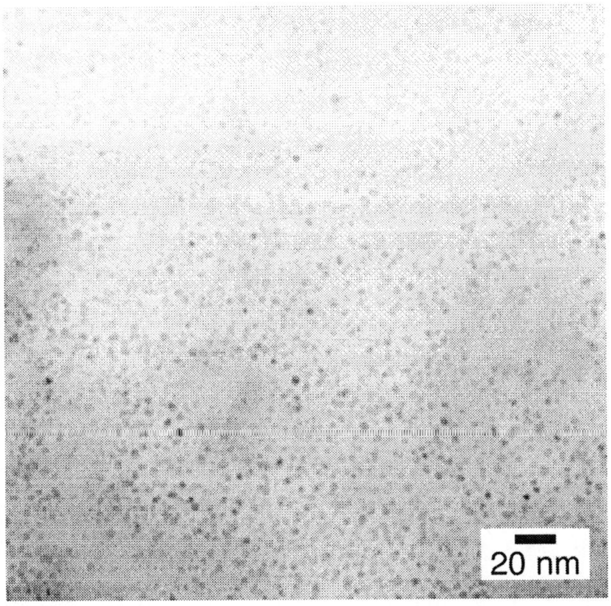

Figure 1. Transmission electron micrograph of the C_{60} dispersion in water

The suspension is very stable, but the stability is strongly affected by salt addition. When NaCl was added to the dispersion at the concentration of 1%, C_{60} precipitated as dark brown precipitates. This suggests that the clusters are stabilized by electrostatic repulsion, as has been reported previously for other aqueous dispersions [12,14].

Solutions of C_{60} in organic solvents and dispersions in water can be distinguished easily by the color of the sample solution. Thus, the solutions of C_{60} are magenta, while the dispersions are yellow [14]. However, the spectroscopic properties of C_{60} are very different when NMP is used as the initial solvent. The saturated solution of C_{60} in NMP is wine red. Even when the solution is injected into excess amount of water, the mixture exhibited the color similar to the NMP solution, but fainter due to lower C_{60} concentration. Figure 2 shows absorption spectra of C_{60} in water before and after dialysis. It should be mentioned that the solution of C_{60} in NMP gives essentially the same spectra as those before dialysis. The absorption spectra of the solution before dialysis, and thus the spectra of the solution in NMP, are significantly different from the typical absorption spectra of C_{60} in solution [21, 22]. Broad absorption starts at around 700 nm, and the intensity increases monotonically down to the shorter wavelengths, with shoulders at around 350 nm and 270 nm. No vibronic fine structure is seen in the spectra. Dialysis of the solution against water for one month altered the spectra slightly. Although the spectra are still dominated by the similar broad absorption, two peaks, which appear as shoulders in the spectra before dialysis, appear clearly at 260 and 331 nm. In addition, the position of these peaks agrees well with that of the peaks observed in the absorption spectra of the aqueous dispersion of C_{60} prepared with THF [14].

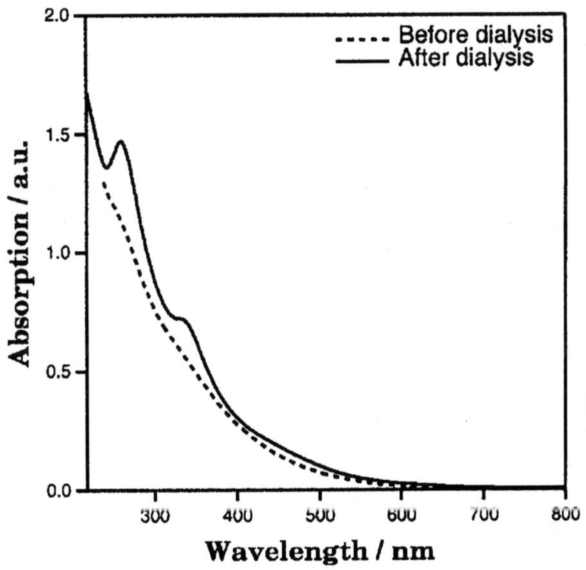

Figure 2. UV-visible absorption spectra of C_{60} in water before (dotted line) and after (solid line). The solution before dialysis contains 1% (v/v) NMP

The very unique absorption spectra of C_{60} in NMP suggest that there is strong interaction between the NMP and the C_{60} molecules. The interaction could be responsible for the unusually high solubility of C_{60} in this solvent. The spectra do not change when the solution is injected into excess amount of water. The similar spectra are observed even after dialysis. We have reported that C_{60} forms clathrate crystals when a solution in THF is injected into excess amount of water [14]. Considering the possible strong interaction between C_{60} and NMP, it is even more likely that clathrate crystals of C_{60} and NMP are formed in the aqueous dispersion. Thus, the guest NMP molecules interact with the C_{60} molecules in the clathrate crystal as they do in the solution, giving rise to the similar absorption spectra as those observed in pure NMP.

The stability of the dispersion seems to be achieved by electrostatic repulsion as was the case for the dispersions prepared by THF. For the clusters of C_{60} prepared with THF, we proposed that electron-transfer from THF to C_{60} is important for the surface charge of the fullerene cluster [14]. The interaction between NMP and C_{60} could also be electron-transfer interaction with NMP being an electron donor.

Another possible explanation for the very different absorption spectra of C_{60} in NMP systems would be that a chemical reaction between C_{60} and NMP takes place. C_{60} is known to react easily with amines by simple mixing at room temperature [23]. However, for C_{60} solubilized in water by PVP, it has been reported that mass spectra of C_{60} with PVP in water are identical to those in toluene [6]. As PVP is a polymeric analogue of NMP, this strongly suggests that C_{60} remains intact in NMP.

On the other hand, reasons for the significantly smaller size of the clusters obtained in this work is not at all clear. The strong interaction between C_{60} and NMP, mentioned above, would be somehow related to the difference in size of the cluster obtained.

4. Conclusion

C_{60} was dispersed in water by injecting a saturated solution of C_{60} in NMP into excess amounts of water, followed by dialysis against water. Although the procedure gives the stable dispersion of C_{60} in water, the spectroscopic properties of the dispersion were remarkably different from the dispersions reported previously. In addition, the size of the C_{60} cluster thus obtained is significantly smaller than those reported previously. The results clearly show that the properties of the aqueous dispersion of C_{60} are strongly affected by the initial organic solvent used for the preparation. Further study of the dispersion, particularly the structure of the C_{60} cluster, is in progress.

Acknowledgement

Professor Koki Horikoshi, JAMSTEC, is acknowledged for his continuous support to this work. Mr. Katsuyuki Uematsu, JAMSTEC, is acknowledged for performing TEM observations.

References

1. Tokuyama H, Yamago S, Nakamura E, J. Am. Chem. Soc, 1993; 115: 7918.
2. Friedman SH, DeCamp DL, Sijbesma RP, Srdanov G, Wudl F, Kenyon GL, J. Am. Chem. Soc, 1993; 115: 6506.
3. Dugan LL, Turetsky DM, Du C, Lobner D, Wheeler M, Almli CR, Shen CK-F, Luh T-Y, Choi DW, Lin T-S, Proc. Natl. Acad. Sci. USA, 1997; 94: 9434.
4. Tabata Y, Murakami Y, Ikada Y, Jpn. J. Cancer. Res., 1997; 88: 1108.
5. Andersson T, Nilsson K, Sundahl M, Westman G, Wennerström O, J. Chem. Soc., Chem. Commun., 1992; 604.
6. Yamakoshi Y, Yagami T, Fukuhara K, Sueyoshi S, Miyata N, J. Chem. Soc., Chem. Commun., 1994; 517.
7. Lai DT, Neumann MA, Matsumoto M, Sunamoto J, Chem. Lett., 2000; 64.
8. Hungerbühler H, Guldi DM, Asmus K-D, J. Am. Chem. Soc, 1993; 115: 3386.
9. Ruoff RS, Tse DS, Malhotra R, Lorents DC, J. Phys. Chem., 1993; 97: 3379.
10. Scrivens WA, Tour JM, J. Am. Chem. Soc, 1994; 116: 4517.
11. Andrievsky GV, Kosevich MV, Vovk OM, Shelkovsky VS, Vashchenko LA, J. Chem. Soc., Chem. Commun., 1995; 1281.
12. Mchedlov-Petrossyan NO, Klochkov VK, Andrievsky GV, J. Chem. Soc., Faraday Trans., 1997; 93: 4343.
13. Prilutski Y, Durov S, Bulavin L, Valerij P, Astashkin Y, Yashchuk V, Ogul'Chanski T, Buzaneva E, Andrievsky G, Mol. Cryst. Liq. Cryst., 1998; 324: 65.
14. Deguchi S, Alargova RG, Tsujii K, J. Am. Chem. Soc, 2001; submitted:
15. Alargova RG, Deguchi S, Tsujii K, J. Am. Chem. Soc, 2001; submitted:
16. Sun Y-P, Bunker CE, Nature, 1993; 365: 398.
17. Ghosh HN, Sapre AV, Mittal JP, J. Phys. Chem., 1996; 100: 9439.
18. Nath S, Pal H, Palit DK, Sapre AV, Mittal JP, J. Phys. Chem. B, 1998; 102: 10158.
19. Mrzel A, Mertelj A, Omerzu A, Copic M, Mihailovic D, J. Phys. Chem. B, 1999; 103: 11256.
20. Dresselhaus MS, Dresselhaus G, Eklund PC, Science of Fullerenes and Carbon Nanotubes, San Diego: Academic Press, 1996.
21. Hare JP, Kroto HW, Taylor R, Chem. Phys. Lett., 1991; 177: 394.
22. Leach S, Vervloet M, Despres A, Breheret E, Hare JP, Dennis TJ, Kroto HW, Taylor R, Walton DRM, Chem. Phys., 1992; 160: 451.
23. Seshadri R, Govindaraj A, Nagarajan R, Pradeep T, Rao CNR, Tetrahedron Lett., 1992; 33: 2069.

Nanoscience & Nanotechnology
eds. E. Balabanova, I. Dragieva, Heron Press, Sofia, 2001

SURFACE MORPHOLOGY AND OPTICAL PROPERTIES OF METAL-POLYMER NANOSTRUCTURED COMPOSITES

S.A. Zavyalov, M.Yu. Yablokov, A.N. Pivkina, A.A. Lushnikov and J. Schoonman[1]

Karpov Institute of Physical Chemistry,Vorontsovo Pole, 10, 103064 Moscow, Russia
[1]Delft University of Technology, Delft Interfaculty Research Center: Renewable Energy, Julianalaan 136, 2628 BL Delft, The Netherlands

Abstract. Nanostructured metal (Pd, Sn, Cu)-polymer (poly-para-xylylene) and metal-oxide-polymer composites reveal synergism of properties of the initial components, which gives rise to specific electrical, mechanical, and optical properties related to an ordered distribution of nanopartices over the matrix volume. Various methods of preparing a metal-polymer nanocomposite exist and from the point of view of homogeneity, methods where polymerisation and formation of the nanoparticles are performed simultaneously are very promising. This paper reports on the experimental set-up for the production of nanocomposites by vacuum co-condensation of the metal and a monomer (paracyclophane) vapours. The morphology, optical, and electrical properties of the deposited composites has been studied.

Keywords: nanostructured materials, metal-polymer nanocomposites, surface morphology, optical absorption

1. Introduction

To date, nanostructured metal-polymer and metal oxide-polymer composites are the subject of increased attention of researchers in different areas of science and technology. Among the perspectives for using these materials are novel types of solar cells and chemical gas sensors. The electrical, mechanical, and optical properties of the nanocomposites are improved by dispersing a metallic phase in a polymeric matrix.

At a particular concentration of nanoparticles close to the percolation threshold the nanocomposite materials exhibit properties specific to the nanoparticles. These properties differ considerably from those for bulk materials and the composite acquires these during composite formation. Using different polymers adds elements of self-organisation to the formation of the composite, for inorganic particles are located in cells of the polymer between grids and layers [1] which gives rise to specific properties of the composite related to an ordered distribution of nanopartices over the matrix volume. Therefore, the application of the nanostructured materials is foreseen in several technical areas – developing of basically new solar cells, new sensors, and catalytic systems, rechargeable batteries, syntheses of materials with new or improved properties, aerosol drags, and memory cells. The nanocomposites combining unique optical and semiconducting properties are of great importance for solar energy conversion.

This paper reports on the production of nanocomposites by vacuum co-condensation of the metal and a monomer vapour. The influence of the metal content on the morphology, optical, and electrical properties of the deposited composites has been studied.

2. Methods

Figure 1 shows the set-up for depositing composite films by co-condensation of metal vapour and an active precursor, which then polymerises. The set-up consists of four principle units: the zone of treating the polymer precursor, the zone of evaporating metals, the reactor and the deposition zone.

The poly-para-xylylene (PPX) matrix is synthesised as follows: in the evaporation zone the precursor is heated up to 380–420 K. The paracyclophane [2] pyrolyses at 930 K giving rise to biradicals of para-xylylene. These molecules then deposit without polymerisation onto a cooled (down to 77 K) substrate. The flux of metal atoms (Pd, Sn, Cu) evaporated from a bulk sample condenses onto the substrate together with the monomer. The condensate consists of nanoparticles of the metal and the monomer. In heating the substrate up to ambient temperature the monomer polymerises to poly-para-xylylene. The structure thus obtained is a porous matrix with dispersed nanoparticles in it.

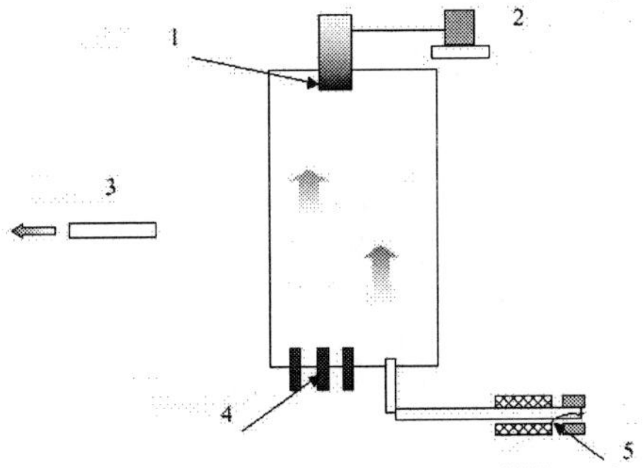

Figure 1. Experimental set-up: 1) cooled substrate; 2) computer data logging; 3) vacuum pump; 4) metal precursor; 5) paracyclophane powder

Composites were deposited on quartz substrates with Pt contacts, connected with an ohmmeter for the *in situ* and *ex situ* measurements of the electrical resistance. The metal content in the composites was calculated by atomic absorption analysis using a *Perkin-Elmer 503* spectrometer. The surface morphology and composite structure were studied by atomic force microscopy (*SOLVER-P47*). The optical spectra of the films were recorded with a spectrophotometer *Shimadzu UV-3100*.

3. Results

Freshly synthesised tin-containing samples are normally of grey colour with a metallic luster. On contacting with the ambient air during 2 minutes the samples became transparent for composites with a Sn concentration below the percolation threshold of 10 Vol.% (Figure 2a), but the samples with a tin content beyond 10 Vol.% do not change their colour during several months (Figure 2b). This effect could be attributed to the fast oxidation of Sn nanoparticles not forming the percolative chains, whereas Sn particle connected into chains do not oxidize.

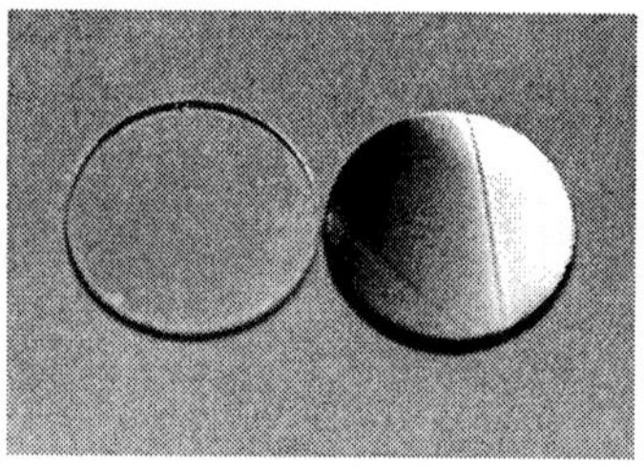

a) b)

Figure 2. Sn-containing samples of nanocomposites after the air exposure: a) 8 Vol.% of Sn; b) 12 Vol.% of Sn

Figure 3 shows the temperature coefficient (the slope of $R(T)/R(20°C)$ *versus* temperature dependency) of the electrical resistance in vacuum of tin and copper containing as-prepared composites for metal contents from 4 Vol.% to 12 Vol.%. The composites with a metal content of 12 Vol.% show the positive temperature-coefficient indicating a conductivity determined by a continuous network with metal-like contacts between the nanoparticles. In contrast, when the metal concentration is 8 Vol.% the temperature-coefficient becomes semiconductor-like. This indicates a loss of metal-like contacts between particles of the metal phase. This is to be expected from percolation behaviour of the composites on the metal filler content. The percolation threshold can be determined by the variation in the temperature dependence of the electrical resistance, which is for the present case 10 Vol.% of metal.

AFM analysis of model Pd composites shows that the surface of the nanocomposites is sufficiently uniform. The spherical polymeric globules of a size up to 200 nm are well distinguishable. The pore size does not exceed 100 nm. In addition, at the spherulite surfaces imperfections are distinguishable deviating their shape from sphere. The phase-contrast regime of the AFM (Figure 4a) allows for identifying these surface imperfections as the particles of the inorganic phase, establishing location, and evaluates their sizes.

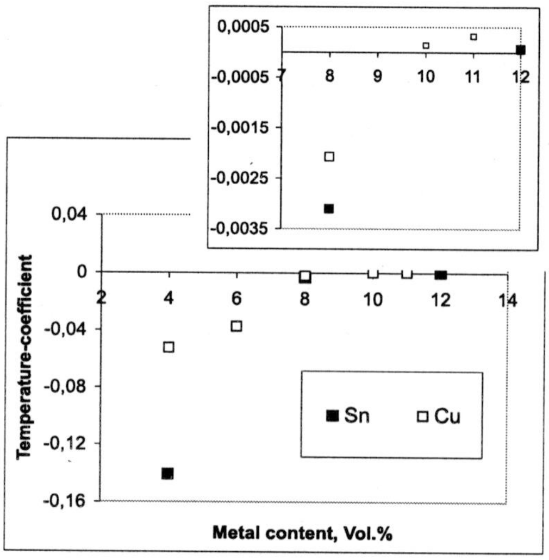

Figure 3. Dependency of temperature-coefficient of resistivity on the metal content

According to Figure 4b, where the cross section of the composition profile is presented, the inorganic particle size is 7–10 nm.

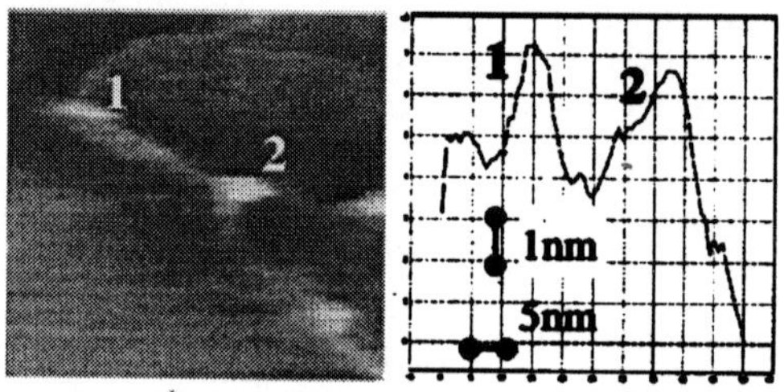

a) b)

Figure 4. a) Phase-contrast AFM image of the composite. Dark regions are polymer spherulites, light spots are Pd nano-particles situated at the boundary between polymer spherulites; b) Cross-section 1–2

According to Figure 4b, where the cross section of the composition profile is presented, the inorganic particle size is 7–10 nm.

The tiny containing composites were used for studying the evolution of the surface morphology with increasing content of the filler from 8 Vol.% to 16 Vol.% (Table 1).

Table 1. Parameters of composites

#	Sn Vol.%	Resistivity in Vacuum	Morphology
1	8	6 MOhm	Isolated particles, localization on the polymer spherulite surface
2	10	880 Ohm	Formation and growth of
3	12	45 Ohm	nanoparticle chains
4	16	13 Ohm	Percolative chains spreading

Figure 5 shows the AFM phase contrast image of a Sn-Polymer composite (Sample #1) containing 8 Vol.% of Sn. Scan size area is 140 × 140 nm. Separated tin particle is localized on the polymer spherulite surface.

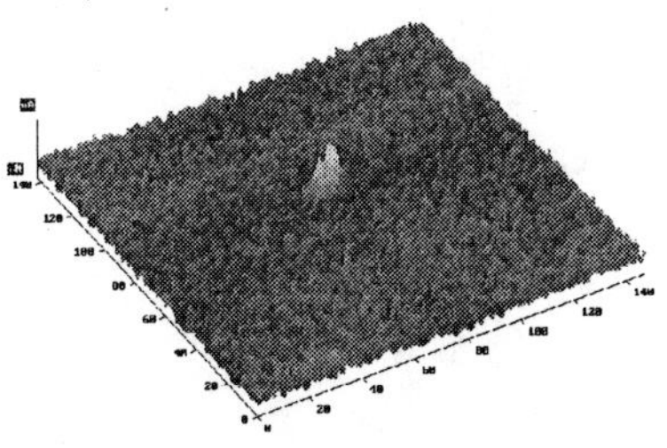

Figure 5. AFM phase contrast image of the sample with 8 Vol.% of Sn

Figure 6 presents the AFM phase contrast image of the Sample #4 containing 16 Vol.% of Sn, scan size 140 × 140 nm. Nanoparticles aggregate connecting into continuous chains. Spreading of connected nanoparticles chains is observed.

Figure 6. AFM phase contrast image of the sample with 16 Vol.% of Sn

Figures 7-8 present optical absorption spectra of Cu and Sn-containing composites, respectively. The metal concentrations in volume per cents are indicated on the plots. The changes in composite structure on a nanometer scale are crucial for optical absorption.

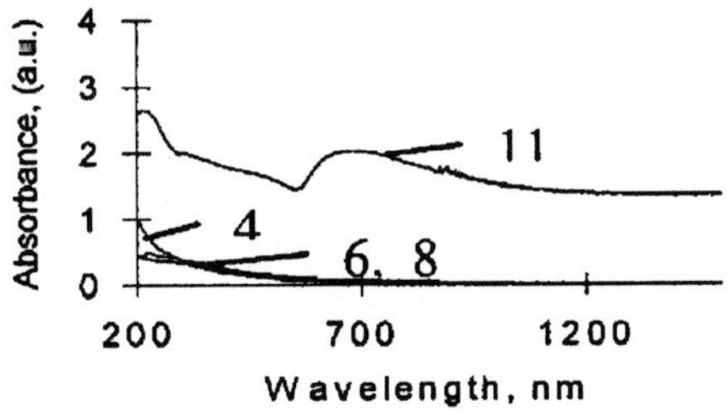

Figure 7. Optical absorbtion spectra of Cu-containing nanocomposites

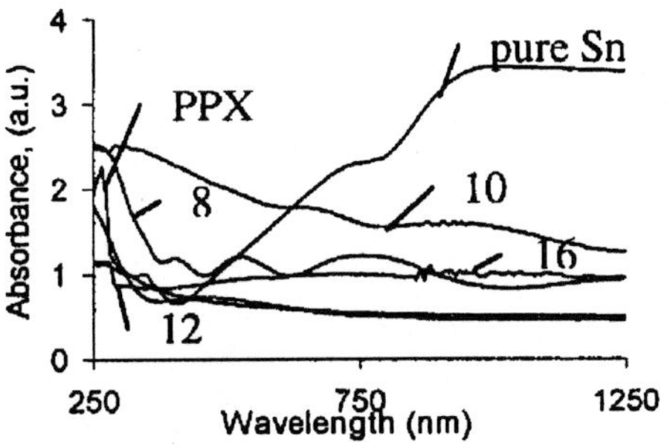

Figure 8. Optical absorbtion spectra of Sn-containing nanocomposites

4. Conclusions

Samples of nanoporous composites of a metal and the polymer PPX have been synthesised in the form of thin films. AFM studies show the metal nanoparticles having a size of 7 to 10 nm. Depending on the metal concentration in as-prepared samples three different types of surface morphology are distinguished to be a function of the metal concentration. The percolation threshold of these systems is evaluated by measuring the temperature dependence of the electrical resistance. The type of Absorbance versus Wavelength dependency for the samples beyond the percolation threshold differs from that of the sample with a nano-metal content below the threshold.

Acknowledgement

The authors are grateful to the Netherlands Organization for Scientific Research (NWO, grant #047-007-008) for financial support of the project.

References

1. Pomogailo A.D., Rozenberg A.S., Ufland U.E. Nanoscale Metal Particles in Polymers, Moscow, "CHIMIYA", 2000, pp. 671.
2. Szwarc, M., Polym. Eng. Sci., 1976, 16, 47.

Nanoscience & Nanotechnology
eds. E. Balabanova, I. Dragieva, Heron Press, Sofia, 2001

DISPERSE STRENGHTENING OF POLYMERS – THEORETICAL CONSIDERATIONS AND EXPERIMENTS WITH UDDP

S. Stavrev, R. Kotsilkova[1], J. Karadjov, Z. Karagyozova

Space Research Institute, BAS, Sofia 1000, Moskovska str. 6
[1]Central Laboratory for Physico-Chemical Mechanics, BAS

Abstract. Infrared spectroscopy shows that shock-wave synthesized ultradisperse diamond powder has a surface, rich in carboxyl gruops which may form strong hydrogen bonds with moderately polar polymers. When used as a filler,ultradisperse diamond powder increases the shock resistance of epoxide resin and leads to a dramatic increase of the wear resistance of polyamide.

Keywords: nanodisperse diamond, epoxy resins, polyamide, composite materials

It is well known that the properties of polymers can be modified by the addition of fillers – powders or fibers. In both cases, the interaction between the surface of the filler particles and the polymer matrix is of crucial importance. Different methods are used to improve the adhesion between the filler and the polymer, especially when both differ in polarity. When powdered inorganic fillers (silica, aluminium oxide, etc.) are used to improve the qualities of polymers, the surface of the particles can be made more hydrophobic by treatment with organic siloxanes. The surface of highly hydrophobic carbon fibers is oxydized to improve the adhesion to moderately polar polymers (epoxy resins, polyamides, etc.) [1]. Here we discuss the use of a new material, which can be used as an additive to polymers.

UDDP is synthesized in our laboratory by the shock wave method [2]. Diamond is formed from the free carbon of the explosive used (a trinitrotoluene/hexogene mixture). After detonation, the solid residue is collected from the camera. The crude product consists of diamond particles ($\sim$ 40% by weight), and finely dispersed graphite. The latter is removed by energic oxydation of the mixture with potassium bichromate/sulfuric acid. After several washings, diamond is obtained in an almost pure form as a gray powder. The pieces of diamond are in the nanometer range (1 to 20 nm, 6 nm average size) with a specific surface of 300–400 m^2/g. For many industrial applications the crude diamond/graphite mixture can be successfully used instead of the more expensive purified UDDP.

Natural and conventional synthetic diamonds have a highly hydrophobic surface, which is used for centuries in the separation of natural diamond from the rock. The surface of shock wave synthesized UDDP is oxydized and much less hydrophobic. IR spectroscopy suggests that it is covered with carboxyl, and probably to a much lesser extent with carbonyl and hydroxyl groups (C=O, 1720 cm^{-1} and C–O, 1120 cm^{-1}). The polarity of the surface of UDDP was estimated experimentally by the affinity of UDDP to different solvents with a density equal to 1 and dielectric constants varying in a wide range. A small quantity of UDDP was placed in 5 ml of solvent, sonificated for 15 min, and allowed to precipitate. In a hydrophobic solvent (carbon tetrachloride/xylene, $\varepsilon = 2.3$), as well as in water ($\varepsilon = 80$), UDDP reaggregates and precipitates very quickly (in the order of minutes). When moderately polar solvents were used (pyridine, $\varepsilon = 12.4$, and especially DMF/DMSO, $\varepsilon = 40$) UDDP forms a colloidal solution which is stable for several months. This finding suggests that UDDP combines the exceptional hardness of other diamond species with a more polar surface which may form strong hydrogen bonds with epoxy resins and polyamides. Both contain polar hydroxy, amino, or amido groups suitable for hydrogen bonding. A hypothetical view of the interaction of the surface of a diamond particle with the polyamide matrix is shown at the picture.

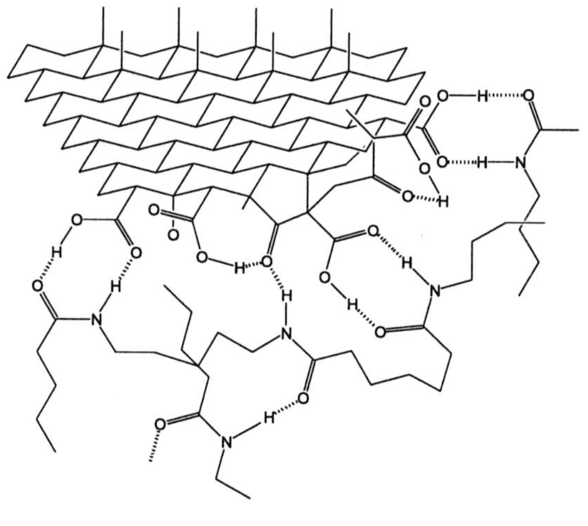

Our first experiments show that an addition of 2 vol.% UDDP to epoxy resin leads to a 30% increase of impact strenght. When added to polyamide (both nylon 6 and nylon 66 were tested) 0.7 vol.% UDDP leads to a decrease of the friction factor and to a dramatic increase (up to 6 fold) of the wear resistance of the polymer. Further experiments on the use of UDDP as an additive to polymers are under way in our laboratory.

References

1. Encyclopedia of polymer science and technology, John Wiley & Sons, 1990.
2. BG patent # 49267, BG patent # 61132.

Nanoscience & Nanotechnology
eds. E. Balabanova, I. Dragieva, Heron Press, Sofia, 2001

THERMAL AND ELECTRIC CONDUCTIVITY OF UDDP COMPACTS FOR THE MICROELECTRONIC INDUSTRY

S. Stavrev, L. Markov, J. Karadjov, A. Petrova

Space Research Institute, BAS, Sofia 1000, Moskovska str. 6

Abstract. Compacts were made from ultradisperse diamond powder (UDDP) alone, and with the addition of 50% AlN, or SiC. Compacts made from pure UDDP are most prom-ising as a non metallic material for the production of pads. They combine good mechanic stability with high ther-mocon-ductivity and a very low electric conductiv-ity, and can be used as pads in microelectronics.

Keywords: nanodisperse diamond, silicon carbide, aluminium nitride

The problem with heat dissipation is crucial in the design of microelectronic devices with high power consumption.

There are three possible ways the heat can be dissipated: thermoconductivity, radiation, and convection. In microelectronics thermoconductivity is the principal mechanism which takes the heat from the crystal and transfers it to the radiator.

Thus microelectronic circuits are separated from the radiator by special pads [1,2]. They are made from materials with good insulating properties, high thermo-conductivity and low price [3].

The pad problem can be solved by two ways:

1) The use of metal pads with high thermoconductiv-ity coated with a layer of materials with good iso-lating properties.

2) The use of pads from non metal material with high thermoconductivity.

The most useful materials with high termoconduc-tivity are given in Table 1.

Table 1

Material	Thermoconductivity W/°K.m
Natural diamond	2000
Synthetic diamond	900
Beryllium oxide	210
Aluminum nitride	120
Silver	450
Copper	400
Gold	310
Aluminum	220
Silicon	160

Highest thermoconductivity is typical for natural diamond and precious metals, especially silver. Their high price either and necessity of electric isolation limit their application only to unique, expensive appurtenances.

A widely used material with good isolating properties is BeO. It has a relatively high thermoconductivity, but its extreme toxicity limits its use.

In the last decade, many attempts were made to solve the heat dissipation problem by the use of pads made from synthetic diamond compacts.

The classic method for synthetic diamond production is not very useful because the diamond obtained is soiled with graphite, which leads to a considerable decrease in its thermoconductivity. Ultradisperse mono-crystal diamond powder (UDDP) with good purity is synthesized in our laboratory. To avoid contamination with graphite, after the shock-wave synthesis diamond is purified by energic oxydation. The so obtained UDDP has a thermoconductivity by 50% higher than natural diamond, an average particle size of 60 Å, and a specific surface of 360–400 m^2/cm^3.

Three different compositions have been com-pacted and investigated:

- UDDP: 100% was compacted at $T = 1450°C$, $P = 9,5$ Gpa, $t = 90$ min
- UDDP + 50% AlN was compacted at $T = 1600°C$, $P = 9,5$ Gpa, $t = 60$ min
- UDDP + 50% SiC was compacted at $T = 1450°C$, $P = 9,5$ Gpa, $t = 60$ min

Figure 1. Pure UDDP

Figure 2. UDDP + AIN

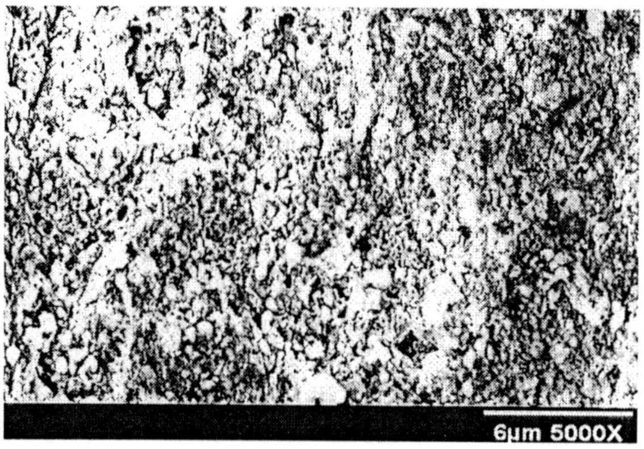

Figure 3. UDDP + SiC

The samples were investigated in Prof. Anderson's laboratory (Rola, Missouri, USA). The structures of the three samples (scanning electron microscopy images) are shown on Figures 1, 2 and 3.

Some results about the reached density of the compacts, their electric conductivity and thermocon-ductivity are presented in Table 2.

Table 2

Material	Density % from the theory	Thermo-conductivity W/m.°K	Electric conductivity S/m
UDDP	87,5	1260	$(6{,}3 + 0{,}5)\times10^{-7}$
UDDP + 50% SiC	89.7	420	3.6 + 0.2
UDDP + 50% AlN	87	700	186 + 2

Compacts made from pure UDDP are most promising as a non metallic material for the production of pads. They combine good mechanic stability with high thermoconductivity and a very low electric conductivity.

Aknowledgments

The authors wish to thank to the research group in TISNUM, Troitsk, Russia and the laboratory of prof. Anderson in the University in Missouri, USA for their assistance and helpful discus-sion.

References

1. Б. Финдайзен, Е. Фридрих, "Спеченньiе и конпозиционньiе материаллы", М., Металургия, 1983
2. Ю. Тармополский, "Пространственно армированньiе материаллы", М., Машиностроение, 1987
3. F. Kimock, "Commercial applications of eon beam deposited diamond-like carbon coating. Surface and coatings technology", vol.56, 1993, pp, 273-279

Nanoscience & Nanotechnology
eds. E. Balabanova, I. Dragieva, Heron Press, Sofia, 2001

THERMAL STABILITY AND DENSITY MEASUREMENT OF A NEW ALLOTROPIC FORM OF CARBON

S. Stavrev, L. Markov, J. Karadjov, A. Petrova

Space Research Institute, BAS, Sofia 1000, Moskovska str. 6

Abstract. A new carbon phase was obtained from ultradisperse diamond powder by the action of a shock wave. Thermal analysis shows that in an inert atmosphere the new material is stable in the tempera-ture range up to 1500°K. Raman and X-ray spectra of the new mate-rial are surprisingly similar, but not identical, to the corresponding spectra of boron carbide. Our hypothesis is that under the action of the shock wave on dia-mond a new carbon phase is formed which contains carbon icosahedres.

Keywords: diamond, carbon

The new material was obtained during experiments with ultradisperse diamond powder (UDDP) [1] in an attempt to increase the size of diamond particles under the action of a shock wave. When a very high pressure was applied, diamond was transformed into a new carbon phase. In our experiments, UDDP was placed axially in a cylindrical explosive charge. The geometry of the charge was optimized to obtain maximal pressure (more than 470 kbar), a reaction time of 9 to 12 ns, and a temperature of 4100°K. After the detonation, the solid residue was collected from the camera. The excess graphite was removed by energic oxydation of the mixture with potassium bichromate and sulfuric acid. After several washings, the material was obtained as black glittering crystals with a size varying from 5 to 200 μm (Figure 1). From 2 grams of UDDP 3.6 grams of material was formed [2]. The weight increase comes from the free carbon of the explosive used (a TNT/hexogene mixture). Chemical analysis shows almost pure carbon with insignificant im-purities of copper and iron (from the reaction vessel) and potassium and chrome (from the oxydizer used). Thermal analysis shows that in an inert atmosphere the new material is stable in the temperature range up to 1500°K. Its hardness is close to this of diamond (100 $\pm$ 20 GPa). The electric conductivity of the material is relatively high which suggests it is a narrow gap semiconductor, but precise measurement, especially in a wide temperature range, requires further purification.

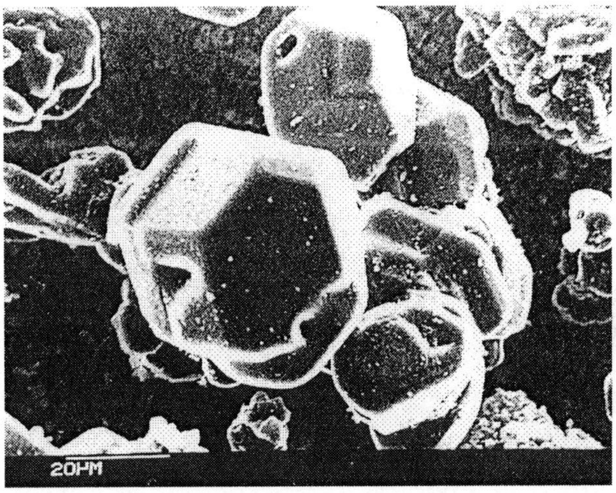

Figure 1. SEM image of the new carbon phase

Raman and X-ray spectra of the new material are surprisingly similar, but not identical, to the corresponding spectra of boron carbide. The crystal structure of the latter is formed by B_{12}^{4-} icosahedres and near-linear C_3^{4+} ions. It is well known that in mild conditions (moderate temperature and normal pressure) carbon may substitute for boron in numerous boron containing chemical species with structures varying from trigonal bipiramid to icosahedre. This is not surprising since carbon atom is isoelectronic and isosteric with boron anion B^-. Our hypothesis is that under the action of the shock wave on diamond a new carbon phase is formed which contains carbon icosahedres.

References

1. BG patent # 49267, BG patent # 61132.
2. BG patent # 104645.

LIST OF CONTRIBUTORS